A Primer on
Smooth Manifolds

A Primer on Smooth Manifolds

Luca Vitagliano

University of Salerno, Italy

World Scientific

EW JERSEY · LONDON · SINGAPORE · BEIJING · SHANGHAI · HONG KONG · TAIPEI · CHENNAI · TOKYO

Published by

World Scientific Publishing Co. Pte. Ltd.
5 Toh Tuck Link, Singapore 596224
USA office: 27 Warren Street, Suite 401-402, Hackensack, NJ 07601
UK office: 57 Shelton Street, Covent Garden, London WC2H 9HE

Library of Congress Cataloging-in-Publication Data
Names: Vitagliano, Luca, author.
Title: A primer on smooth manifolds / Luca Vitagliano, Universsity of Salerno, Italy.
Description: New Jersey : World Scientific, [2024] | Includes bibliographical references and index.
Identifiers: LCCN 2023043228 | ISBN 9789811283949 (hardcover) |
    ISBN 9789811283956 (ebook) | ISBN 9789811283963 (ebook other)
Subjects: LCSH: Manifolds (Mathematics) | Geometry, Differential.
Classification: LCC QA613 .V47 2024 | DDC 516.3/6--dc23/eng20230106
LC record available at https://lccn.loc.gov/2023043228

British Library Cataloguing-in-Publication Data
A catalogue record for this book is available from the British Library.

For any available supplementary material, please visit
https://www.worldscientific.com/worldscibooks/10.1142/13612#t=suppl

Desk Editors: Nimal Koliyat/Nijia Liu

Typeset by Stallion Press
Email: enquiries@stallionpress.com

Preface

Differential Geometry is one of the big branches of modern Geometry. It studies geometric problems with tools from Calculus, Linear Algebra, and Commutative Algebra. Standard topics in Differential Geometry are smooth curves and surfaces in the standard Euclidean space $\mathbb{R}^n$. However, nowadays those notions are encompassed by the abstract definition of a *smooth manifold*. Smooth manifolds are the main objects in Differential Geometry, and these notes provide a short introduction to them and to calculus on them. Historically, the definition of a smooth manifold has two main motivations:

Defining a coordinate invariant calculus: In undergraduate Analysis courses, we learn calculus on $\mathbb{R}^n$ and its open subsets. The standard Euclidean space is equipped with canonical coordinates $t^1, \ldots, t^n$, and we traditionally build calculus in terms of those coordinates. However, it is often useful to change coordinates. For instance, polar coordinates in $\mathbb{R}^3$ are better suited to study the motion of a particle moving under the action of a central potential with center located at the origin. We are thus led to the question of how to translate the main constructions in calculus (derivatives, integrals, etc.) from standard coordinates to polar coordinates or, more generally, any coordinate system. There is an even more fundamental question: Does calculus have any *coordinate invariant meaning*, i.e., is there anything in calculus which is independent of the choice of coordinates? Note that this question is particularly relevant in view of the fact that coordinates are actually a *human artifact*: There are no coordinates in nature! For instance, in the physical 3-dimensional space (or, better to say, 4-dimensional space-time) where we live, there are no distinguished coordinates, and using Cartesian coordinates, as we often do, is an arbitrary human choice, with some advantages in some cases, but still a human choice. In some sense, Differential Geometry is *what remains of calculus after removing all the dependences on the arbitrary choice of a coordinate system.*

Defining a calculus on more general spaces: As we already stressed, in undergraduate Analysis courses, we learn calculus on $\mathbb{R}^n$ and its open subsets. However, there are more general spaces than that where it is useful to perform calculus. For instance, when describing the motion of a particle on the surface of the 2-dimensional sphere, it would be useful to be able to compute derivatives of curves and functions *on the sphere*, independently on how the sphere embeds into $\mathbb{R}^3$. Even more generally, there are physical systems whose configuration space (space of positions) cannot be embedded in the standard Euclidean space in any physically meaningful way. The typical example is the rigid body. Let B be a rigid body moving in our 3-dimensional physical Euclidean space E. We choose Cartesian coordinates on E and think of it as $\mathbb{R}^3$. In particular, it is equipped with a standard orthonormal frame $\mathcal{R}_{can}$. Now let G be the center of mass of B, and let $\mathcal{R}_B$ be an orthonormal reference frame attached to B and defining the same orientation as $\mathcal{R}_{can}$. Then, the positions of B in E are parameterized by the position of G, given by its coordinates, together with the position of $\mathcal{R}_B$ with respect to $\mathcal{R}_{can}$, given by a special orthogonal matrix. Summarizing, the configuration space of B is (isomorphic to) $\mathbb{R}^3 \times SO(3)$, and a natural question is as follows: Is it possible to define a calculus on such space (so to study the motion of B)? Actually, $\mathbb{R}^3 \times SO(3)$ is an example of a smooth manifold, and calculus on smooth manifolds is a part of Differential Geometry.

So Differential Geometry was born to provide a coordinate invariant calculus on one side and to define a calculus on more general spaces on the other side. Later, mostly for its deep conceptual contents, it became an important field of research with connections to several other branches of Mathematics including Algebraic Topology, Algebraic Geometry, Mathematical Physics, and Partial Differential Equations.

Notation and Conventions

If A, B are sets, $C \subseteq A$, $D \subseteq B$ are subsets, and $F : A \to B$ is a map such that $F(C) \subseteq D$, we will usually denote again by $F : C \to D$ the restriction of F to C *in the domain* and to D *in the codomain*. Despite this notation might seem confusing, it is very useful in practice. Note that indicating explicitly the domain and the codomain in $F : A \to B$, and $F : C \to D$, should actually remove the ambiguity.

The identity map of a set A is denoted by $\mathrm{id}_A : A \to A$ or simply $\mathrm{id} : A \to A$. The inclusion of a subset $A \subseteq B$ in a set B is usually denoted by $i_A : A \hookrightarrow B$. It is the restriction to A of the identity map $\mathrm{id}_B : B \to B$.

A *cover* of a set A is a family $\mathcal{F} = \{B\}$ of parts such that $A = \bigcup_{B \in \mathcal{F}} B$. We also say that $\mathcal{F}$ *covers* A.

A *countable set* is a set which is either finite or in bijection with integers.

We denote by $\mathbb{R}$ the field of real numbers. Let n be a non-negative integer and let $\mathbb{R}^n$ be the standard n-dimensional Euclidean space. We will usually denote by $t^1, \ldots, t^n$ the standard coordinates on $\mathbb{R}^n$ interpreting them as real valued functions on $\mathbb{R}^n$. As such, they can be seen as the components of the identity map $\mathrm{id}_{\mathbb{R}^n} : \mathbb{R}^n \to \mathbb{R}^n$. The standard coordinates $t^1, \ldots, t^n$ restrict to any open subset $U \subseteq \mathbb{R}^n$, and we denote with the same symbols $t^1, \ldots, t^n$ the restrictions. When $n = 1$, we usually denote by t (or x) the only coordinate. When $n = 2$ (resp. 3), we usually write x, y (resp. x, y, z) instead of t^1, t^2 (resp. t^1, t^2, t^3). Occasionally, we will deviate from these conventions.

Let n, m be non-negative integers, and let $U \subseteq \mathbb{R}^n$ and $V \subseteq \mathbb{R}^m$ be open subsets. By definition, a real function $f : U \to \mathbb{R}$ is C^∞ or *smooth* (at some point $P \in U$) if it possesses partial derivatives of arbitrarily high order (at P). Given smooth functions $f, g : U \to \mathbb{R}$, the *sum*

$$f + g : U \to \mathbb{R}, \quad P \mapsto (f + g)(P) := f(P) + g(P)$$

and the *product*

$$fg : U \to \mathbb{R}, \quad P \mapsto (fg)(P) := f(P)g(P)$$

are smooth functions as well. A map $F = (F^1, \ldots, F^m) : U \to V$ is C^∞ or *smooth* (at some point $P \in U$) if all its components F^i, $i = 1, \ldots, m$, are smooth (at P). The identity $\mathrm{id}_U : U \to U$ is a smooth map, and the composition of smooth maps is smooth. A smooth map $\Phi : U \to V$ is a *diffeomorphism* if it is additionally invertible, and its inverse $\Phi^{-1} : V \to U$ is smooth as well. In this case, Φ^{-1} is also a diffeomorphism. The identity is a diffeomorphism and the composition of diffeomorphisms is a diffeomorphism. As both a diffeomorphism and its inverse are continuous, a diffeomorphism $\Phi : U \to V$ transforms open subsets into open subsets, i.e., $U_0 \subseteq U$ is an open subset if and only if so is $\Phi(U_0) \subseteq V$.

Let $U \subseteq \mathbb{R}^n$ be an open subset, with standard coordinates $t^1, \ldots, t^n$, let $P \in U$ be a point, and let $f : U \to \mathbb{R}$ be a smooth function. The partial derivatives of f at P will be denoted as follows:

$$\frac{\partial}{\partial t^i}\Big|_{(t^1, \ldots, t^n) = P} f(t^1, \ldots, t^n) \quad \text{or simply} \quad \frac{\partial}{\partial t^i}\Big|_P f \quad \text{or} \quad \frac{\partial f}{\partial t^i}(P).$$

The partial derivative functions will be denoted as follows:

$$\frac{\partial f}{\partial t^i}(t^1, \ldots, t^n) \quad \text{or simply} \quad \frac{\partial f}{\partial t^i}.$$

If $n = 1$, the only derivative will be denoted as $\frac{d}{dt}$.

Let $M(n, m; \mathbb{R})$ be the space of $n \times m$ real matrices. For a matrix $A = (a^i_j)^{i=1,\ldots,n}_{j=1,\ldots,m} \in M(n, m; \mathbb{R})$, the upper index i will usually range over the rows, and the lower index j will range over the columns. For instance, if $B = (b^j_k)^{j=1,\ldots,m}_{k=1,\ldots,q} \in M(m, q; \mathbb{R})$ is another matrix, then the product $A \cdot B \in M(n, q; \mathbb{R})$ is given by

$$A \cdot B = \left(\sum_{j=1}^{m} a^i_j b^j_k \right)^{i=1,\ldots,n}_{k=1,\ldots,q}.$$

About the Author

Luca Vitagliano is a Full Professor of Geometry at the University of Salerno, Italy, where he regularly teaches Geometry courses to bachelor's, master's, and PhD students. His main research interests lie in Differential Geometry, with a slight emphasis on those aspects that are inspired by Mathematical Physics. He is the (co)author of around 40 scientific papers in the field published in distinguished international mathematical journals and has been invited to present his results to several international conferences.

Contents

Chapter 1

Charts, Atlases, and Smooth Manifolds

In this chapter we introduce the main objects of study of these lecture notes, namely *smooth manifolds*. Roughly, a smooth manifold is a space which *looks like* some standard Euclidean space $\mathbb{R}^n$ near by every point. In order to make this definition precise, we have to learn how to put coordinates on a space in such a way that the *transition* from one coordinate frame to the other is a *smooth* enough process.

1.1 Charts and Atlases

We begin defining *charts* on a set. So, let M be a set, and let n be a nonnegative integer.

Definition 1.1 (Chart). An *n-dimensional chart* on M is a map $\varphi : U \to \widehat{U}$, where

- $U \subseteq M$ is a subset,
- $\widehat{U} \subseteq \mathbb{R}^n$ is a non-empty open subset,
- φ is a bijection.

A chart will be often denoted by (U, φ) and with $\widehat{U}$ we always indicate the range of φ. Then U is called a *coordinate domain* and φ the *coordinate map*. The coordinate map is a vector valued map with components usually denoted as $x^1, \ldots, x^n$ (or $y^1, \ldots, y^n$, etc.) and called *coordinates* on U determined by the chart (U, φ). Accordingly, we will often write $(U, \varphi = (x^1, \ldots, x^n))$.

A chart on M *around a point* $p \in M$ is a chart (U, φ) such that $U \ni p$. A chart (U, φ) around a point $p \in M$ is *centered at* p if $\widehat{U} = \varphi(U) \ni 0$ and $\varphi(p) = 0$.

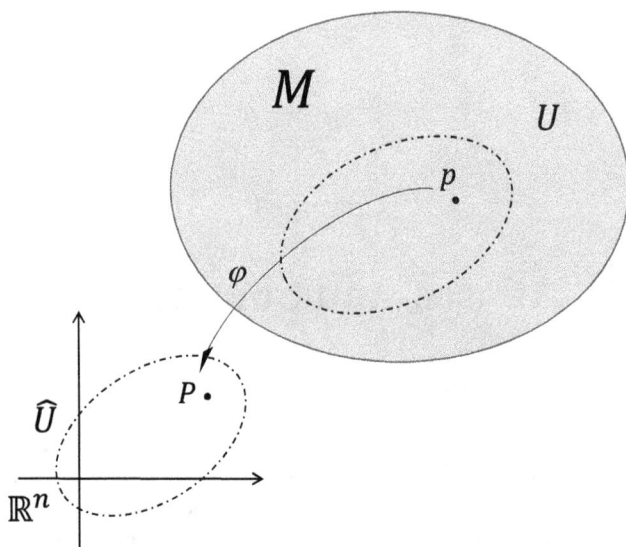

Figure 1.1.   A chart.

See Figure 1.1 for a pictorial representation of a chart. We now make some examples of charts on sets.

Example 1.2 (The Standard Chart on $\mathbb{R}^n$). The identity map id : $\mathbb{R}^n \to \mathbb{R}^n$ is an n-dimensional chart on $\mathbb{R}^n$ whose associated coordinates are the standard coordinates $t^1, \ldots, t^n$. The chart $(\mathbb{R}^n, \mathrm{id})$ is also called the *standard chart* on $\mathbb{R}^n$. The standard chart on $\mathbb{R}^n$ is centered at 0.   ♦

Example 1.3. Example 1.2 can be slightly generalized considering any n-dimensional real vector space V and any frame $\mathcal{R} = (e_1, \ldots, e_n)$ of V. Recall that the *coordinate map* associated with $\mathcal{R}$ is the vector space isomorphism

$$\varphi_{\mathcal{R}} : V \to \mathbb{R}^n, \quad v \mapsto \varphi_{\mathcal{R}}(v) := (x^1(v), \ldots, x^n(v))$$

mapping a vector v to its components $(x^1(v), \ldots, x^n(v))$ in the frame $\mathcal{R}$. In their turn, $(x^1(v), \ldots, x^n(v))$ are implicitly given by $v = \sum_{i=1}^{n} x^i(v)e_i$. It is clear that $(V, \varphi_{\mathcal{R}})$ is an n-dimensional chart on V centered at 0, for any frame $\mathcal{R}$.   ♦

Example 1.4 (The Stereographic Charts on the Sphere). Consider the $(n+1)$-dimensional standard Euclidean space $\mathbb{R}^{n+1}$ with standard coordinates $t^1, \ldots, t^{n+1}$. The *n-dimensional sphere* is the subset $S^n \subseteq \mathbb{R}^{n+1}$

defined by

$$S^n = \left\{ P \in \mathbb{R}^{n+1} : \|P\| = 1 \right\},$$

where $\|-\|$ is the standard *Euclidean norm*: for $P = (P^1, \ldots, P^{n+1})$,

$$\|P\| := \sqrt{(P^1)^2 + \cdots + (P^{n+1})^2}.$$

We now define two charts on S^n: the *stereographic charts*. To do this, we preliminarily denote by

$$P_+ := (0, \ldots, 0, 1) \quad \text{and} \quad P_- = (0, \ldots, 0, -1)$$

the intersections of the sphere with the t^{n+1}-axis and call them the *north pole* and the *south pole*, respectively. We also denote

$$U_+ := S^n \smallsetminus \{P_+\} \quad \text{and} \quad U_- := S^n \smallsetminus \{P_-\}.$$

For any point $P \in U_+$, the line through P and P_+ intersects the hyperplane $t^{n+1} = 0$ in a point with coordinates denoted $(X_+^1(P), \ldots, X_+^n(P), 0)$. The map

$$\varphi_+ : U_+ \to \mathbb{R}^n, \quad P \mapsto \varphi_+(P) := (X_+^1(P), \ldots, X_+^n(P))$$

is called the *stereographic projection* (from the north, see Figure 1.2).

It is easy to see that (U_+, φ_+) is an n-dimensional chart on S^n (see Exercise 1.1) called the *stereographic chart* (from the north). Replacing P_+ with P_- and U_+ with U_-, we can define another chart (U_-, φ_-) on S^n:

$$\varphi_- : U_- \to \mathbb{R}^n, \quad P \mapsto \varphi_-(P) := (X_-^1(P), \ldots, X_-^n(P)).$$

The coordinate map φ_- is called the *stereographic projection from the south*, and the chart (U_-, φ_-) is the *stereographic chart from the south*. Note that the stereographic chart from the north (U_+, φ_+) is centered at the south pole P_- and, similarly, the stereographic chart from the south (U_-, φ_-) is centered at the north pole P_+. ◆

Exercise 1.1. Consider the stereographic charts

$$(U_+, \varphi_+ = (X_+^1, \ldots, X_+^n)) \quad \text{and} \quad (U_-, \varphi_- = (X_-^1, \ldots, X_-^n))$$

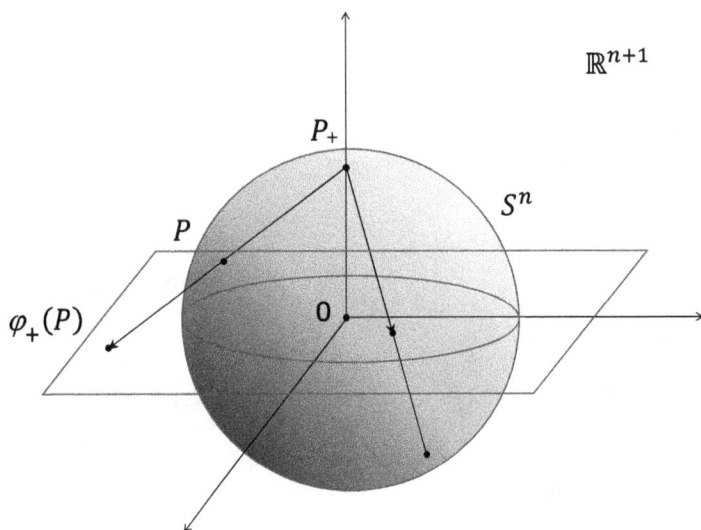

Figure 1.2. Stereographic projection from the north.

on the sphere, from Example 1.4. Show that they are indeed charts and that the stereographic projections are explicitly given by

$$X^i_+(P) = \frac{P^i}{1 - P^{n+1}} \quad \text{and} \quad X^i_-(P) = \frac{P^i}{1 + P^{n+1}},$$

for $P = (P^1, \ldots, P^{n+1}) \in S^n$.

Example 1.5 (More Charts on the Sphere). In this example, we define $2(n+1)$ more charts on S^n. We begin considering the *standard n-dimensional open disk*, i.e., the open subset $D^n \subseteq \mathbb{R}^n$ defined by

$$D^n := \{P \in \mathbb{R}^n : \|P\| < 1\}.$$

Now, for all $i = 1, \ldots, n+1$, let $U_{i,\pm} \subseteq S^n$ be the subset defined by

$$U_{i,\pm} := \left\{(P^1, \ldots, P^{n+1}) \in S^n : \pm P^i > 0\right\}.$$

Consider the *orthogonal projection* (onto the hyperplane $t^i = 0$)

$$\pi_i : \mathbb{R}^{n+1} \to \mathbb{R}^n,$$

$$(P^1, \ldots, P^n) \mapsto \pi_i(P^1, \ldots, P^n) := (P^1, \ldots, \widehat{P^i}, \ldots, P^{n+1}),$$

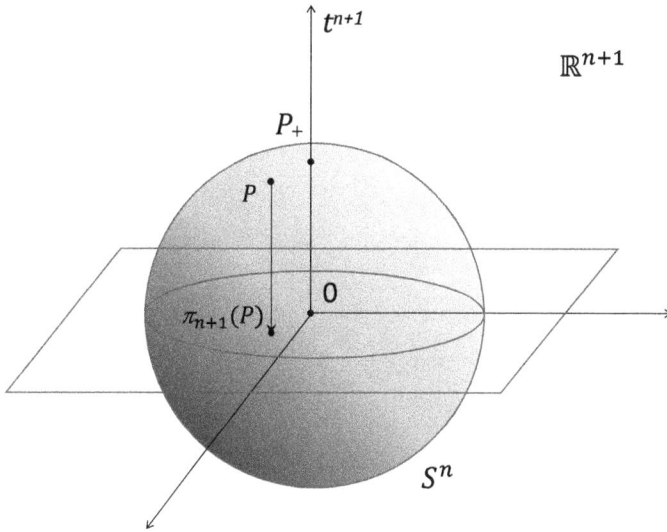

Figure 1.3. Orthogonal projection on the hyperplane $t^{n+1} = 0$.

where a hat "$\widehat{}$" denotes omission (see Figure 1.3 for the case $(U_{n+1,+}, \pi_{n+1})$). The projection π_i maps $U_{i,\pm}$ to D^n, hence we can consider the restriction:

$$\pi_i : U_{i,\pm} \to D^n.$$

It is clear that the $(U_{i,\pm}, \pi_i)$ are all n-dimensional charts on the sphere. Note that $(U_{i,\pm}, \pi_i)$ is a chart centered at

$$(0, \ldots, \underbrace{\pm 1}_{i\text{-th place}}, \ldots, 0).$$

$\blacklozenge$

Example 1.6 (Affine Charts on the Projective Space). On $\mathbb{R}^{n+1} \smallsetminus \{0\}$ with standard coordinates $t^0, \ldots, t^n$, consider the equivalence relation $\sim$ given by

$$P \sim Q \Leftrightarrow \text{there exists a non-zero real number } \lambda \text{ such that } P = \lambda Q.$$

The *n-dimensional (real) projective space* is the quotient

$$\mathbb{R}P^n := \mathbb{R}^{n+1} \smallsetminus \{0\} / \sim$$

of $\mathbb{R}^{n+1} \smallsetminus \{0\}$ under the equivalence relation $\sim$. Equivalently, $\mathbb{R}P^n$ is the space of lines through the origin in $\mathbb{R}^{n+1}$. The equivalence class $[P]$ of the point $P = (P^0, \ldots, P^n) \in \mathbb{R}^{n+1} \smallsetminus \{0\}$ will be denoted by

$$[P^0 : \cdots : P^n],$$

and the real numbers $P^0, \ldots, P^n$ are called the *homogeneous coordinates* of $[P] \in \mathbb{R}P^n$. In this example, we define n-dimensional charts on $\mathbb{R}P^n$. First of all, for every $i = 0, \ldots, n$, let $U_i \subseteq \mathbb{R}P^n$ be the subset defined by

$$U_i := \Big\{ [P^0 : \cdots : P^n] : P^i \neq 0 \Big\}.$$

The map

$$\varphi_i : U_i \to \mathbb{R}^n, \quad [P] = [P^0 : \cdots : P^n] \mapsto \varphi_i([P]) := \left(\frac{P^0}{P^i}, \ldots, \frac{\widehat{P^i}}{P^i}, \ldots, \frac{P^n}{P^i} \right),$$

is well defined (here, as in Example 1.5, "$\widehat{}$" denotes omission). Additionally, it is invertible with the inverse given by

$$\varphi_i^{-1} : \mathbb{R}^n \to U_i,$$
$$Q = (Q^1, \ldots, Q^n) \mapsto \varphi_i^{-1}(Q) := [Q^1 : \cdots : \underbrace{1}_{i\text{-th place}} : \cdots : Q^n].$$

Hence, the (U_i, φ_i) are all n-dimensional charts on the n-dimensional projective space. They are called the *affine charts* on $\mathbb{R}P^n$. The i-th affine chart (U_i, φ_i) is centered at

$$[0 : \cdots : \underbrace{1}_{i\text{-th place}} : \cdots : 0].$$

◆

Given an n-dimensional chart (U, φ) on a set M, we can use the coordinate map φ to identify the coordinate domain U with $\widehat{U} := \varphi(U)$. In turn, we can use this identification to transfer notions in calculus (smoothness, derivatives, etc.) from $\widehat{U}$ to U. For instance, we could declare that a real function $f : U \to \mathbb{R}$ is smooth at a point $p \in U$ if the composition $f \circ \varphi^{-1} : \widehat{U} \to \mathbb{R}$ is smooth at $\varphi(p)$. If we want to do the same for the whole M, we need to have, at least, a family of charts $\mathcal{A} = \{(U, \varphi)\}$ *covering* M in the sense that $M = \bigcup_{(U,\varphi) \in \mathcal{A}} U$. Suppose we have such a family $\mathcal{A}$, let $f : M \to \mathbb{R}$ be a (real valued) function on M, and let $p \in M$. Then, it is natural to declare that f is smooth at p if the composition $f \circ \varphi^{-1} : U \to \mathbb{R}$ is smooth, for some chart $(U, \varphi) \in \mathcal{A}$ around p. However, the latter condition may depend on (U, φ), unless $\mathcal{A}$ satisfies suitable additional conditions.

The above discussion motivates the following two definitions.

Definition 1.7 (Compatible Charts). Two n-dimensional charts $(U, \varphi = (x^1, \ldots, x^n)), (V, \psi = (y^1, \ldots, y^n))$ on a set M are *compatible* if either

- $U \cap V = \emptyset$ or
- $U \cap V \neq \emptyset$ and, additionally, the following two conditions are satisfied:

    (1) $\varphi(U \cap V) \subseteq \hat{U}$ and $\psi(U \cap V) \subseteq \hat{V}$ are (non-empty) open subsets and
    (2) $\psi \circ \varphi^{-1} : \varphi(U \cap V) \to \psi(U \cap V)$ is a diffeomorphism (hence $\varphi \circ \psi^{-1} : \psi(U \cap V) \to \varphi(U \cap V)$ is a diffeomorphism as well).

The map $\psi \circ \varphi^{-1} : \varphi(U \cap V) \to \psi(U \cap V)$ is called the *transition map* between the charts (U, φ) and (V, ψ) (or between the coordinates $(x^1, \ldots, x^n)$ and $(y^1, \ldots, y^n)$, see Figure 1.4).

Note that the compatibility between charts is a reflexive and symmetric relation, but it is not transitive (can you provide a counter-example?).

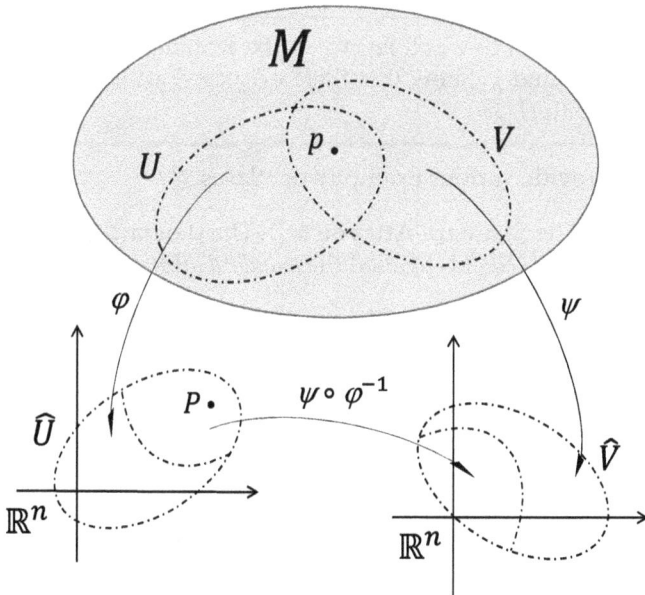

Figure 1.4.   Compatible charts.

Definition 1.8 (Atlas). An *n-dimensional smooth atlas* (or, simply, an *atlas*) on a set M is a family $\mathcal{A} = \{(U, \varphi)\}$ of n-dimensional charts such that

- $\mathcal{A}$ *covers* M, i.e., $M = \bigcup_{(U,\varphi) \in \mathcal{A}} U$ and
- any two charts in $\mathcal{A}$ are compatible.

Before providing examples of atlases, we want to stress that there are various ways to produce a compatible chart from a given chart.

Example 1.9 (Subcharts). Let M be a set and let (U, φ) be a chart on M. Choose any non-empty open subset $\widehat{V} \subseteq \widehat{U}$ in $\widehat{U} = \varphi(U)$ and put $V := \varphi^{-1}(\widehat{V})$. In particular, we can restrict φ to obtain a new map $\varphi : V \to \widehat{V}$. It is clear that (V, φ) is again a chart, called a *subchart* of (U, φ). Additionally, the transition map between (U, φ) and (V, φ) is simply the identity id : $\widehat{V} \to \widehat{V}$. Hence any subchart of (U, φ) is compatible with (U, φ). ◆

Example 1.10. Let M be a set and let (U, φ) be an n-dimensional chart on M. Choose any diffeomorphism $\Psi : \widehat{U} \to \widehat{W}$ onto an(other) open subset $\widehat{W} \subseteq \mathbb{R}^n$. The composition $\Psi \circ \varphi : U \to \widehat{W}$ is bijective, hence $(U, \Psi \circ \varphi)$ is a chart. Additionally, the transition map between (U, φ) and $(U, \Psi \circ \varphi)$ is precisely $\Psi : \widehat{U} \to \widehat{W}$. Hence (U, φ) and $(U, \Psi \circ \varphi)$ are compatible. ◆

> **Exercise 1.2.** Let M be a set, let $p \in M$ be a point, and let (U, φ) be a chart on M around p. Show that there exists a chart centered at p and compatible with (U, φ).

We now provide various examples of atlases.

Example 1.11 (The Standard Atlas on $\mathbb{R}^n$). The standard chart $(\mathbb{R}^n, \mathrm{id})$ on $\mathbb{R}^n$ forms an atlas $\{(\mathbb{R}^n, \mathrm{id})\}$ called the *standard atlas* of $\mathbb{R}^n$. ◆

Example 1.12. Let V be an n-dimensional real vector space and let $\mathcal{R}$ be a frame of V. The chart $(V, \varphi_\mathcal{R})$ from Example 1.3 forms an atlas $\{(V, \varphi_\mathcal{R})\}$. ◆

Example 1.13 (Two Atlases on the Sphere). Consider the n-dimensional sphere $S^n \subseteq \mathbb{R}^{n+1}$. The stereographic projections (U_+, φ_+), (U_-, φ_-) (Example 1.4) cover S^n and are compatible (see Exercise 1.3). Hence they form an atlas on the sphere, called the *stereographic atlas*.

Similarly, the charts $(U_{i,\pm}, \pi_i)$ from Example 1.5 cover S^n and are pairwise compatible. Hence they form another atlas $\{(U_{i,\pm}, \pi_i)\}_{i=1,\ldots,n+1}$ on the sphere. Finally, note that the charts $(U_{i,\pm}, \pi_i)$ are also compatible with the stereographic projections (Exercise 1.3). ◆

Exercise 1.3. Consider the n-dimensional sphere $S^n \subseteq \mathbb{R}^{n+1}$. Show that

(1) the stereographic charts (Example 1.4) form an n-dimensional atlas,
(2) the charts $(U_{i,\pm}, \pi_i)$ from Example 1.5 form an n-dimensional atlas,
(3) the charts $(U_{i,\pm}, \pi_i)$ are compatible with the stereographic projections.

Example 1.14 (The Affine Atlas on the Projective Space). Consider the n-dimensional projective space $\mathbb{R}P^n$. The affine charts (U_i, φ_i) (Example 1.6) cover $\mathbb{R}P^n$ and are compatible (see Exercise 1.4 below). Hence they form an atlas, called the *affine atlas* on the projective space. ◆

Exercise 1.4. Consider the n-dimensional projective space $\mathbb{R}P^n$. Show that the affine charts (Example 1.6) form an n-dimensional atlas on it.

We will show in the following chapter that the notion of *smoothness* of a function $f : M \to \mathbb{R}$ (at a point $p \in M$) is independent of the choice of a chart (around p) in a fixed atlas. However, it may still depend on the atlas, unless we restrict ourselves to a class of *compatible atlases*.

Definition 1.15 (Compatible Atlases). Two atlases $\mathcal{A}, \mathcal{B}$ on a set M are *compatible* if the charts of $\mathcal{A}$ are compatible with the charts of $\mathcal{B}$ or, equivalently, if $\mathcal{A} \cup \mathcal{B}$ is an atlas as well.

Example 1.16. Let V be an n-dimensional real vector space and let $\mathcal{R}, \mathcal{S}$ be frames of V. The charts $(V, \varphi_{\mathcal{R}})$ and $(V, \varphi_{\mathcal{S}})$ are compatible. Indeed the transition map is

$$\varphi_{\mathcal{S}} \circ \varphi_{\mathcal{R}}^{-1} : \mathbb{R}^n \to \mathbb{R}^n$$

which is a vector space isomorphism, hence a diffeomorphism. It follows that the atlases

$$\{(V, \varphi_{\mathcal{R}})\} \quad \text{and} \quad \{(V, \varphi_{\mathcal{S}})\}$$

are compatible. ◆

Example 1.17. According to Exercise 1.3, the stereographic atlas and the atlas $\{(U_{i,\pm}, \pi_i)\}_{i=1,\ldots,n+1}$ consisting of the orthogonal projections are compatible atlases on the n-dimensional sphere. ◆

Proposition 1.18. *The compatibility between atlases on a set is an equivalence relation.*

Proof.   The compatibility between atlases is clearly reflexive and symmetric. It remains to prove that it is transitive. So, let $\mathcal{A}, \mathcal{B}, \mathcal{C}$ be atlases on a set M and suppose that $\mathcal{A}$ is compatible with $\mathcal{B}$, and $\mathcal{B}$ is compatible with $\mathcal{C}$. We want to show that $\mathcal{A}$ is compatible with $\mathcal{C}$. To do this, consider a chart (U, φ) in $\mathcal{A}$ and a chart (W, χ) in $\mathcal{C}$ and show that they are compatible. If $U \cap W = \varnothing$, there is nothing to prove. So let $U \cap W \neq \varnothing$ and look at the transition map:

$$\chi \circ \varphi^{-1} : \varphi(U \cap W) \to \chi(U \cap W). \tag{1.1}$$

We remark, preliminarily, that (1.1) is invertible. Now, $\varphi(U \cap W) \subseteq \widehat{U} = \varphi(U)$ is an open subset. Indeed, let $P \in \varphi(U \cap W)$, and let $p = \varphi^{-1}(P)$. As $\mathcal{B}$ is an atlas, there is a chart (V, ψ) in $\mathcal{B}$ around p. Additionally, (V, ψ) and (U, φ) are compatible, meaning that

(1) $\psi(U \cap V) \subseteq \widehat{V}$ and $\varphi(U \cap V) \subseteq \widehat{U}$ are open subsets and
(2) the transition map

$$\varphi \circ \psi^{-1} : \psi(U \cap V) \to \varphi(U \cap V)$$

is a diffeomorphism.

But (V, ψ) and (W, χ) are also compatible. In particular, $\psi(V \cap W) \subseteq \widehat{V}$ is an open subset as well. As the intersection of two open subsets is an open subset, it follows that

$$\psi(U \cap V) \cap \psi(V \cap W) = \psi(U \cap V \cap W) \subseteq \widehat{V}$$

is an open subset. Hence

$$\varphi(U \cap V \cap W) = (\varphi \circ \psi^{-1})(\psi(U \cap V \cap W)) \subseteq \widehat{U}$$

is an open subset. More precisely, it is an open neighborhood of P in $\varphi(U \cap W)$. It follows from the arbitrariness of P that $\varphi(U \cap W) \subseteq \widehat{U}$ is an open subset (see Figure 1.5). Exchanging the roles of (U, φ) and (W, χ), we immediately see that $\chi(U \cap W) \subseteq \widehat{W}$ is also an open subset.

Next, we show that the transition map (1.1) is smooth. It is enough to show that it is *smooth around every point*, i.e., for every $P \in \varphi(U \cap W)$, there

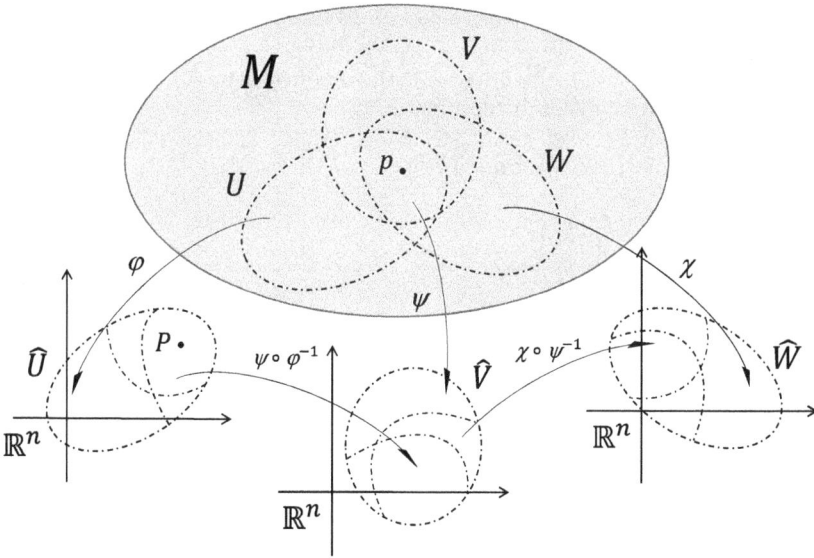

Figure 1.5. Compatibility of atlases is transitive.

exists an open neighborhood U_0 of P in $\varphi(U \cap W)$ such that the restriction

$$(\chi \circ \varphi^{-1})|_{U_0} : U_0 \to \chi(U \cap W)$$

is smooth. This is indeed the case, e.g., we can choose $U_0 = \varphi(U \cap V \cap W)$, where (V, ψ) is as in the first part of the proof. Then

$$(\chi \circ \varphi^{-1})|_{U_0} = (\chi \circ \psi^{-1}) \circ (\psi \circ \varphi^{-1})|_{U_0},$$

which is the smooth composition of two smooth maps. Finally, exchanging again the roles of (U, φ) and (W, χ) reveals that $\varphi \circ \chi^{-1}$ is also smooth around every point, hence it is smooth. This concludes the proof. □

Given an atlas $\mathcal{A}$ on a set M, it is often useful to consider all charts on M that are compatible with all charts in $\mathcal{A}$. With this in mind, we give the following

Definition 1.19 (Smooth Structure). An atlas $\mathcal{A}$ on a set M is *maximal* if it is not properly contained in any other (compatible) atlas or, equivalently, if it already contains all charts compatible with its charts. A maximal n-dimensional atlas on a set M is also called an n-dimensional *smooth structure* on M.

Yet in other words, an atlas $\mathcal{A}$ is maximal if, whenever a chart (U, φ) is compatible with all the charts in $\mathcal{A}$, we have $(U, \varphi) \in \mathcal{A}$. For instance, a maximal atlas contains all subcharts of its charts, and all charts obtained composing a chart with a diffeomorphism.

Proposition 1.20. *Let M be a set. Then*

(1) *every atlas $\mathcal{A}$ on M is contained in a unique maximal atlas $\mathcal{A}_{max}$, necessarily compatible with $\mathcal{A}$,*
(2) *two atlases are contained in the same maximal atlas if and only if they are compatible.*

Proof. Left as Exercise 1.5. □

Exercise 1.5. Prove Proposition 1.20. (**Hint:** *For the first part define*

$\mathcal{A}_{max} = \{\text{charts } (U, \varphi) \text{ on } M : (U, \varphi) \text{ is compatible with every chart in } \mathcal{A}\}.$

For the second part, use Proposition 1.18.)

For instance, the charts on a real vector space determined by frames are all in the same maximal atlas. Similarly, the stereographic atlas on the sphere and the atlas of orthogonal projections are contained in the same maximal atlas.

Exercise 1.6 (The Standard Smooth Structure on $\mathbb{R}^n$). Consider the standard one-chart atlas $\{(\mathbb{R}^n, \mathrm{id})\}$ on $\mathbb{R}^n$. According to Proposition 1.20, it is contained in a unique maximal atlas $\mathcal{A}_{can}$ called the *standard smooth structure* on $\mathbb{R}^n$. Show that $\mathcal{A}_{can}$ consists of charts (U, φ), where $U \subseteq \mathbb{R}^n$ is any open subset and $\varphi : U \to \widehat{U}$ is any diffeomorphism between open subsets of $\mathbb{R}^n$.

An atlas $\mathcal{A}$ on a set M determines a distinguished family of subsets of M according to the following

Definition 1.21 (Open Subset with Respect to an Atlas). A subset $\mathcal{U} \subseteq M$ is $\mathcal{A}$-open (or, simply, *open* if this does not lead to confusion) if, for every point $p \in \mathcal{U}$, there exists a chart $(U, \varphi) \in \mathcal{A}$ around p, such that $\varphi(\mathcal{U} \cap U) \subseteq \widehat{U}$ is an open subset.

Proposition 1.22. *A subset $\mathcal{U} \subseteq M$ is $\mathcal{A}$-open if and only if, for every chart $(V, \psi) \in \mathcal{A}$, $\psi(\mathcal{U} \cap V) \subseteq \widehat{V}$ is an open subset.*

Proof. The "if" part of the statement is obvious. Let's prove the "only if" part. So let $\mathcal{U} \subseteq M$ be an $\mathcal{A}$-open subset, and let (V, ψ) be a chart in $\mathcal{A}$. We want to show that $\psi(\mathcal{U} \cap V) \subseteq \widehat{V}$ is an open subset. If $\mathcal{U} \cap V = \varnothing$, then $\psi(\mathcal{U} \cap V) = \varnothing \subseteq \widehat{V}$ is an open subset. On the other hand, let $\mathcal{U} \cap V \neq \varnothing$, pick a point $P \in \psi(\mathcal{U} \cap V)$, and let $p = \psi^{-1}(P)$. As $\mathcal{U}$ is open, there is a chart $(U, \varphi) \in \mathcal{A}$ around p such that $\varphi(\mathcal{U} \cap U)$ is open. As (V, ψ) and (U, φ) are in the same atlas, they are compatible, hence the transition map

$$\psi \circ \varphi^{-1} : \varphi(U \cap V) \to \psi(U \cap V)$$

is a diffeomorphism between open subsets of $\mathbb{R}^n$. It follows that

$$\varphi(\mathcal{U} \cap U \cap V) = \varphi(\mathcal{U} \cap U) \cap \varphi(U \cap V) \subseteq \widehat{U}$$

is an open subset, and

$$\psi(\mathcal{U} \cap U \cap V) = (\psi \circ \varphi^{-1})(\varphi(\mathcal{U} \cap U \cap V)) \subseteq \psi(U \cap V)$$

is an open subset as well. From the arbitrariness of P, $\psi(\mathcal{U} \cap V) \subseteq \widehat{V}$ is an open subset. □

We denote by $\tau_{\mathcal{A}}$ the family of $\mathcal{A}$-open subsets in M. The main properties of $\tau_{\mathcal{A}}$ are summarized in the following

Proposition 1.23 (Topology Determined by a Smooth Structure). *Let $\mathcal{A}$ be an atlas on a set M, and let $\tau_{\mathcal{A}}$ be the family of $\mathcal{A}$-open subsets of M. First of all, $\tau_{\mathcal{A}}$ does only depend on the smooth structure containing $\mathcal{A}$. In other words, any two compatible atlases $\mathcal{A}, \mathcal{B}$ determine the same family of open subsets: $\tau_{\mathcal{A}} = \tau_{\mathcal{B}}$. Additionally,*

(1) *$\varnothing, M$ are in $\tau_{\mathcal{A}}$,*
(2) *for every subfamily $\mathcal{F} \subseteq \tau_{\mathcal{A}}$, the union $\bigcup_{\mathcal{U} \in \mathcal{F}} \mathcal{U}$ is in $\tau_{\mathcal{A}}$,*
(3) *for every finitely many $\mathcal{U}_1, \ldots, \mathcal{U}_k \in \tau_{\mathcal{A}}$, the intersection $\mathcal{U}_1 \cap \cdots \cap \mathcal{U}_k$ is in $\tau_{\mathcal{A}}$.*

Exercise 1.7. Prove Proposition 1.23. (**Hint:** *Use Proposition 1.22. For Properties (1), (2), (3), use also the analogous properties enjoyed by the open subsets of $\mathbb{R}^n$.*)

Properties (1), (2), (3) in Proposition 1.23 can be axiomatized to get what is called a *topology* on a set. A set equipped with a topology is called a *topological space,* and the branch of Mathematics studying topological spaces is *Topology.* In the following section, we briefly recall those fundamentals of Topology that will be useful in the rest of these notes.

1.2   Topological Spaces

Let X be a set.

Definition 1.24 (Topological Space). A *topology* on X is a family $\tau = \{\mathcal{U}\}$ of subsets of X such that

(1) $\varnothing, X \in \tau$,
(2) for every subfamily $\mathcal{F} \subseteq \tau$, the union $\bigcup_{\mathcal{U} \in \mathcal{F}} \mathcal{U}$ is in τ,
(3) for every finitely many $\mathcal{U}_1, \ldots, \mathcal{U}_k \in \tau$, the intersection $\mathcal{U}_1 \cap \cdots \cap \mathcal{U}_k$ is in τ.

A set X equipped with a topology τ is a *topological space*. The elements of τ are called *open subsets*. A subset $C \subseteq X$ is *closed* if its complement $X \smallsetminus C$ is open.

Example 1.25 (Atlas Topology). Let $(M, \mathcal{A})$ be a set equipped with an atlas. The family $\tau_{\mathcal{A}}$ of $\mathcal{A}$-open subsets in M is a topology, called the *atlas topology*, which only depends on the smooth structure $\mathcal{A}_{\max}$ containing $\mathcal{A}$: $\tau_{\mathcal{A}} = \tau_{\mathcal{A}_{\max}}$. Accordingly, M is a topological space, whose open subsets are $\mathcal{A}$-open subsets. In what follows, we will call them just open subsets if there is no risk of confusion. ♦

Example 1.26. The open subsets in $\mathbb{R}^n$ form a topology denoted τ_{can} and called the *standard topology* of $\mathbb{R}^n$. On the other hand, the standard atlas $\mathcal{A} = \{(\mathbb{R}^n, \mathrm{id})\}$ induces its own atlas topology $\tau_{\mathcal{A}}$. It is easy to see that, actually, $\tau_{\mathcal{A}} = \tau_{\mathrm{can}}$. Indeed a subset $\mathcal{U} \subseteq \mathbb{R}^n$ is $\mathcal{A}$-open if and only if $\mathrm{id}(\mathcal{U} \cap \mathbb{R}^n)$ is open, i.e., $\mathcal{U} \in \tau_{\mathrm{can}}$. In the following, we will always interpret $\mathbb{R}^n$ as a topological space with the standard topology. ♦

Example 1.27. On any set X, there are at least two canonical topologies:

- the smallest possible one, consisting of $\varnothing$ and X only, called the *concrete topology* and denoted τ_{con},
- the largest possible one, consisting of all subsets in X, called the *discrete topology* and denoted τ_{dis}.

A topological space (X, τ) is *concrete* if $\tau = \tau_{\mathrm{con}}$, and it is *discrete* if $\tau = \tau_{\mathrm{dis}}$. Note that a topological space (X, τ) is discrete if and only if every singleton $\{p\} \subseteq X$, with $p \in X$, is an open subset. It immediately follows from this remark that a 0-dimensional atlas induces the discrete topology. ♦

Remark 1.28. Let (X, τ) be a topological space, and let σ be the family of closed subsets of X. It is easy to see that

(1) $\varnothing, X \in \sigma$,
(2) for every subfamily $\mathcal{G} \subseteq \sigma$, the intersection $\bigcap_{C \in \mathcal{G}} C$ is in σ,
(3) for every finitely many $C_1, \ldots, C_k \in \sigma$, the union $C_1 \cup \cdots \cup C_k$ is in σ.

Moreover a family σ of subsets of a set X enjoys Properties (1), (2), (3), if and only if it is the family of closed subset with respect to some (necessarily unique) topology on X. $\qquad\qquad\diamond$

Every subset $Y \subseteq X$ in a topological space X is a topological space in a canonical way. Namely, define

$$\tau_Y := \{Y \cap \mathcal{U} : \mathcal{U} \subseteq X \text{ is an open subset}\}.$$

Then τ_Y is a topology on Y, called the *subspace topology*. The subset $Y \subseteq X$ equipped with the subspace topology is called a *subspace*. Note that

- if Y itself is open in X, then its subspace topology consists of open subsets of X contained in Y, and, similarly,
- if Y is closed, then closed subsets with respect to the subspace topology are closed subsets of X contained in Y.

Example 1.29. $\mathbb{Z}^n \subseteq \mathbb{R}^n$ is a closed and discrete subspace. Every finite subspace in $\mathbb{R}^n$ is closed and discrete. $\qquad\qquad\blacklozenge$

Example 1.30. Being a subset in $\mathbb{R}^{n+1}$, the n-dimensional sphere S^n is a subspace. Denote by τ_{S^n} the subspace topology. On the other hand, the stereographic atlas $\mathcal{A}$, or, equivalently, the atlas of orthogonal projections, induces its own atlas topology $\tau_{\mathcal{A}}$. We will show in the following section (see Example 1.36) that, actually, $\tau_{S^n} = \tau_{\mathcal{A}}$. $\qquad\qquad\blacklozenge$

The notion of topology allows us to talk about *continuity*. Before giving the main definition, we recall that, given open subsets $\mathcal{U} \subseteq \mathbb{R}^m$ and $\mathcal{V} \subseteq \mathbb{R}^n$, the continuity of a map $F : \mathcal{U} \to \mathcal{V}$ can be expressed purely in terms of the (subspace) topologies of $\mathcal{U}$ and $\mathcal{V}$. Namely, a map $F : \mathcal{U} \to \mathcal{V}$ is continuous if and only if, for every open subset $V \subseteq \mathcal{V}$, the preimage $F^{-1}(V) \subseteq \mathcal{U}$ is an open subset. This suggests the following

Definition 1.31 (Continuous Map). A map $F : X \to Y$ between topological spaces is *continuous* if, for every open subset $V \subseteq Y$, the preimage $F^{-1}(V) \subseteq X$ is an open subset in X.

Let (X, τ_X) and (Y, τ_Y) be topological spaces and let $F : X \to Y$ be a map. We write $F : (X, \tau_X) \to Y$, or $F : X \to (Y, \tau_Y)$, or $F : (X, \tau_X) \to (Y, \tau_Y)$, if we want to insist that X and Y are topological spaces with the topologies τ_X and τ_Y respectively (and not different ones).

For instance, every map $F : X \to (Y, \tau_{\text{con}})$ to a concrete space is continuous. Every map $F : (X, \tau_{\text{dis}}) \to Y$ from a discrete space is continuous. The inclusion $i_A : A \hookrightarrow X$ of a subspace $A \subseteq X$ is also continuous. Indeed, for any subset $B \subseteq X$, the preimage $i_Y^{-1}(B)$ is simply $A \cap B$, so, if B is open, $i_Y^{-1}(B)$ is open (with respect to the subspace topology).

For any topological space (X, τ), the identity map $\text{id} : (X, \tau) \to (X, \tau)$ is continuous. However, if the domain and the codomain are equipped with different topologies, say τ and τ', then $\text{id} : (X, \tau) \to (X, \tau')$ needs not to be continuous. For instance, if X possesses more than one element, then the concrete topology τ_{con} and the discrete topology τ_{dis} on X are different, and, while the map $\text{id} : (X, \tau_{\text{dis}}) \to (X, \tau_{\text{con}})$ is continuous, the map $\text{id} : (X, \tau_{\text{con}}) \to (X, \tau_{\text{dis}})$ is not. Finally, it is easy to see that the composition of continuous maps is continuous. It follows that the restriction $F : A \to Y$ of a continuous map $F : X \to Y$ to a subspace $A \subseteq X$ is also continuous. Indeed, it is the composition of the continuous map $i_A : A \hookrightarrow X$ followed by $F : X \to Y$.

Example 1.32. Let $G : X \to Y$ be a (not necessarily continuous) map between topological spaces, let $B \subseteq Y$ be a subspace, and suppose that G takes values in B. In particular, we can consider the restriction $G : X \to B$ of $G : X \to Y$ to B in the codomain, and $G : X \to B$ is continuous if and only if so is $G : X \to Y$. Indeed, let $G : X \to B$ be continuous, then $G : X \to Y$ is continuous as well because it is the composition of the continuous map $G : X \to B$ followed by the inclusion $i_B : B \hookrightarrow Y$. Conversely, let $G : X \to Y$ be continuous, and let $U \subseteq B$ be an open subset. To avoid confusion, denote $G_B : X \to B$ the restriction of G. By definition of subspace topology, there is an open subset $\mathcal{U} \subseteq Y$ such that $U = \mathcal{U} \cap B = i_B^{-1}(\mathcal{U})$. Hence

$$G_B^{-1}(U) = G_B^{-1}(i_B^{-1}(\mathcal{U})) = (i_B \circ G_B)^{-1}(\mathcal{U}) = G^{-1}(\mathcal{U})$$

which is open because $G : X \to Y$ is continuous.   ◆

Definition 1.33 (Homeomorphism). A map $\Phi : X \to Y$ between topological spaces is a *homeomorphism* if it is continuous, bijective, and Φ^{-1} is also continuous. Two topological spaces X, Y are *homeomorphic* if there is a homeomorphism $\Phi : X \to Y$ connecting them.

Note that, as the example provided by the maps $\text{id} : (X, \tau_{\text{dis}}) \to (X, \tau_{\text{con}})$ and $\text{id} : (X, \tau_{\text{con}}) \to (X, \tau_{\text{dis}})$ shows, a continuous, bijective map needs not to be a homeomorphism, hence we cannot remove the last condition from the definition. On the other hand, the inverse of a homeomorphism is clearly always a homeomorphism. The identity $\text{id} : (X, \tau) \to$

(X, τ) is a homeomorphism and the composition $\Psi \circ \Phi : X \to Z$ of homeomorphisms $\Phi : X \to Y$ and $\Psi : Y \to Z$ is a homeomorphism whose inverse homeomorphism is $\Phi^{-1} \circ \Psi^{-1} : Z \to X$. Finally, diffeomorphisms between open subsets of $\mathbb{R}^n$ are, in particular, homeomorphisms.

Let $\Phi : X \to Y$ be a homeomorphism between topological spaces. It immediately follows from the definition that a subset $\mathcal{U} \subseteq X$ is open if and only if $\Phi(\mathcal{U}) \subseteq Y$ is an open subset. In other words, the homeomorphism Φ identify both X and Y and their topologies. So, homeomorphic topological spaces should be regarded as "the same topological space".

The notion of topology is very general, and a generic topological space may be rather wild. For this reason, one usually considers topological spaces with particularly nice properties only. In this section, we define *Hausdorff* and *II-countable* topological spaces. Before explaining these notions, we stress that we will not use them much in these notes, however, we briefly discuss them here for completeness, especially because they are technically unavoidable to prove important, more advanced properties of *smooth manifolds*.

We begin with the Hausdorff property. Let X be a topological space, and let $p \in X$ be a point. First of all, an *open neighborhood* of p is an open subset $U \subseteq X$ containing p. Now, X is a *Hausdorff space* if any two different points $p, q \in X$ are *separated* by open subsets, i.e., there exist an open neighborhood U of p and an open neighborhood V of q such that $U \cap V = \varnothing$. For instance, $\mathbb{R}^n$ is a Hausdorff space, and any discrete space is Hausdorff. Every subspace of a Hausdorff space is Hausdorff (see Exercise 1.8), so the n-dimensional sphere, with the subspace topology, is a Hausdorff space.

> **Exercise 1.8.** Show that every subspace of a Hausdorff space is Hausdorff.

A Hausdorff space enjoys the nice property that its one-point subspaces are closed. Indeed let X be a Hausdorff space, and let $p \in X$ be a point. We want to show that $X \smallsetminus \{p\}$ is open. To do this, take $q \in X \smallsetminus \{p\}$, i.e., $q \neq p$. From the Hausdorff property, there is, in particular, an open neighborhood V_q of q not containing p, i.e., $V_q \subseteq X \smallsetminus \{p\}$. Hence

$$X \smallsetminus \{p\} = \bigcup_{q \in X \smallsetminus \{p\}} V_q$$

is open.

Not all topological spaces are Hausdorff. For instance, a concrete space with more than one element is *not* Hausdorff.

We now come to II-countability. First, we need to discuss the new notion of *basis for a topology*. Let (X, τ) be a topological space. A *basis* for the topology τ is a subfamily $\mathcal{B} \subseteq \tau$ such that every open subset can be written as a union of elements of $\mathcal{B}$. For instance, the open disks form a basis for the standard topology of $\mathbb{R}^n$. The open disks with rational centers and radii do also form a basis. Finally, *multi-rectangles*

$$I_1 \times \cdots \times I_n \in \mathbb{R}^n,$$

with $I_1, \ldots, I_n \subseteq \mathbb{R}$ open intervals, do also form a basis (and we can even restrict to multi-rectangles with rational vertices). Note that any basis for a discrete topology should contain all one-point subsets.

Let (X, τ) be a topological space, and let $\mathcal{B}$ be a basis for τ. It immediately follows from the definition that $\mathcal{B}$ is an *open cover* of X, i.e., it consists of open subsets covering X: $X = \cup_{\mathcal{U} \in \mathcal{B}} \mathcal{U}$. Additionally, for all $\mathcal{U}, \mathcal{V} \in \mathcal{B}$, the intersection $\mathcal{U} \cap \mathcal{V}$ is a union of open subsets in $\mathcal{B}$. Conversely, if X is a set, and $\mathcal{B}$ is a family of subsets covering X and such that for all $\mathcal{U}, \mathcal{V} \in \mathcal{B}$, the intersection $\mathcal{U} \cap \mathcal{V}$ is a union of subsets in $\mathcal{B}$, then $\mathcal{B}$ is a basis for a, necessarily unique, topology $\tau_{\mathcal{B}}$. Specifically, $\tau_{\mathcal{B}}$ consists of subsets of the form

$$\bigcup_{\mathcal{U} \in \mathcal{F}} \mathcal{U}$$

for some subfamily $\mathcal{F} \subseteq \mathcal{B}$. The proof of this fact is easy and it is left as Exercise 1.9.

Exercise 1.9. Let X be a set, and let $\mathcal{B}$ be a family of subsets covering X and such that for all $\mathcal{U}, \mathcal{V} \in \mathcal{B}$, the intersection $\mathcal{U} \cap \mathcal{V}$ is a union of subsets in $\mathcal{B}$. Show that

$$\tau_{\mathcal{B}} := \{\cup_{\mathcal{U} \in \mathcal{F}} \mathcal{U} : \mathcal{F} \subseteq \mathcal{B}\}$$

is a topology and $\mathcal{B}$ is a basis for it.

Given a topological space (X, τ), it is often useful to have a basis $\mathcal{B}$ for the topology τ. For instance, a map $F : Y \to X$ from another topological space Y is continuous provided only $F^{-1}(B) \subseteq Y$ is open for any $B \in \mathcal{B}$ (show it as an exercise!).

A topological space (X, τ) is *II-countable* if there is a countable basis for the topology τ. For instance, $\mathbb{R}^n$ is II-countable (open disks with rational centers and radii are countably many), and any concrete space is II-countable. Every subspace of a II-countable space is II-countable (see Exercise 1.10), so the n-dimensional sphere, with the subspace topology, is II-countable.

> **Exercise 1.10.** Show that every subspace of a II-countable space is II-countable. (**Hint:** *Let* (X, τ) *be a topological space and let* $Y \subseteq X$ *be a subspace. Denote by* τ_Y *the subspace topology of* Y. *Show, preliminarily, that, given a basis* $\mathcal{B}$ *for* τ, *the family of subsets* $\mathcal{B}_Y := \{B \cap Y : B \in \mathcal{B}\}$ *is a basis for* τ_Y. *Now, choose* $\mathcal{B}$ *to be countable.*)

Not all topological spaces are II-countable. For instance, a discrete space with uncountably many points is not II-countable.

We conclude this section discussing *products of topological spaces*. Let $X_1, \ldots, X_k$ be topological spaces. In the Cartesian product $X_1 \times \cdots \times X_k$, consider the following family of subsets:

$$\mathcal{B} := \{\mathcal{U}_1 \times \cdots \times \mathcal{U}_k : \mathcal{U}_i \subseteq X_i \text{ is an open subset for all } i = 1, \ldots, k\}.$$

It is easy to see that $\mathcal{B}$ enjoys the *properties of the basis*, i.e., the properties stated in Exercise 1.9 (show it as an exercise). It follows that $\mathcal{B}$ is a basis for a topology on $X_1 \times \cdots \times X_k$ called the *product topology*. The Cartesian product $X_1 \times \cdots \times X_k$ equipped with the product topology is called the *product space*. Note that, if, for all i, $\mathcal{B}_i$ is a basis for the topology of X_i, then

$$\mathcal{B}^\times := \{B_1 \times \cdots \times B_k : B_i \in \mathcal{B}_i \text{ for all } i = 1, \ldots, k\} \tag{1.2}$$

is a basis for the product topology (can you show it?).

The projections onto the factors

$$\mathrm{pr}_i : X_1 \times \cdots \times X_k \to X_i,$$

$$(p_1, \ldots, p_k) \mapsto \mathrm{pr}_i(p_1, \ldots, p_k) := p_i, \quad i = 1, \ldots, k$$

are continuous map. Indeed, if $\mathcal{U} \subseteq X_i$ is an open subset, then

$$\mathrm{pr}_i^{-1}(\mathcal{U}) = X_1 \times \cdots \times \underbrace{\mathcal{U}}_{i\text{-th place}} \times \cdots \times X_k$$

which is an open subset in the product.

Example 1.34 (The Standard Topology on $\mathbb{R}^n$ is the Product Topology). Let $n, n_1, \ldots, n_k$ be non-negative integers such that $n_1 + \cdots + n_k = n$. The map

$$\Phi : \mathbb{R}^{n_1} \times \cdots \times \mathbb{R}^{n_k} \to \mathbb{R}^n$$

given by

$$((P^1, \ldots, P^{n_1}), \ldots, (P^{n-n_k+1}, \ldots, P^n)) \mapsto (P^1, \ldots, P^n)$$

is a homeomorphism. Indeed, first remember that the open multi-rectangles form a basis for the topology of $\mathbb{R}^k$ (for all k). Hence products of open multi-rectangles form a basis for the topology of $\mathbb{R}^{n_1} \times \cdots \times \mathbb{R}^{n_k}$. Now, the preimage under Φ of an open multi-rectangle in $\mathbb{R}^n$ is the product of open multi-rectangles, showing that Φ is continuous. Additionally, Φ is clearly invertible, and the preimage under Φ^{-1} of a product of multi-rectangles is a multi-rectangle, showing that Φ^{-1} is continuous as well. In the following, we will often use Φ to identify the topological spaces $\mathbb{R}^{n_1} \times \cdots \times \mathbb{R}^{n_k}$ and $\mathbb{R}^n$. ◆

Exercise 1.11. Let $X, X_1, \ldots, X_k$ be topological spaces. Consider a map

$$F : X \to X_1 \times \cdots \times X_k.$$

The i-th *component* of F is, by definition, the composition $F_i = \mathrm{pr}_i \circ F : X \to X_i$ of F followed by the projection $\mathrm{pr}_i : X_1 \times \cdots \times X_k \to X_i$ onto the i-th factor, $i = 1, \ldots, k$. Prove that F is continuous if and only if so are its components.

Exercise 1.12. Let X be a topological space. Consider the product $X \times X$. The *diagonal* is the subspace

$$\Delta := \{(p, p) \in X \times X : p \in X\} \subseteq X \times X.$$

Prove that X is a Hausdorff space if and only if Δ is a closed subspace in $X \times X$.

The product of Hausdorff spaces is Hausdorff, and the product of II-countable spaces is II-countable. Indeed, let $X_1, \ldots, X_k$ be Hausdorff spaces. Take two different points $p = (p_1, \ldots, p_k)$ and $q = (q_1, \ldots, q_k)$ in the product space $X_1 \times \cdots \times X_k$. This means that there exists $i \in \{1, \ldots, k\}$ such that $p_i \neq q_i$. As X_i is Hausdorff, p_i, q_i are separated by open subsets U_i, V_i. But then p, q are separated by the open subsets $\mathrm{pr}_i^{-1}(U_i), \mathrm{pr}_i^{-1}(V_i)$. For the II-countability, it is enough to note that, if $\mathcal{B}_1, \ldots, \mathcal{B}_k$ are countable bases of $X_1, \ldots, X_k$ then $\mathcal{B}^\times$ (see (1.2)) is a countable basis of $X_1 \times \cdots \times X_k$.

We conclude with a short discussion on *compactness* and *connectedness* in Topology. Begin with compactness. A topological space (X, τ) is *compact* if for any open cover $\mathcal{C} = \{\mathcal{U}\}$ of X there are finitely many open subsets $\mathcal{U}_1, \ldots, \mathcal{U}_k \in \mathcal{C}$ such that $\{\mathcal{U}_1, \ldots, \mathcal{U}_k\}$ is again a(n open) cover of X, i.e., $X = \mathcal{U}_1 \cup \cdots \cup \mathcal{U}_k$. A subset Y of a topological space X is *compact*, if it is a compact topological space when equipped with the subspace topology; in other words, for any family $\mathcal{C} = \{\mathcal{U}\}$ of open subsets of X such that $Y \subseteq \bigcup_{\mathcal{U} \in \mathcal{C}} \mathcal{U}$, there are finitely many open subsets $\mathcal{U}_1, \ldots, \mathcal{U}_k \in \mathcal{C}$ such that

$Y \subseteq \mathcal{U}_1 \cup \cdots \cup \mathcal{U}_k$. Concrete spaces and finite discrete spaces are compact. An infinite discrete space is not compact. The Euclidean space $\mathbb{R}^n$ is not compact (cover $\mathbb{R}^n$ with open disks of finite radius and note that there are no finitely many open disks covering it). Actually, one can show that a subspace Y of $\mathbb{R}^n$ is compact if and only if it is closed and *bounded*, i.e., there exists a disk of finite radius containing it. This agrees with the notion of compactness usually covered in standard Calculus classes. For instance, the sphere $S^n \subseteq \mathbb{R}^{n+1}$ is a compact subspace.

> **Exercise 1.13.** Prove that a closed subset in a compact topological space is compact. Prove also that a compact subspace in a Hausdorff space is closed.

Finally, a topological space X is *connected* if the only subsets of X which are simultaneously open and closed are the empty subset $\varnothing$ and X itself. In other words, X is the only non-empty subset which is simultaneously open and closed. A subset Y of a topological space X is *connected* if it is a connected topological space when equipped with the subspace topology. Concrete spaces are connected and a discrete space is connected if and only if it consists of just one point. The Euclidean space $\mathbb{R}^n$ is connected, and every convex subset $S \subseteq \mathbb{R}^n$ is connected, in particular every interval $I \subseteq \mathbb{R}$ is connected.

For much more on Topology and topological spaces, the reader may consult, e.g., Kosniowski (1980), Lee (2011), and Manetti (2015).

1.3 Smooth Manifolds

We begin characterizing the atlas topology of a set equipped with an atlas.

Proposition 1.35 (Properties of the Atlas Topology). *Let $(M, \mathcal{A})$ be a set equipped with an n-dimensional atlas. The atlas topology $\tau_{\mathcal{A}}$ induced on M by $\mathcal{A}$ is the unique topology such that, for every chart $(U, \varphi) \in \mathcal{A}$,*

(1) the coordinate domain $U \subseteq M$ is open and
(2) the coordinate map $\varphi : U \to \widehat{U} \subseteq \mathbb{R}^n$ is a homeomorphism.

Here both U and $\widehat{U}$ are equipped with the subspace topology.

Proof. First of all, we show that the atlas topology $\tau_{\mathcal{A}}$ possesses both Properties (1) and (2) in the statement. Property (1) is clear. For Property (2), we have to show that both $\varphi : U \to \widehat{U}$ and $\varphi^{-1} : \widehat{U} \to U$

are continuous. Equivalently, we have to show that for every open sub-sets $V \subseteq U$ and $\widehat{V} \subseteq \widehat{U}$, both $\varphi^{-1}(\widehat{V})$ and $\varphi(V) = (\varphi^{-1})^{-1}(V)$ are open. So let $\widehat{V} \subseteq \widehat{U}$ be an open subset, and consider $\varphi^{-1}(\widehat{V}) \subseteq U$. We have $\varphi(\varphi^{-1}(\widehat{V}) \cap U) = \varphi(\varphi^{-1}(\widehat{V})) = \widehat{V}$, which is open. This shows that $\varphi^{-1}(\widehat{V}) \in \tau_A$. So it is an open subset of M contained in the open subspace U, hence it is open in the subspace U. Now, let $V \subseteq U$ be an open subset. Being an open subset in an open subspace, it belongs to τ_A. In particular, $\varphi(V \cap U) = \varphi(V) \subseteq \widehat{U}$ is an open subset.

Now we prove uniqueness. Let τ, τ' be topologies on M with both Properties (1) and (2), and prove that $\tau = \tau'$. First, for any chart $(U, \varphi) \in A$, denote by τ_U the subspace topology on U induced by τ and by τ'_U the subspace topology induced by τ'. Take an element $\mathcal{U} \in \tau$. Hence $U \cap \mathcal{U} \in \tau_U$. As $\varphi : (U, \tau_U) \to \widehat{U}$ is a homeomorphism, $\varphi(U \cap \mathcal{U}) \subseteq \widehat{U}$ is an open subset. But $\varphi : (U, \tau'_U) \to \widehat{U}$ is a homeomorphism as well, hence $U \cap \mathcal{U} \in \tau'_U \subseteq \tau'$. Finally,

$$\mathcal{U} = \bigcup_{(U, \varphi) \in A} U \cap \mathcal{U}$$

is the open union of open subsets of (M, τ'). This shows that $\tau \subseteq \tau'$. Exchanging the role of τ and τ', we see that $\tau' \subseteq \tau$, and this concludes the proof.   □

Example 1.36 (The Atlas Topology on the Sphere is the Subspace Topology). Consider the n-dimensional sphere $S^n \subseteq \mathbb{R}^{n+1}$ and its stereographic atlas A. In this example, we use Proposition 1.35 to show that the subspace topology τ_{S^n} and the atlas topology τ_A coincide. According to Proposition 1.35, it is enough to show that τ_{S^n} possesses Properties (1) and (2) in the statement. So, consider the stereographic charts $(U_+, \varphi_+), (U_-, \varphi_-)$. The coordinate domain U_+ is

$$U_+ = S^n \setminus \{P_+\} = S^n \cap (\mathbb{R}^{n+1} \setminus \{P_+\}).$$

As the second factor is open in $\mathbb{R}^n$, it follows that U_+ is open in the subspace topology of S^n and similarly for U_-. For Property (2), note that $\varphi_+ : U_+ \to \mathbb{R}^n$ is the restriction to the subspace U_+ of the continuous map:

$$F : \mathbb{R}^{n+1} \setminus \{t^{n+1} = 1\} \to \mathbb{R}^n$$

$$P = (P^1, \ldots, P^{n+1}) \mapsto F(P) = (F^1(P), \ldots, F^n(P)),$$

where

$$F^i(P) = \frac{P^i}{1 - P^{n+1}}, \quad i = 1, \ldots, n.$$

Hence, φ_+ is continuous with respect to the subspace topology. It is easy to see that its inverse $\varphi_+^{-1} : \mathbb{R}^n \to U_+$ is the restriction to the subspace $U_+ \subseteq \mathbb{R}^{n+1}$ in the codomain of the continuous map

$$G : \mathbb{R}^n \to \mathbb{R}^{n+1}$$

$$Q = (Q^1, \ldots, Q^n) \mapsto G(Q) = (G^1(Q), \ldots, G^{n+1}(Q)),$$

where

$$G^i(Q) = \frac{2Q^i}{\|Q\|^2 + 1}, \quad \text{for } i = 1, \ldots, n,$$

and

$$G^{n+1}(Q) = \frac{\|Q\|^2 - 1}{\|Q\|^2 + 1}.$$

Hence, from Example 1.32, $\varphi_+^{-1} : \mathbb{R}^n \to U_+$ is also continuous with respect to the subspace topology. We conclude that $\varphi_+ : U_+ \to \mathbb{R}^n$ is a homeomorphism with respect to the subspace topology. Similarly for φ_-. Summarizing, τ_{S^n} shares both Properties (1) and (2) in the statement of Proposition 1.35 and coincides with τ_A. It immediately follows that (S^n, τ_A) is a Hausdorff, II-countable topological space. ♦

Having a smooth structure $\mathcal{A}$ on a set M is enough to define a *differential calculus* on M. However, it is often useful in practice to restrict to the case when the atlas topology τ_A is particularly nice. Specifically, it is usually required to (M, τ_A) to be a *Hausdorff* and *II-countable* topological space. The following definition introduces the key object of this chapter and these lecture notes.

Definition 1.37 (Smooth Manifold). An *n-dimensional smooth manifold* is a set M equipped with an n-dimensional smooth structure $\mathcal{A} = \mathcal{A}_{\max}$ such that (M, τ_A) is a Hausdorff and II-countable topological space. The non-negative integer n is also called the *dimension* of M and denoted $\dim M$.

Example 1.38 (Standard Euclidean Spaces and Spheres are Manifolds). The standard Euclidean space $\mathbb{R}^n$ and the n-dimensional sphere are n-dimensional smooth manifolds. ♦

Remark 1.39. There is another (equivalent) definition of a smooth manifold M, where the topology is pre-existing on M rather than induced by an atlas. Such alternative definition involves the notion of a *topological manifold* (see, e.g., Lee, 2013). ◇

Let $(M, \mathcal{A})$ be a set equipped with a smooth structure. From now on, by a *chart* (resp. an *atlas*) on M we will always mean a chart in $\mathcal{A}$ (resp. an atlas contained in $\mathcal{A}$) unless otherwise stated.

Remark 1.40. For every open subset $\mathcal{U} \subseteq M$ and every $p \in \mathcal{U}$, there is a chart (U, φ) around p such that $U \subseteq \mathcal{U}$. In other words, the coordinate domains of a maximal atlas $\mathcal{A}$ form a basis for the topology $\tau_{\mathcal{A}}$. Indeed, let (U', φ) be any chart around p, and put $U := U' \cap \mathcal{U}$. Then U is an open subset contained in U' and (U, φ) is a subchart such that $U \subseteq \mathcal{U}$. But, as maximal atlases contain all subcharts of its charts, we have $(U, \varphi) \in \mathcal{A}$. $\Diamond$

There are simple criteria to check whether $(M, \mathcal{A})$ is a smooth manifold or not, i.e., whether $(M, \tau_{\mathcal{A}})$ is Hausdorff and II-countable or not.

Proposition 1.41. *Let $(M, \mathcal{A})$ be a set equipped with a smooth structure. Then*

- $(M, \tau_{\mathcal{A}})$ *is Hausdorff if and only if, for any two points $p, q \in M$, either p, q belong to the same coordinate domain or they belong to disjoint coordinate domains;*
- $(M, \tau_{\mathcal{A}})$ *is II-countable if and only if it possesses a countable atlas.*

Proof.   Let p, q be distinct points of M. If M is Hausdorff, then p, q are separated by open subsets $\mathcal{U}, \mathcal{V}$. As coordinate domains form a basis for the topology of M, then p, q are also separated by coordinate domains. Conversely, let p, q belong to disjoint coordinate domains U, V, then $U, V \subseteq M$ are open subsets separating them. On the other hand, if p, q belong to the coordinate domain U of the same chart (U, φ), then $\varphi(p), \varphi(q) \in \widehat{U}$ are distinct point. As $\widehat{U}$ is Hausdorff, there are open subsets $V, W \subseteq \widehat{U}$ separating them. Hence $\varphi^{-1}(V), \varphi^{-1}(W)$ are open subsets of U (hence of M) separating p, q.

For the second part of the statement, let $\mathcal{B} \subseteq \tau_{\mathcal{A}}$ be a countable basis. Denote by $\mathcal{B}_{\mathcal{A}} \subseteq \mathcal{B}$ the subfamily consisting of open subsets $B \in \mathcal{B}$ with the property that there exists $(U, \varphi) \in \mathcal{A}$ such that $B \subseteq U$. As the coordinate domains cover M, and every coordinate domain is covered by open subsets in $\mathcal{B}_{\mathcal{A}}$, it follows that $\mathcal{B}_{\mathcal{A}}$ covers M. Additionally, being an open subset in a coordinate domain, every $B \in \mathcal{B}_{\mathcal{A}}$ can be seen itself as the coordinate domain of a (sub)chart (B, φ). Choose just one such chart (B, φ_B) for every B and note that $\{(B, \varphi_B) : B \in \mathcal{B}_{\mathcal{A}}\}$ is a countable atlas on M. Conversely, let $\mathcal{A}_0 \subseteq \mathcal{A}$ be a countable atlas. For every $(U, \varphi) \in \mathcal{A}_0$, let $\mathcal{F}_U$ be the family of open subsets of U of the form $\varphi^{-1}(\mathcal{D})$, where $\mathcal{D} \subseteq \widehat{U}$ is an open disk with rational center and radius. It is clear that

$$\bigcup_{(U,\varphi)\in\mathcal{A}_0} \mathcal{F}_U$$

is a countable basis for $\tau_{\mathcal{A}_0} = \tau_{\mathcal{A}}$ (can you prove it in details?). $\qquad\square$

Example 1.42. Let V be an n-dimensional real vector space. We know that all the charts $(V, \varphi_\mathcal{R})$ determined by frames $\mathcal{R}$ of V belong to the same maximal atlas $\mathcal{A}_{\max}$. Then $(V, \mathcal{A}_{\max})$ is an n-dimensional smooth manifold. $\qquad\blacklozenge$

Example 1.43 (Real Projective Spaces are Manifolds). The n-dimensional real projective space $\mathbb{R}P^n$ equipped with its affine atlas is an n-dimensional smooth manifold. Indeed, let $[\bar{P}] = [\bar{P}^0 : \cdots : \bar{P}^n], [\bar{Q}] = [\bar{Q}^0 : \cdots : \bar{Q}^n]$ be distinct points of $\mathbb{R}P^n$. Then either they belong to the same coordinate domain or they do not belong simultaneously to any of the affine coordinate domains; in other words, whenever $P^i \neq 0$, we must have $Q^i = 0$ and vice versa. In this second case, without loosing much generality, we assume that $\bar{P}^0, \ldots, \bar{P}^{k-1} \neq 0$, $\bar{P}^k = \cdots = \bar{P}^n = 0$ for some $0 < k \leq n$. Then we have $\bar{Q}^0 = \cdots = \bar{Q}^{k-1} = 0$ and $\bar{Q}^k, \ldots, \bar{Q}^n \neq 0$, and we can even choose $\bar{P}^0 = \bar{Q}^k = 1$. Now, let δ, ϵ be positive real numbers and, in $\mathbb{R}^n$, consider the following open subsets:

$$\widehat{U}'_0 := \{(A^1, \ldots, A^n) : |A^{k-1} - \bar{P}^{k-1}| < \delta\}$$

and

$$\widehat{U}'_k := \{(B^1, \ldots, B^n) : |B^1| < \epsilon\}$$

and their preimages $U'_0 = \varphi_0^{-1}(\widehat{U}'_0) \subseteq U_0$ and $U'_k = \varphi_k^{-1}(\widehat{U}'_k) \subseteq U_k$. It is clear that $(U'_0, \varphi_0), (U'_k, \varphi_k)$ are (sub)charts around $[\bar{P}], [\bar{Q}]$. It remains to show that δ, ϵ can be chosen so that $U'_0 \cap U'_k = \varnothing$. To do this, recall that a point $[P] \in U'_0 \cap U'_k$ has homogeneous coordinates of the form

$$(1, A^1, \ldots, A^n), \quad \text{with } |A^{k-1} - \bar{P}^{k-1}| < \delta, \tag{1.3}$$

but it also has homogeneous coordinates of the form

$$(B^1, \ldots, \underbrace{1}_{k\text{-th place}}, \ldots, B^n) \quad \text{with } |B^1| < \epsilon. \tag{1.4}$$

It follows that $B^1 \neq 0$ and

$$A^{k-1} = \frac{1}{B^1}. \tag{1.5}$$

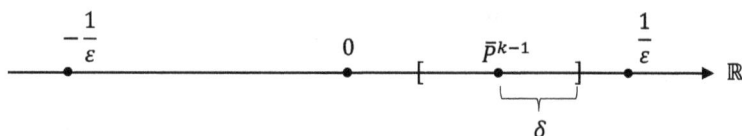

Figure 1.6.   Choosing ϵ, δ in an appropriate way.

We conclude noting that (1.3)–(1.5) cannot be simultaneously satisfied if we choose

$$\frac{1}{\epsilon} > \max\left\{ |\bar{P}^{k-1} - \delta|, |\bar{P}^{k-1} + \delta| \right\}$$

(see Figure 1.6). II-countability is clear.   ◆

Finally, we show that products of smooth manifolds are smooth manifolds, and open subsets in a smooth manifold are smooth manifolds in a natural way. This will enrich significantly our list of examples.

Let $M_1, \ldots, M_k$ be smooth manifolds of dimension $n_1, \ldots, n_k$ and let $\mathcal{A}_1, \ldots, \mathcal{A}_k$ be their smooth structures. Put $n := n_1 + \cdots + n_k$. First of all, we can construct an n-dimensional atlas on the product $M_1 \times \cdots \times M_k$ as follows. For every $i = 1, \ldots, k$, let $(U_i, \varphi_i) \in \mathcal{A}_i$ be a chart on M_i. Consider the product $U_1 \times \cdots \times U_k \subseteq M_1 \times \cdots \times M_k$ and the map

$$\varphi_1 \times \cdots \times \varphi_k : U_1 \times \cdots \times U_k \rightarrow \widehat{U}_1 \times \cdots \times \widehat{U}_k \subseteq \mathbb{R}^n$$

given by

$$(p_1, \ldots, p_k) \mapsto (\varphi_1(p_1), \ldots, \varphi_k(p_k)).$$

It is clear that $(U_1 \times \cdots \times U_k, \varphi_1 \times \cdots \times \varphi_k)$ is a chart on $M_1 \times \cdots \times M_k$. Every such chart is called a *product chart*. It is easy to see that the coordinate domains of product charts cover $M_1 \times \cdots \times M_k$ and every two product charts are compatible (see Exercise 1.14). Hence product charts form an n-dimensional atlas on $M_1 \times \cdots \times M_k$, called the *product atlas*, and denoted $\mathcal{A}^\times$. Additionally, the atlas topology induced on $M_1 \times \cdots \times M_k$ by the product atlas coincides with the product topology.

Exercise 1.14. Prove that product charts on $M_1 \times \cdots \times M_k$ form an atlas, the *product atlas*. Prove also that the atlas topology induced by the product atlas coincides with the product topology. (**Hint:** *Prove preliminarily that if $\varphi_i : X_i \to Y_i$ are homeomorphisms between topological spaces, $i = 1, \ldots, k$, then the map*

$$\varphi_1 \times \cdots \times \varphi_k : X_1 \times \cdots \times X_k \to Y_1 \times \cdots \times Y_k$$

given by

$$(p_1, \ldots, p_k) \mapsto (\varphi_1(p_1), \ldots, \varphi_k(p_k))$$

is a homeomorphism between the product spaces. Finally, use this fact and Proposition 1.35 to prove the claim.)

It follows that $(M_1 \times \cdots \times M_k, \tau_{\mathcal{A}^\times})$ is a Hausdorff and II-countable topological space. Hence $M_1 \times \cdots \times M_k$ is an n-dimensional manifold.

Example 1.44 (The Torus). The n-dimensional manifold

$$T^n := \underbrace{S^1 \times \cdots \times S^1}_{n \text{ times}}$$

is called the *n-dimensional torus*. ♦

We now come to open subsets in a manifold. So, let M be a manifold, let $\mathcal{A}$ be its smooth structure, and let $\mathcal{U} \subseteq M$ be an open subset. Every chart $(U, \varphi) \in \mathcal{A}$ such that $U \subseteq \mathcal{U}$ is clearly a chart on $\mathcal{U}$. Additionally, the coordinate domains of such charts cover $\mathcal{U}$. In other words,

$$\mathcal{A}_{\mathcal{U}} := \{(U, \varphi) \in \mathcal{A} : U \subseteq \mathcal{U}\}$$

is an atlas on $\mathcal{U}$. It is almost immediate to see that the atlas topology induced by $\mathcal{A}_{\mathcal{U}}$ coincides with the subspace topology (do you see it?). Hence with the atlas topology, $\mathcal{U}$ is Hausdorff and II-countable. We conclude that $\mathcal{U}$ is a manifold of the same dimension as M. Every such manifold is called an *open submanifold* of M.

Example 1.45 (The General Linear Group). Consider the vector space $M(n, \mathbb{R})$ of $n \times n$ real matrices. The *general linear group* $\mathrm{GL}(n, \mathbb{R}) \subseteq$

$M(n, \mathbb{R})$ consists, by definition, of invertible matrices or, equivalently, non-zero determinant matrices. As the determinant map

$$\det : M(n, \mathbb{R}) \to \mathbb{R}$$

is continuous (even polynomial), and $\mathbb{R} \smallsetminus \{0\} \subseteq \mathbb{R}$ is an open subset, it follows that $\mathrm{GL}(n, \mathbb{R}) = \det^{-1}(\mathbb{R} \smallsetminus \{0\}) \subseteq M(n, \mathbb{R})$ is an open subset, hence an open submanifold of $M(n, \mathbb{R})$.

More generally, let V be an n-dimensional real vector space and let End V the n^2-dimensional vector space of endomorphisms $h : V \to V$ of V. The *general linear group* $\mathrm{GL}(V)$ of V consists of invertible endomorphisms and (similarly as for $\mathrm{GL}(n, \mathbb{R})$) it is an open submanifold in End V.   ♦

Chapter 2

Smooth Maps and Submanifolds

In this chapter, we show how the smooth structure on a smooth manifold allows us to define the notion of *smoothness*. In particular, we will define *smooth real valued functions on a manifold*, *smooth maps between manifolds*, and *smooth subsets of a manifold*, called *submanifolds*. These notions are the first building blocks of a *differential calculus on manifolds*, in particular a *coordinate-free* calculus.

2.1 Smooth Functions

Let M be an n-dimensional manifold.

Definition 2.1 (Smooth Function). A *smooth function* on M is a real valued function $f : M \to \mathbb{R}$ such that, for every $p \in M$, there exists a chart (U, φ) on M such that the real valued function $f \circ \varphi^{-1} : \widehat{U} \to \mathbb{R}$ is smooth.

Given a real valued function $f : M \to \mathbb{R}$, a point $p \in M$, and a chart $(U, \varphi = (x^1, \ldots, x^n))$ on M around p, the function $f \circ \varphi^{-1} : \widehat{U} \to \mathbb{R}$ is called a *coordinate representation of f around p* and will be denoted by $\widehat{f}_U$, or simply $\widehat{f}$, if this does not lead to confusion. So Definition 2.1 can be restated saying that a function $f : M \to \mathbb{R}$ is smooth if *it has a smooth coordinate representation around every point*. We will sometimes consider the restriction $f : U \to \mathbb{R}$ and write $f = f(x^1, \ldots, x^n)$ to express that, restricted to U, f can be seen as a function of the coordinates $x^1, \ldots, x^n$.

Proposition 2.2. *A function $f : M \to R$ is smooth if and only if all its coordinate representations are smooth.*

Proof.   The "if part" of the statement is obvious. Let's prove the "only if part". So let $f : M \to \mathbb{R}$ be a smooth function, and let (V, ψ) be any chart on M. We want to show that the coordinate representation $\widehat{f}_V = f \circ \psi^{-1} : \widehat{V} \to \mathbb{R}$ is smooth. It is enough to show that $\widehat{f}_V$ is smooth around every point. To do this, take $Q \in \widehat{V}$, and let $p = \psi^{-1}(Q)$. As f is smooth, there is a chart (U, φ) around p such that the coordinate representation $\widehat{f} = f \circ \varphi^{-1} : \widehat{U} \to \mathbb{R}$ is smooth. Hence, so is the restriction $\widehat{f} : \varphi(U \cap V) \to \mathbb{R}$ of $\widehat{f}$ to the open subset $\varphi(U \cap V)$. Consider the open subset $\psi(U \cap V) \subseteq \widehat{V}$ and note that the restriction $\widehat{f}_V : \psi(V \cap U) \to \mathbb{R}$ is given by the following composition (see Figure 2.1):

$$\widehat{f}_V = \widehat{f} \circ (\varphi \circ \psi^{-1}).$$

As both the transition map $\varphi \circ \psi^{-1}$ and the coordinate representation $\widehat{f}$ are smooth, so is $\widehat{f}_V : \psi(V \cap U) \to \mathbb{R}$. From the arbitrariness of Q, the coordinate representation $\widehat{f}_V : \widehat{V} \to \mathbb{R}$ is smooth as well.   □

Note that a real function $f : \mathcal{U} \to \mathbb{R}$ on an open subset $\mathcal{U}$ of $\mathbb{R}^n$ is smooth in the sense of Definition 2.1 if and only if it is smooth in the standard sense. All constant functions

$$\mathrm{const} : M \to \mathbb{R}$$

have constant coordinate representations, hence they are smooth.

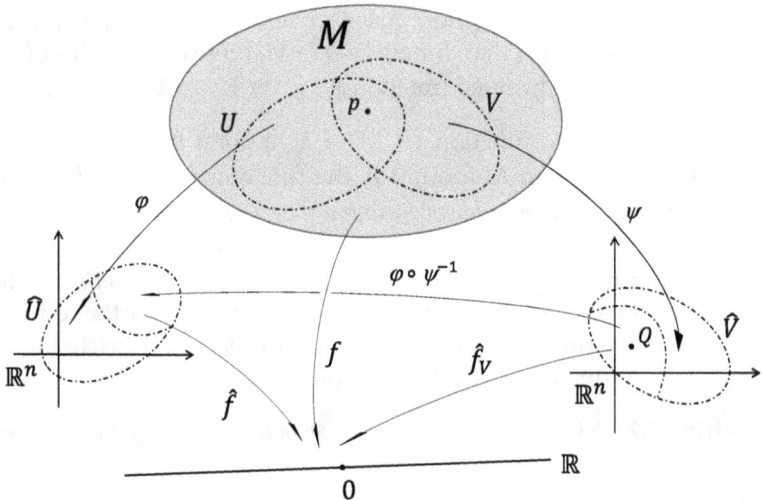

Figure 2.1.   All coordinate representations of a smooth function are smooth.

Exercise 2.1. Let $Q(t^0, \ldots, t^n), R(t^0, \ldots, t^n)$ be homogeneous polynomials of the same degree in the real indeterminates $t^0, \ldots, t^n$. Assume, additionally, that $R(t^0, \ldots, t^n)$ does only vanish at the origin (e.g., this is the case if $R(t^0, \ldots, t^n) = (t^0)^2 + \cdots + (t^n)^2$). Prove that the real valued function

$$f : \mathbb{R}P^n \to \mathbb{R}, \quad [P] = [P^0 : \cdots : P^n] \mapsto f([P]) := \frac{Q(P^0, \ldots, P^n)}{R(P^0, \ldots, P^n)}$$

is well defined and smooth.

It is clear that a function $f : M \to \mathbb{R}$ is smooth if and only if it is *locally smooth*, i.e., for every point $p \in M$, there exists an open neighborhood $\mathcal{U} \subseteq M$ of p such that the restriction $f|_{\mathcal{U}} : \mathcal{U} \to \mathbb{R}$ of f to $\mathcal{U}$ is smooth, where we interpret $\mathcal{U}$ as a manifold with the *open submanifold structure*. The claim easily follows from the fact that the charts in M whose coordinate domain is contained in $\mathcal{U}$ form an atlas on $\mathcal{U}$ (check the details as an exercise!). Otherwise stated, it holds the following

Lemma 2.3 (Gluing Lemma). *Let M be a manifold, and let $C = \{\mathcal{U}\}$ be an open cover of M (i.e., every element $\mathcal{U} \in C$ is an open subset $\mathcal{U} \subseteq M$, and $M = \bigcup_{\mathcal{U} \in C} \mathcal{U}$). For every $\mathcal{U} \in C$, let $f_{\mathcal{U}} : \mathcal{U} \to \mathbb{R}$ be a smooth function. If*

$$f_{\mathcal{U}}|_{\mathcal{U} \cap \mathcal{V}} = f_{\mathcal{V}}|_{\mathcal{U} \cap \mathcal{V}}$$

for every $\mathcal{U}, \mathcal{V} \in C$, then there exists a unique smooth function $f : M \to \mathbb{R}$ such that

$$f|_{\mathcal{U}} = f_{\mathcal{U}}$$

for all $\mathcal{U} \in C$.

There are a lot of smooth functions on a manifold, as the following theorem shows.

Theorem 2.4 (Bump Function Theorem). *Let M be a smooth manifold. For every point $p \in M$, and every open neighborhood $\mathcal{U} \subseteq M$ of p, there exists a smooth function $\beta : M \to \mathbb{R}$ on M such that $\beta = 1$ around p (i.e., there is an open neighborhood $V \subseteq M$ of p such that $\beta|_V = 1$), and the support of β is contained in $\mathcal{U}$.*

The statement of Theorem 2.4 needs some explanations. First of all, let X be a topological space, and let $A \subseteq X$ be a subset. The *closure of A in X* is the smallest closed subset $\overline{A} \subseteq X$ containing A. Equivalently, $\overline{A}$ is the (closed) intersection of all closed subsets of X containing A. Given a real valued function $f : X \to \mathbb{R}$, the *support* of f is the closure

$$\operatorname{supp} f := \overline{\{p \in X : f(p) \neq 0\}}.$$

Note that, if f is a continuous function, then the subset $\{p \in X : f(p) \neq 0\}$ is open (while the support of f is always closed by definition).

We postpone the proof of Theorem 2.4 to the end of this section. We only stress here that the Hausdorff property of M is a key ingredient of the proof (and the theorem does not hold true in general for a set M with a smooth structure $\mathcal{A}$ such that $(M, \tau_{\mathcal{A}})$ is not Hausdorff). A smooth function β like in the statement of Theorem 2.4 is called a *bump function* (relative to the data $(p, \mathcal{U})$). Note that, as a real-valued function on M vanishes outside its support, given a bump function β relative to the data $(p, \mathcal{U})$, in particular we have $\beta|_{M \smallsetminus \mathcal{U}} = 0$. In Example 2.8, we show that the Bump Function Theorem holds true when $M = \mathbb{R}^n$. The general case can be proved from this one (via charts, the Hausdorff property of manifolds, and a compactness argument, see the proof at the end of the present section). Before doing this, we present some algebraic structures on the space of smooth functions on a manifold. We begin defining *real algebras*.

Definition 2.5 (Algebra). A *real, associative, commutative algebra with unit*, or simply an $\mathbb{R}$-*algebra*, or just an *algebra*, is a real vector space $(A, +, \cdot)$ equipped with an additional interior operation $\star : A \times A \to A$ such that

- $\star$ is $\mathbb{R}$-bilinear,
- $\star$ is associative.

In particular, $(A, +, \star)$ is a ring. Additionally, we require that the ring $(A, +, \star)$ is commutative and possesses a unit. A *subalgebra* of an algebra A is a vector subspace $B \subseteq A$ which is additionally a subring containing 1.

Note that a subalgebra B of an algebra A is also an algebra with the restricted operations. As for rings and vector spaces, the symbols $\star, \cdot$ are usually omitted in products, and we write $a\alpha$ and $\alpha\beta$ instead of $a \cdot \alpha$ and $\alpha \star \beta$, for $a \in \mathbb{R}$, and α, β elements in an algebra. As the first trivial example, note that $\mathbb{R}$ is an algebra in the obvious way. Complex numbers do also form a real algebra. More generally, for any algebra A, the map $i : \mathbb{R} \to A$, $a \mapsto a \cdot 1$, is an injection identifying $\mathbb{R}$ with a subalgebra of A. In the following, we will often identify $\mathbb{R}$ with its image $i(\mathbb{R})$ under i. This shows that an algebra A can be also seen as a *ring extension* $\mathbb{R} \subseteq A$.

Example 2.6 (Algebra of Real-Valued Functions). Let X be a set. The space $\mathcal{F}(X, \mathbb{R})$ of real valued functions $f : X \to \mathbb{R}$ is an algebra with the following *point-wise* operations:

$$+ : (f, g) \mapsto f + g, \quad (f + g)(p) := f(p) + g(p),$$
$$\star : (f, g) \mapsto fg, \quad (fg)(p) := f(p)g(p),$$
$$\cdot : (a, f) \mapsto af, \quad (af)(p) := af(p),$$

where $f, g \in \mathcal{F}(X, \mathbb{R})$ and $a \in \mathbb{R}$. The zero in $\mathcal{F}(X, \mathbb{R})$ is the constant function whose constant value is 0, the 1 in $\mathcal{F}(X, \mathbb{R})$ is the constant function whose constant value is 1, and the inclusion $i : \mathbb{R} \hookrightarrow \mathcal{F}(X, \mathbb{R})$ identifies a real number a with the constant function, also denoted a, whose constant value is a itself. ♦

Proposition 2.7 (Algebra of Smooth Functions). *Let M be a manifold. Then smooth functions on M form a subalgebra of the algebra $\mathcal{F}(M, \mathbb{R})$ of all real valued functions $f : M \to \mathbb{R}$. In other words, sum, products, and products by real scalars of smooth functions are smooth functions.*

Proof. Left as Exercise 2.2. □

Exercise 2.2. Prove Proposition 2.7. (**Hint:** *Show that, given smooth functions f, g, a real number a, and a chart (U, φ), we have the following relations between coordinate representations:*

$$\widehat{f + g} = \widehat{f} + \widehat{g},$$
$$\widehat{f \cdot g} = \widehat{f} \cdot \widehat{g},$$
$$\widehat{af} = a\widehat{f}.$$

Finally, use the fact that sum, products, and products by real scalars of smooth functions on $\widehat{U}$ are smooth functions.)

The algebra of smooth functions on a manifold M will be denoted by $C^\infty(M)$.

Example 2.8 (Bump Functions on $\mathbb{R}^n$). Now, we prove the Bump Function Theorem in the special case $M = \mathbb{R}^n$. From this case, the general case can be proved with a little more Topology. We begin considering the real valued function $f : \mathbb{R} \to \mathbb{R}$ given by

$$f(t) := \begin{cases} 0 & \text{if } t < 0 \\ e^{-\frac{1}{t}} & \text{if } t \geq 0 \end{cases}, \quad t \in \mathbb{R},$$

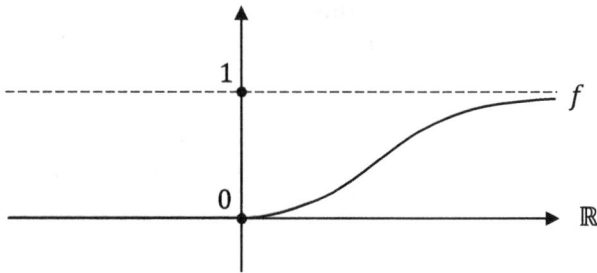

Figure 2.2.   The function $e^{-1/t}$.

see Figure 2.2. By studying the derivatives of f at the origin, one can prove that f is a smooth function. Second, for every two real numbers r, R such that $r < R$, we define a smooth function $h : \mathbb{R} \to \mathbb{R}$ by putting

$$h(t) = \frac{f(R-t)}{f(R-t) + f(t-r)}, \quad t \in \mathbb{R},$$

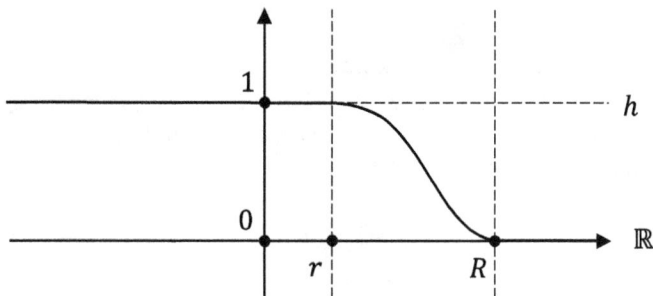

Figure 2.3.   Cut-off function.

see Figure 2.3. Clearly, $h(t) = 1$ for all $t \leq r$, and $h(t) = 0$ for all $t \geq R$. Additionally, $0 < h(t) < 1$ for $r < t < R$. The function h is called a *cut-off function*. Finally, let $P_0 \in \mathbb{R}^n$, and let $\mathcal{U} \subseteq \mathbb{R}^n$ be any open neighborhood of P_0. There are positive real numbers r, R such that $r < R$ and, additionally, the closed disks D_r, D_R centered at P_0 with radii r, R are contained in $\mathcal{U}$. It follows that the smooth function

$$\beta : \mathbb{R}^n \to \mathbb{R}, \quad P \mapsto \beta(P) := h\left(\|P - P_0\|\right)$$

is a bump function relative to the data $(P_0, \mathcal{U})$. Indeed $\beta|_{D_r} = 1$, in particular, $\beta = 1$ in an open neighborhood of P_0, and $\operatorname{supp}\beta = D_R \subseteq \mathcal{U}$ (see Figure 2.4).

♦

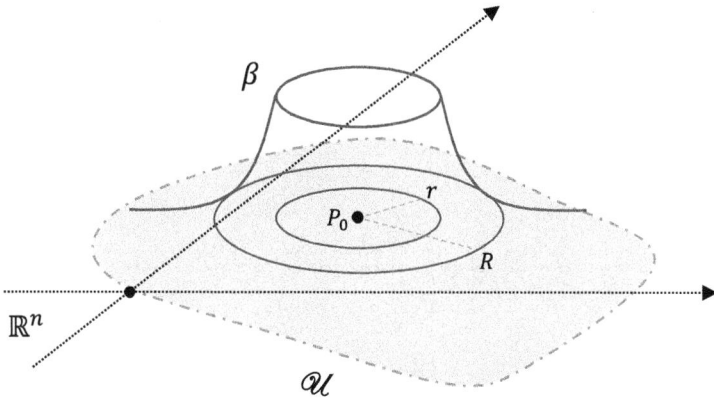

Figure 2.4. Bump function on $\mathbb{R}^n$.

Proof of Theorem 2.4. Let M, $\mathcal{U}$, and p be as in the statement. Choose a chart (U, φ) around p such that $U \subseteq \mathcal{U}$, and choose a bump function $\widehat{\beta} \in C^\infty(\mathbb{R}^n)$ relative to the data $(\widehat{U}, \varphi(p))$. Do this as in Example 2.8 so that $\operatorname{supp} \beta$ is a closed disk $D \subseteq \mathbb{R}^n$ (centered at $\varphi(p)$). In particular, D is a compact subspace in $\mathbb{R}^n$ (it is closed and bounded). The function

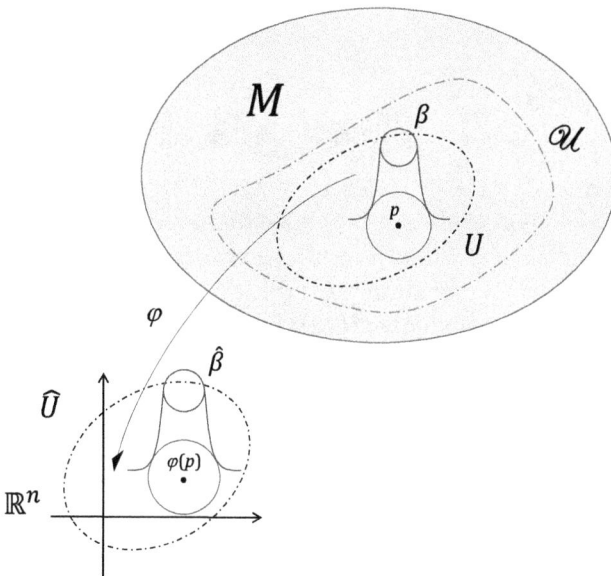

Figure 2.5. The proof of the Bump Function Theorem.

$\beta' : U \to \mathbb{R}$ defined as the composition $\beta' := \widehat{\beta} \circ \varphi$ is clearly smooth. Moreover, $\operatorname{supp} \beta' = \varphi^{-1}(D)$. But $\varphi^{-1} : \widehat{U} \to U$ is a homeomorphism, and homeomorphisms preserve the compactness of subspaces (do you see it? If not, prove it as an exercise), hence $\operatorname{supp} \beta' \subseteq U \subseteq \mathcal{U} \subseteq M$ is a compact subspace containing p. As M is a Hausdorff space, it follows that $\operatorname{supp} \beta'$ is also a closed subspace in M (see Exercise 1.13), not just a closed subset of U. Finally, consider the function $\beta : M \to \mathbb{R}$ defined by

$$\beta(p) := \begin{cases} \beta'(p) & \text{if } p \in U \\ 0 & \text{if } p \in M \smallsetminus \operatorname{supp} \beta' \end{cases},$$

see Figure 2.5. Clearly, β is well defined, and it is smooth by the Gluing Lemma 2.3. Moreover, $\operatorname{supp} \beta = \operatorname{supp} \beta'$. We conclude that β is a bump function relative to the data $(\mathcal{U}, p)$ as desired. □

2.2 Smooth Maps

We can compare two smooth manifolds via *smooth maps* between them. Let M and N be smooth manifolds, with $\dim M = m$ and $\dim N = n$.

Definition 2.9 (Smooth Map). A *smooth map* between M and N is a map $F : M \to N$ with the following property: for every $p \in M$, there exist a chart (U, φ) on M around p and a chart (V, ψ) on N around $F(p)$, such that

(1) $F(U) \subseteq V$ and
(2) the composition $\psi \circ F \circ \varphi^{-1} : \widehat{U} \to \widehat{V}$ is smooth.

Given a map $F : M \to N$, a point $p \in M$, a chart (U, φ) on M around p, and a chart (V, ψ) on N around $F(p)$ such that $F(U) \subseteq V$, the composition $\psi \circ F \circ \varphi^{-1} : \widehat{U} \to \widehat{V}$ is called a *coordinate representation of F around p* and will be denoted by $\widehat{F}_{U,V}$ or simply $\widehat{F}$ if this does not lead to confusion (see Figure 2.6). If two such charts (U, φ), (V, ψ) exist for every $p \in M$, we say that F *has coordinate representations* (around every point). So Definition 2.9 can be restated saying that a map $F : M \to N$ is smooth if *it has a smooth coordinate representation around every point*. Let $F : M \to N$ be a map having coordinate representations, and let $(U, \varphi = (x^1, \ldots, x^m))$ be a chart on M and $(V, \psi = (y^1, \ldots, y^n))$ a chart on N such that $F(U) \subseteq V$. In this situation, we will sometimes consider the vector-valued map $\psi \circ F : U \to \widehat{V}$. Its components will be denoted $F^1, \ldots, F^n$: $\psi \circ F = (F^1, \ldots, F^n)$ and we will write $F^a = F^a(x^1, \ldots, x^m)$, $a = 1, \ldots, n$, to express the fact that the F^a can be seen as functions of the coordinates $x^1, \ldots, x^m$. Note that $F^a = y^a \circ$

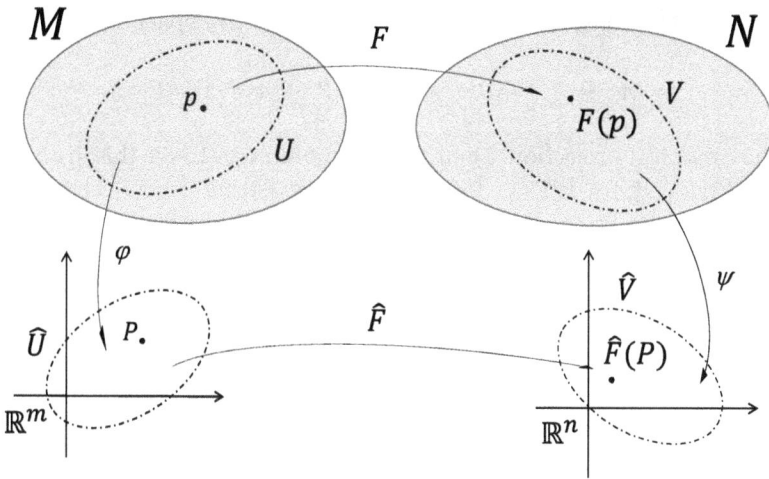

Figure 2.6. Coordinate representation of a map between manifolds.

$F : U \to \mathbb{R}$ for all $a = 1, \ldots, n$, and their coordinate representations $\widehat{F}^a$ with respect to the chart (U, φ) are exactly the components of the coordinate representation $\widehat{F}$: $\widehat{F} = (\widehat{F}^1, \ldots, \widehat{F}^n)$. Accordingly, F is smooth if and only if it has a coordinate representation $\widehat{F}_{U,V}$ around every point such that the F^a are smooth.

Proposition 2.10. *A map $F : M \to N$ is smooth if and only if it has a coordinate representation around every point $p \in M$, and all its coordinate representations are smooth.*

Proof. Left as Exercise 2.3. □

Exercise 2.3. Prove Proposition 2.10.

We now provide a list of examples of smooth maps. A map $F : \mathcal{U} \to \mathcal{V}$ between open subsets of standard Euclidean spaces $\mathcal{U} \subseteq \mathbb{R}^m$ and $\mathcal{V} \subseteq \mathbb{R}^n$ is smooth in the sense of Definition 2.9 if and only if it is smooth in the standard sense. All constant maps const : $M \to N$ have constant coordinate representations, hence they are smooth. A map $f : M \to \mathbb{R}$ is smooth if and only if it is a smooth function in the sense of Definition 2.1. Finally, the identity id : $M \to M$ is a smooth map. More generally, the inclusion $i_{\mathcal{U}} : \mathcal{U} \hookrightarrow M$ of an open submanifold $\mathcal{U} \subseteq M$ has coordinate representations, and they can be chosen to be simply the identity (choose the same chart in $\mathcal{U}$ and in M). Hence $i_{\mathcal{U}}$ is a smooth map.

Example 2.11. Consider the n-dimensional projective space $\mathbb{R}P^n$, and let

$$\pi : \mathbb{R}^{n+1} \smallsetminus \{0\} \to \mathbb{R}P^n, \quad P \mapsto \pi(P) := [P]$$

be the canonical projection. Then π is a smooth map. To see this, fix a point $P_0 = (P_0^0, \ldots, P_0^n) \in \mathbb{R}^{n+1} \smallsetminus \{0\}$. Then there exists $i \in \{0, \ldots, n\}$ such that $P_0^i \neq 0$. Denote $V_i := \{t^i \neq 0\} \subseteq \mathbb{R}^{n+1} \smallsetminus \{0\}$. It is an open subset and (V_i, id) is a subchart of the chart $(\mathbb{R}^{n+1} \smallsetminus \{0\}, \mathrm{id})$ on $\mathbb{R}^{n+1} \smallsetminus \{0\}$ around P_0. Additionally, $\pi(V_i) \subseteq U_i$, the coordinate domain of the i-th affine chart (U_i, φ_i) on $\mathbb{R}P^n$. This shows that π has coordinate representations around every point P_0. Now, we compute

$$\widehat{\pi} = \varphi_i \circ \pi \circ \mathrm{id}^{-1} = \varphi_i \circ \pi : \widehat{V_i} = V_i \to \mathbb{R}^n.$$

For every $P = (P^0, \ldots, P^n) \in V_i$, we have

$$\widehat{\pi}(P) = \varphi_i(\pi(P)) = \varphi_i([P]) = \left(\frac{P^0}{P^i}, \ldots, \frac{\widehat{P^i}}{P^i}, \ldots, \frac{P^n}{P^i} \right).$$

So $\widehat{\pi}$ is a smooth map for all i (it has smooth components). We conclude that π is a smooth map. ◆

Exercise 2.4. Prove that the inclusion $i_{S^n} : S^n \hookrightarrow \mathbb{R}^{n+1}$ of the n-dimensional sphere in the $(n+1)$-dimensional standard Euclidean space is a smooth map.

Exercise 2.5. Let m, n be non-negative integers, and let $Q^0(t^0, \ldots, t^m)$, $\ldots, Q^n(t^0, \ldots, t^m)$ be homogeneous polynomials in the indeterminate $t^0, \ldots, t^m$, all of the same degree, and such that they do only vanish simultaneously at the origin. Prove that the map

$$F : \mathbb{R}P^m \to \mathbb{R}P^n, \quad [P] \mapsto F([P]) := [Q^0(P) : \cdots : Q^n(P)]$$

is well defined and smooth.

Example 2.12 (Projections onto Factors are Smooth). Let $M_1, \ldots, M_k$ be smooth manifolds of dimensions $n_1, \ldots, n_k$, respectively. Consider the product manifold $M_1 \times \cdots \times M_k$. The projection onto the i-th factor

$$\mathrm{pr}_i : M_1 \times \cdots \times M_k \to M_i, \quad (p_1, \ldots, p_k) \mapsto \mathrm{pr}_i(p_1, \ldots, p_k) := p_i$$

is a smooth map for every $i = 1, \ldots, k$. Indeed, let $(\bar{p}_1, \ldots, \bar{p}_k) \in M_1 \times \cdots \times M_k$ and, for every $j = 1, \ldots, k$, let (U_j, φ_j) be a chart on M_j around $\bar{p}_j$. Consider the product chart

$$(U_1 \times \cdots \times U_k, \varphi_1 \times \cdots \times \varphi_k).$$

Clearly, $\mathrm{pr}_i(U_1 \times \cdots \times U_k) \subseteq U_i$, showing that pr_i has coordinate representation

$$\widehat{\mathrm{pr}}_i : \widehat{U}_1 \times \cdots \times \widehat{U}_k \to \widehat{U}_i.$$

Additionally, denote by $(t^1_{(j)}, \ldots, t^{n_j}_{(j)})$ the standard coordinates on $\widehat{U}_j$. Then $\widehat{\mathrm{pr}}_i$ is given by

$$\widehat{\mathrm{pr}}_i(t^1_{(1)}, \ldots, t^{n_1}_{(1)}, \ldots, t^1_{(k)}, \ldots, t^{n_k}_{(k)}) = (t^1_{(i)}, \ldots, t^{n_i}_{(i)}),$$

which is smooth. ♦

The proof of the following lemma should be clear (is it?).

Lemma 2.13 (Gluing Lemma for Smooth Maps). *Let M, N be manifolds, and let $C = \{U\}$ be an open cover of M. For every $U \in C$, let $F_U : U \to N$ be a smooth map. If*

$$F_U|_{U \cap V} = F_V|_{U \cap V}$$

for every $U, V \in C$, then there exists a unique smooth map $F : M \to N$ such that

$$F|_U = F_U$$

for all $U \in C$.

Let M, N be manifolds, and let $F : M \to N$ be a map. In order to check whether or not F is smooth, we have to check, first of all, whether or not F has coordinate representations. Note that continuous maps $F : M \to N$ do always have coordinate representations around every point. Indeed, let $p \in M$, and let (V, ψ) be a chart on N around $F(p)$. The preimage $F^{-1}(V) \subseteq M$ is an open neighborhood of p. Hence there is a chart (U, φ) around p such that $U \subseteq F^{-1}(V)$ (the coordinate domains of charts in the maximal atlas form a basis for the topology). It follows that $F(U) \subseteq V$. Actually, a map $F : M \to N$ is smooth if and only if it is continuous, and its coordinate representations (they exist by continuity) are smooth. The only non-obvious thing in this claim is that a smooth map is continuous. This is the content of the following

Proposition 2.14. *Smooth maps are continuous.*

Proof.　Let $F : M \to N$ be a smooth map between manifolds, and let $\mathcal{V} \subseteq N$ be an open subset. We want to show that the preimage $F^{-1}(\mathcal{V}) \subseteq M$ is an open subset. So, let $p \in F^{-1}(\mathcal{V})$ be a point, and let (U, φ) be a chart on M around p and (V, ψ) a chart on N around $F(p)$ such that $F(U) \subseteq V$ (they exist because F is smooth). Consider also the coordinate representation

$$\widehat{F} = \psi \circ F \circ \varphi^{-1} : \widehat{U} \to \widehat{V}.$$

As p is arbitrary, it is enough to show that $\varphi(U \cap F^{-1}(\mathcal{V})) \subseteq \widehat{U}$ is an open subset. But

$$\varphi(U \cap F^{-1}(\mathcal{V})) = \widehat{F}^{-1}(\psi(V \cap \mathcal{V}))$$

(do you see it? If not, prove it as Exercise 2.6, see also Figure 2.7). As $\mathcal{V} \subseteq N$ is an open subset, $\psi(V \cap \mathcal{V}) \subseteq \widehat{V}$ is an open subset, and since $\widehat{F}$ is smooth, hence continuous, $\widehat{F}^{-1}(\psi(V \cap \mathcal{V})) \subseteq \widehat{U}$ is an open subset as well. So $\varphi(U \cap F^{-1}(\mathcal{V}))$ is an open subset and this concludes the proof. □

Exercise 2.6. Complete the proof of Proposition 2.14 showing that for any smooth map $F : M \to N$, any chart (U, φ) on M, any chart (V, ψ) on N such that $F(U) \subseteq V$, and any open subset $\mathcal{V}$ (intersecting V), we have

$$\varphi(U \cap F^{-1}(\mathcal{V})) = \widehat{F}^{-1}(\psi(V \cap \mathcal{V})).$$

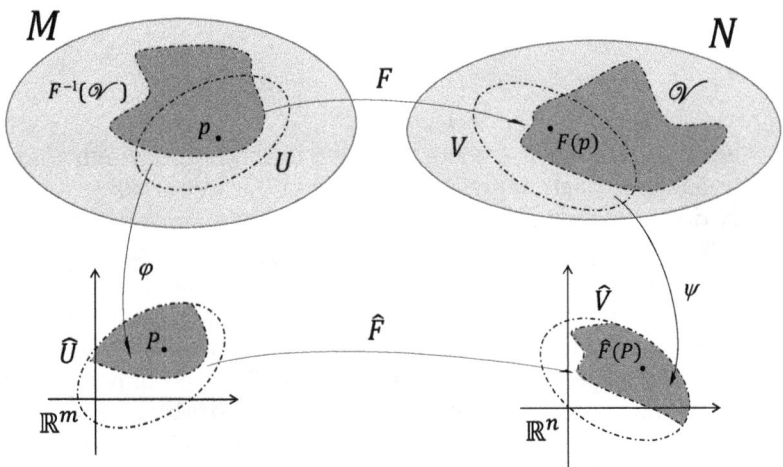

Figure 2.7.　Smooth maps are continuous.

Proposition 2.15. *The composition of smooth maps is smooth.*

Proof. Let $F : M \to N$ and $G : N \to Q$ be smooth maps between manifolds. We want to show that $G \circ F : M \to Q$ is a smooth map. So take a point $p \in M$. As G is smooth, there is a chart (V, ψ) on N around $F(p)$ and a chart (W, χ) on Q around $G(F(p))$ such that $G(V) \subseteq W$ and the coordinate representation $\widehat{G} : \widehat{V} \to \widehat{W}$ is smooth. Since F is continuous, $F^{-1}(V) \subseteq M$ is an open neighborhood of p. Hence there is a chart (U, φ) on M around p such that $U \subseteq F^{-1}(V)$, i.e., $F(U) \subseteq V$. Denote by $\widehat{F} : \widehat{U} \to \widehat{V}$ the coordinate representation of F with respect to (U, φ) and (V, ψ). All coordinate representations of a smooth map are smooth, so $\widehat{F}$ is smooth. Collecting all previous remarks we get that $\widehat{G} \circ \widehat{F} : \widehat{U} \to \widehat{W}$ is a smooth coordinate representation of $G \circ F$ around p (see Figure 2.8). From the arbitrariness of p, we get that $G \circ F$ is smooth. $\qquad\square$

Exercise 2.7. Let $M, M_1, \ldots, M_k$ be manifolds. Consider a map

$$F : M \to M_1 \times \cdots \times M_k.$$

The i-th *component* of F is, by definition, the composition $F_i = \mathrm{pr}_i \circ F : M \to M_i$ of F followed by the projection $\mathrm{pr}_i : M_1 \times \cdots \times M_k \to M_i$ onto the i-th factor, $i = 1, \ldots, k$. Prove that F is smooth if and only if so are its components.

Let M, N be smooth manifolds.

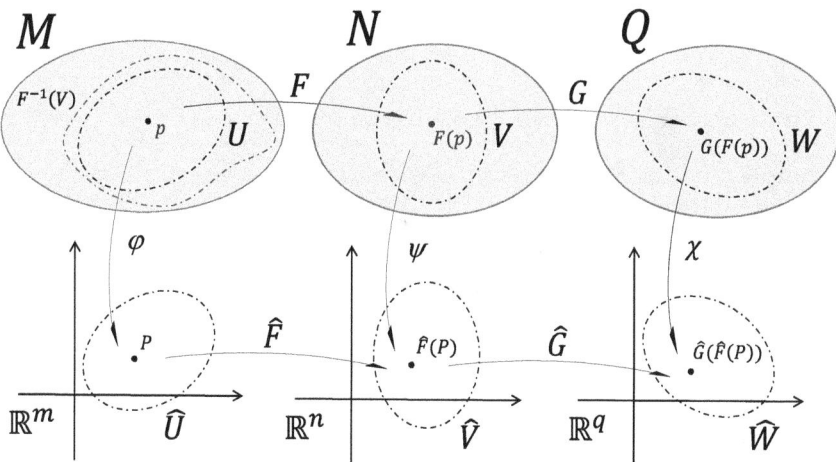

Figure 2.8. The composition of smooth maps is smooth.

Definition 2.16 (Diffeomorphism). A *diffeomorphism* between M and N is a map $\Phi : M \to N$ such that

(1) Φ is smooth,
(2) Φ is invertible,
(3) Φ^{-1} is smooth.

The manifolds M and N are *diffeomorphic* if there is a diffeomorphism $\Phi : M \to N$ connecting them.

A map $\Phi : \mathcal{U} \to \mathcal{V}$ between open subsets of $\mathbb{R}^n$ is a diffeomorphism in the sense of Definition 2.16 if and only if it is a diffeomorphism in the standard sense. Note that Property (3) cannot be dropped from Definition 2.16. For instance, $F : \mathbb{R} \to \mathbb{R}$, $t \mapsto t^3$ is smooth and invertible, but its inverse $F^{-1} : \mathbb{R} \to \mathbb{R}$, $s \mapsto s^{1/3}$ is not smooth. Every diffeomorphism is a homeomorphism. Hence diffeomorphic manifolds are also homeomorphic. The identity is a diffeomorphism, the composition of diffeomorphisms is a diffeomorphism, and the inverse of a diffeomorphism is a diffeomorphism. It follows that *being diffeomorphic* is an equivalence relation. Additionally, diffeomorphisms from a manifold M to itself form a group under composition, called the *group of diffeomorphisms* of M, and denoted $\mathrm{Diffeo}(M)$.

Example 2.17. Consider the homeomorphism

$$\Phi : \mathbb{R}^{n_1} \times \cdots \times \mathbb{R}^{n_k} \to \mathbb{R}^n$$

of Example 1.34. It is easy to see that it is actually a diffeomorphism. ◆

Example 2.18. The map

$$R : S^n \to S^n, \quad P \mapsto -P$$

is a diffeomorphism with $R^{-1} = R$ (do you see it?). ◆

Example 2.19. Let M be a smooth manifold, and let k be a positive integer. Consider the product

$$M^{\times k} := \underbrace{M \times \cdots \times M}_{k \text{ times}}.$$

For every permutation $\sigma \in S_k$, the map

$$\Phi_\sigma : M^{\times k} \to M^{\times k}, \quad (p_1, \ldots, p_k) \mapsto \Phi_\sigma(p_1, \ldots, p_k) := (p_{\sigma(1)}, \ldots, p_{\sigma(k)}),$$

is a diffeomorphism with $\Phi_\sigma^{-1} = \Phi_{\sigma^{-1}}$ (do you see it?). ◆

Exercise 2.8. Consider the correspondence

$$\Phi : \mathbb{R}P^1 \to S^1,$$

$$[P^0 : P^1] \mapsto \Phi([P^0 : P^1]) := \left(\frac{2P^0 P^1}{(P^1)^2 + (P^0)^2}, \frac{(P^1)^2 - (P^0)^2}{(P^1)^2 + (P^0)^2} \right).$$

Prove that it is a well-defined diffeomorphism. (**Hint:** *Consider the affine charts* $(U_0, \varphi_0), (U_1, \varphi_1)$ *on the projective line* $\mathbb{R}P^1$ *and the stereographic charts* $(U_+, \varphi_+), (U_-, \varphi_-)$ *on the circle. Prove that* $\Phi(U_0) \subseteq U_+$ *and* $\Phi(U_1) \subseteq U_-$, *and, more precisely,*

$$\Phi([P]) = \begin{cases} (\varphi_+^{-1} \circ \varphi_0)([P]) & \text{if } [P] \in U_0 \\ (\varphi_-^{-1} \circ \varphi_1)([P]) & \text{if } [P] \in U_1 \end{cases}. \tag{2.1}$$

Use (2.1) to complete the proof.)

Proposition 2.20. *Let M be an n-dimensional manifold and let (U, φ) be a chart on M. Then the coordinate map*

$$\varphi : U \to \widehat{U}$$

is a diffeomorphism (between the open submanifold U of M and the open submanifold $\widehat{U}$ of the Euclidean space $\mathbb{R}^n$).

Proof.   Left as Exercise 2.9.   □

Exercise 2.9. Prove Proposition 2.20.

Diffeomorphisms identify the smooth structures in the following sense. Let $\Phi : M \to N$ be a diffeomorphism. Given a chart (U, φ) on M, the pair $(\Phi(U), \varphi \circ \Phi^{-1})$ is a chart on N. Additionally, the correspondence $(U, \varphi) \mapsto (\Phi(U), \varphi \circ \Phi^{-1})$ is a bijection between the smooth structures on M and N. Accordingly, two diffeomorphic manifolds should be identified.

Our next aim is to show that a smooth map $F : M \to N$ determines an *algebra homomorphism* $F^* : C^\infty(N) \to C^\infty(M)$. We begin defining algebra homomorphisms. So let A, B be (associative, commutative) $\mathbb{R}$-algebras (with unit).

Definition 2.21 (Algebra Homomorphism). An *algebra homomorphism* between A and B is an $\mathbb{R}$-linear map $H : A \to B$ such that

(1) *H preserves the product*, i.e., for every $\alpha, \beta \in A$,

$$H(\alpha\beta) = H(\alpha)H(\beta),$$

(2) *H preserves the unit,* i.e.,

$$H(1) = 1.$$

An *algebra isomorphism* is an invertible algebra homomorphism.

Note that an algebra homomorphism is, in particular, a ring homomorphism. The identity id : $A \to A$ is an algebra homomorphism. More generally, the inclusion of a subalgebra is an algebra homomorphism. The composition of algebra homomorphisms is an algebra homomorphism and the inverse of an algebra isomorphism is an algebra isomorphism. Finally, the composition of algebra isomorphisms is an algebra isomorphism.

Example 2.22 (Pull-Back of Real Valued Functions). Let X, Y be sets and let $F : X \to Y$ be a map. Consider the algebras of real-valued functions $\mathcal{F}(X, \mathbb{R}), \mathcal{F}(Y, \mathbb{R})$. It is easy to see that the map

$$F^* : \mathcal{F}(Y, \mathbb{R}) \to \mathcal{F}(X, \mathbb{R}), \quad g \mapsto F^*(g) := g \circ F$$

is an algebra homomorphism, called the *pull-back along F*. It is clear that the pull-back along the identity id : $X \to X$ is the identity id : $\mathcal{F}(X, \mathbb{R}) \to \mathcal{F}(X, \mathbb{R})$. If Z is another set and $G : Y \to Z$ is another map, then

$$(G \circ F)^* = F^* \circ G^*.$$

It follows that the pull-back $\Phi^* : \mathcal{F}(Y, \mathbb{R}) \to \mathcal{F}(X, \mathbb{R})$ along a bijection $\Phi : X \to Y$ is an algebra isomorphism with inverse given by the pull-back $(\Phi^{-1})^* : \mathcal{F}(X, \mathbb{R}) \to \mathcal{F}(Y, \mathbb{R})$ along the inverse map $\Phi^{-1} : Y \to X$, i.e.,

$$(\Phi^*)^{-1} = (\Phi^{-1})^*.$$

We leave to the reader to check all the claims in this example as Exercise 2.10. ◆

> **Exercise 2.10.** Prove all unproved claims in Example 2.22.

Now, let $F : M \to N$ be a smooth map between smooth manifolds. It follows from Proposition 2.15 that the pull-back $F^* : \mathcal{F}(N, \mathbb{R}) \to \mathcal{F}(M, \mathbb{R})$ takes smooth functions to smooth functions: $F^*(C^\infty(N)) \subseteq C^\infty(M)$.

The restriction $F^* : C^\infty(N) \to C^\infty(M)$ is also an algebra homomorphism, and we also call it the *pull-back along F*. Analogous remarks as for maps of sets hold in the case of smooth maps of manifolds: the pull-back along the identity id : $M \to M$ is the identity id : $C^\infty(M) \to C^\infty(M)$. If Q

is another smooth manifold and $G : N \to Q$ is another smooth map, then $(G \circ F)^* = F^* \circ G^*$. Finally, the pull-back $\Phi^* : C^\infty(N) \to C^\infty(M)$ along a diffeomorphism $\Phi : M \to N$ is an algebra isomorphism with inverse given by the pull-back $(\Phi^{-1})^* : C^\infty(M) \to C^\infty(N)$ along the inverse diffeomorphism $\Phi^{-1} : N \to M$.

Remark 2.23. We actually have a much stronger result that we state without proof in the following

Theorem 2.24. *Let M, N be smooth manifolds:*

(1) *A map $F : M \to N$ is smooth if and only if the pull-back F^* takes smooth functions to smooth functions.*
(2) *Every algebra homomorphism $H : C^\infty(N) \to C^\infty(M)$ is the pull-back along a unique smooth map $F : M \to N$, i.e., for every algebra homomorphism $H : C^\infty(N) \to C^\infty(M)$ there is a unique smooth map $F : M \to N$ such that $H = F^*$.*

In other words, the correspondence $F \mapsto F^*$ is a(n identity preserving and composition reversing) bijection between smooth maps $M \to N$ and algebra homomorphisms $C^\infty(N) \to C^\infty(M)$.

Proof. Omitted (for a detailed proof see, e.g., Nestruev, 2020). □

◇

2.3 Submanifolds

In this section, we present *submanifolds*. They are particularly nice subsets of a manifold M. They are so nice that they inherit a smooth structure from that of M. We will make a systematic use of the *Implicit Function Theorem*. Therefore, we begin recalling the appropriate version of it. First of all, we define *regular systems*. So, let d, n be non-negative integers such that $d \leq n$. Consider the standard Euclidean space $\mathbb{R}^n$ with standard coordinates $(t^1, \ldots, t^n)$, let $\mathcal{U} \subseteq \mathbb{R}^n$ be an open subset, and let

$$F = (F^1, \ldots, F^{n-d}) : \mathcal{U} \to \mathbb{R}^{n-d}$$

be a smooth map. The *zero locus* of F is the closed subset

$$Z(F) := \{P \in \mathcal{U} : F(P) = 0\} \subseteq \mathcal{U}.$$

In other words, $Z(F)$ is the space of solutions of the system

$$S_F : \begin{cases} F^1(t^1, \ldots, t^n) = 0 \\ \quad \vdots \\ F^{n-d}(t^1, \ldots, t^n) = 0 \end{cases}.$$

Denote by $M(n-d, n; \mathbb{R})$ the real vector space of real $(n-d) \times n$ matrices and recall that the *Jacobian matrix* of F is the matrix-valued smooth map

$$J_F : \mathcal{U} \to M(n-d, n; \mathbb{R}), \quad P \mapsto J_F(P) := \left(\frac{\partial F^a}{\partial t^i}(P) \right)_{i=1,\dots,n}^{a=1,\dots,n-d}.$$

Here the index a ranges over the rows, and the index i ranges over the columns. We say that 0 is a *regular value* of the smooth map F, or that the system S_F is *regular*, if

$$\operatorname{rank} J_F(P) = n - d \quad \text{for all } P \in Z(F).$$

Example 2.25. Let $F : \mathbb{R}^n \to \mathbb{R}^{n-d}$ be a surjective affine map. Then F is of the form

$$F(P) = A \cdot P + P_0, \quad P \in \mathbb{R}^n,$$

for some $P_0 \in \mathbb{R}^{n-d}$ and some $(n-d) \times n$ matrix A such that $\operatorname{rank} A = n - d$. Clearly, $J_F = A$ constantly. In particular,

$$S_F : \left\{ F(t^1, \dots, t^n) = 0 \right.$$

is a regular system. This shows that every *reduced linear system* is regular.
♦

Example 2.26. Let $U \subseteq \mathbb{R}^d$ and $V \subseteq \mathbb{R}^{n-d}$ be open subsets, let $f = (f^1, \dots, f^{n-d}) : U \to V$ be a smooth map, and let $F : U \times V \to \mathbb{R}^{n-d}$ be the smooth map given by

$$F : U \times V \to \mathbb{R}^{n-d}, \quad (Q, R) \mapsto F(Q, R) := R - f(Q).$$

So

$$S_F : \begin{cases} t^{d+1} - f^1(t^1, \dots, t^d) = 0 \\ \qquad \vdots \\ t^n - f^{n-d}(t^1, \dots, t^d) = 0 \end{cases},$$

and the zero locus of F is

$$Z(F) = \left\{ (Q, f(Q)) : Q \in U \subseteq \mathbb{R}^d \right\} \subseteq U \times V :$$

the *graph* of the smooth map f. The Jacobian matrix of F is

$$
J_F = \begin{pmatrix}
-\frac{\partial f^1}{\partial t^1} & \cdots & -\frac{\partial f^1}{\partial t^d} & 1 & \cdots & 0 \\
\vdots & \ddots & \vdots & \vdots & \ddots & \vdots \\
-\frac{\partial f^{n-d}}{\partial t^1} & \cdots & -\frac{\partial f^{n-d}}{\partial t^d} & 0 & \cdots & 1
\end{pmatrix},
$$

whose rank is $n - d$ at all points of $U \times \mathbb{R}^{n-d}$, in particular all points of $Z(F)$. Hence S_F is a regular system. ◆

Example 2.27. Let $F : \mathbb{R}^{n+1} \smallsetminus \{0\} \to \mathbb{R}$ be the smooth map given by

$$
F(P) := \|P\|^2 - 1, \quad P \in \mathbb{R}^{n+1}.
$$

In other words,

$$
F(t^1, \ldots, t^n) = (t^1)^2 + \cdots + (t^{n+1})^2 - 1.
$$

The Jacobian matrix of F is the single row

$$
J_F = (2t^1, \ldots, 2t^{n+1})
$$

whose rank is 0 at the origin and 1 elsewhere. As $Z(F) = S^n$ is the n-dimensional sphere, and $0 \notin S^n$, we conclude that 0 is a regular value of F, and the system

$$
S_F : \left\{ (t^1)^2 + \cdots + (t^{n+1})^2 - 1 = 0 \right.
$$

is regular. ◆

Example 2.28. On $\mathbb{R}^2$ with coordinates (x, y), consider the smooth map $F : \mathbb{R}^2 \to \mathbb{R}$ given by

$$
F(x, y) := xy.
$$

The Jacobian matrix of F is the row

$$
J_F = (y, x)
$$

whose rank is 0 at the origin and 1 elsewhere. However, the origin is in $Z(F)$, so the system

$$
S_F : \{ xy = 0
$$

is *not* regular. ◆

Note that, if rank $J_F(P_0) = n - d$ for some $P_0 \in \mathcal{U}$, then there exists an $(n - d) \times (n - d)$ minor $\mathbb{M}$ of J_F which is non-zero at P_0. Hence, by continuity, $\mathbb{M}$ is non-zero in a full open neighborhood $\mathcal{V} \subseteq \mathcal{U}$ of P_0 (in particular, rank $J_F(P) = n - d$ for all $P \in \mathcal{V}$). It follows from this discussion that, if S_F is a regular system, then, for every $P_0 \in Z(F)$, there exists an $(n - d) \times (n - d)$ minor $\mathbb{M}$ of J_F which is non-zero in an open neighborhood $\mathcal{V} \subseteq \mathcal{U}$ of P_0. Let $\mathbb{M}$ be the minor identified by (all the rows and) the columns

$$i_1, \ldots, i_{n-d}$$

of J_F. In this situation, we denote by $(v^1, \ldots, v^{n-d})$ the coordinates $(t^{i_1}, \ldots, t^{i_{n-d}})$ respectively and by $(u^1, \ldots, u^d)$ the remaining coordinates, in their order. So, the n-tuple $(u^1, \ldots, u^d, v^1, \ldots, v^{n-d})$ differs from $(t^1, \ldots, t^n)$ only by a permutation. Denote it σ. In the following, we also denote $u = (u^1, \ldots, u^d)$ and $v = (v^1, \ldots, v^{n-d})$. Given open subsets $V \subseteq \mathbb{R}^{n-d}$ and $U \subseteq \mathbb{R}^d$, we denote by $U \times_\sigma V \subseteq \mathbb{R}^n$ the open subset defined by

$$U \times_\sigma V := \{P \in \mathbb{R}^n : u(P) \in U, \text{ and } v(P) \in V\}.$$

Proposition 2.29. *Let $\mathcal{U}, \mathcal{V} \subseteq \mathbb{R}^n$ be open subsets and let $\Phi : \mathcal{V} \to \mathcal{U}$ be a diffeomorphism. Additionally, let $F : \mathcal{U} \to \mathbb{R}^{n-d}$ be a smooth map. Then 0 is a regular value of F if and only if it is a regular value of $F \circ \Phi$.*

Exercise 2.11. Prove Proposition 2.29 (**Hint:** *First of all, let $\mathcal{W} \subseteq \mathbb{R}^m$ be an open subset and let $G : \mathcal{W} \to \mathcal{U}$ be a smooth map. Use the chain rule for the derivatives of a composition of smooth maps to show that*

$$J_{F \circ G}(Q) = J_F(G(Q)) \cdot J_G(Q) \tag{2.2}$$

for all $Q \in \mathcal{W}$ (see also Chapter 4). Now apply (2.2) to the case $F = \Phi^{-1}$ and $G = \Phi$ to show that the Jacobian matrix J_Φ of Φ is invertible (the inverse being $J_{\Phi^{-1}} \circ \Phi$). Finally, apply (2.2) to the case $G = \Phi$.)

We are now ready to state the *Implicit Function Theorem*.

Theorem 2.30 (Implicit Function Theorem). *Let d, n be non-negative integers such that $d \leq n$, let $\mathcal{U} \subseteq \mathbb{R}^n$ be an open subset, and let*

$$F : \mathcal{U} \to \mathbb{R}^{n-d}$$

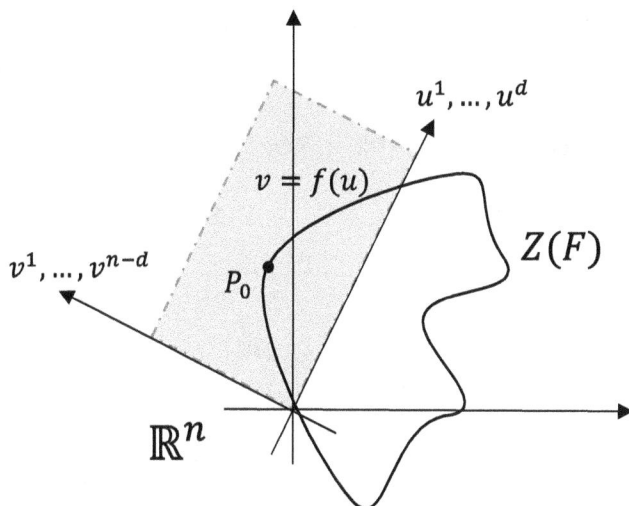

Figure 2.9. The Implicit Function Theorem.

be a smooth map. If 0 is a regular value of F, then, locally, around every point, $Z(F)$ is the graph of a smooth map, *i.e., for every point $P_0 \in Z(F)$, there are*

(1) *an open neighborhood $U \subseteq \mathbb{R}^d$ of $u(P_0)$,*
(2) *an open neighborhood $V \subseteq \mathbb{R}^{n-d}$ of $v(P_0)$, and*
(3) *a smooth map $f = (f^1, \ldots, f^{n-d}) : U \to V$*

such that $U \times_\sigma V \subseteq \mathcal{U}$ and, in $U \times_\sigma V$, the system S_F is equivalent to the system

$$
\begin{cases}
v^1 = f^1(u^1, \ldots, u^d) \\
\quad\vdots \\
v^{n-d} = f^{n-d}(u^1, \ldots, u^d)
\end{cases}
\tag{2.3}
$$

i.e., $Z(F) \cap U \times_\sigma V$ is exactly the space of solutions of (2.3) (here (u, v) are the tuples of coordinates defined in the discussion preceding Exercise 2.11), see Figure 2.9.

For more details on the Implicit Function Theorem 2.30, including the proof, see, e.g., Lang (1997).

Remark 2.31. In the situation described in the statement of the Implicit Function Theorem, we also say that, locally, around the point $P_0 \in Z(F)$, we can *solve the system S_F with respect to the variables $(v^1, \ldots, v^{n-d})$.*

This can be done even globally if F is an affine map (see Example 2.25) and, in this case, the proof of the Implicit Function Theorem boils down to the standard *Gauss elimination method*. ◇

Example 2.32. Let $F : \mathbb{R}^{n+1} \setminus \{0\} \to \mathbb{R}$ be as in Example 2.27, and let

$$P_0 = (P_0^1, \ldots, P_0^{n+1}) \in Z(F) = S^n.$$

Then

$$J_F(P_0) = (2P_0^1, \ldots, 2P_0^{n+1}) \neq 0$$

and there is $i \in \{1, \ldots, n+1\}$ such that $P_0^i \neq 0$. Assume for simplicity that $P_0^i > 0$. In this case, we can choose $v = t^i$ and $u = (t^1, \ldots, \widehat{t^i}, \ldots, t^{n+1})$ (where, as usual, a hat denotes omission). Additionally, we can choose

(1) $U \subseteq \mathbb{R}^n$ the unit open disk,
(2) $V \subseteq \mathbb{R}$ the open interval $(0, +\infty)$, and
(3) $f(u) = \sqrt{1 - \|u\|^2}$

so that

$$U \times_\sigma V = \left\{ \sum_{j \neq i} (t^j)^2 < 1, \text{ and } t^i > 0 \right\}$$

and

$$Z(F) \cap U \times_\sigma V = \{P = (P^1, \ldots, P^n) \in S^n : P^i > 0\}$$

is the space of solution of the system

$$\left\{ v = \sqrt{1 - \|u\|^2} \quad \Leftrightarrow \quad \left\{ t^i = \sqrt{1 - \sum_{j \neq i} (t^j)^2} \right. \right.$$

Yet in other words, (locally around P_0) we were able to solve the equation

$$(t^1)^2 + \cdots + (t^{n+1})^2 - 1 = 0$$

with respect to t^i. The case $P_0^i < 0$ can be discussed in a similar (and obvious) way (do you see it?). ♦

Exercise 2.12. Given the open subset $\mathcal{U} \subseteq \mathbb{R}^n$ and the smooth map $F : \mathcal{U} \to \mathbb{R}^{n-d}$, prove that 0 is a regular value of F and, for every point $P_0 \in Z(F)$, find U, V, f as in the statement of the Implicit Function Theorem, in the following two cases:

- in $\mathbb{R}^3$ with coordinates (x, y, z), $\mathcal{U} = \mathbb{R}^3 \setminus \{x = y = 0\}$, and $F : \mathcal{U} \to \mathbb{R}$ given by

$$F(x, y, z) = \left(\sqrt{x^2 + y^2} - R \right)^2 + z^2 - r^2,$$

where r, R are positive real numbers with $r < R$;
- in $\mathbb{R}^4$ with coordinates (w, x, y, z), $\mathcal{U} = \mathbb{R}^4$, and $F : \mathbb{R}^4 \to \mathbb{R}^2$ given by

$$F(w, x, y, z) = (w^2 - x^2 - 1, y^2 + z^2 - 1),$$

so that

$$S_F : \begin{cases} w^2 - x^2 - 1 = 0 \\ y^2 + z^2 - 1 = 0 \end{cases}.$$

We finally come to the main definition of this section. Let M be an n-dimensional manifold.

Definition 2.33 (Submanifold). A *d-dimensional submanifold* of M is a subset $S \subseteq M$ such that, for every point $p_0 \in S$, there is a chart (U, φ) on M around p_0 such that $\varphi(S \cap U)$ is the solution space of a regular system, i.e., there is a smooth map $F : \widehat{U} \to \mathbb{R}^{n-d}$ such that

(1) 0 is a regular value of F and
(2) $\varphi(S \cap U) = Z(F)$.

A submanifold is a *smooth subset* in the sense that, in local coordinates, it looks like the solution space of a regular system, hence, by the Implicit Function Theorem, as the graph of a smooth map.

Proposition 2.34. *Let $S \subseteq M$ be a submanifold. Then, for every point $p_0 \in S$, and every chart (V, ψ) around p_0, there is a subchart (V_0, ψ) around p_0, such that $\psi(S \cap V_0)$ is the solution space of a regular system.*

Proof. Let (U, φ) be a chart around p_0 such that $\varphi(S \cap U)$ is the solution space of a regular system. Put $V_0 = U \cap V$. Then $\varphi(S \cap V_0)$ is the solution space of a regular system (defined on the open subset $\varphi(V_0) \subseteq \widehat{U}$). We want to show that $\psi(S \cap V_0)$ is also the solution space of a regular system. To do this, note that

(1) $\varphi(V_0)$ and $\psi(V_0)$ are related by the transition map $\varphi \circ \psi^{-1} : \psi(V_0) \to \varphi(V_0)$ and
(2) the transition map $\varphi \circ \psi^{-1}$ identifies $\psi(S \cap V_0)$ and $\varphi(S \cap V_0)$.

Finally, use Proposition 2.29.   □

Example 2.35. A d-dimensional affine subspace A in a vector space V is a d-dimensional submanifold. Indeed, in linear coordinates, A can be always written as the solution space of a reduced linear system (see also Example 2.25).   ◆

Example 2.36 (Graph of a Smooth Map). Let $F : M \to N$ be a smooth map between manifolds, $\dim M = n$. By definition, the *graph* of F is the subset graph $F \subseteq M \times N$ in the product manifold defined by

$$\text{graph}\, F := \{(p, F(p)) : p \in M\}.$$

Then graph $F \subseteq M \times N$ is an n-dimensional submanifold. Indeed, pick a point $p \in M$, and let (U, φ) be a chart on M around p and (V, ψ) a chart on N around $F(p)$ such that $F(U) \subseteq V$. Denote by $(U \times V, \varphi \times \psi)$ the product chart on $M \times N$. It is easy to see that $\varphi \times \psi(\text{graph}\, F)$ is the graph of the coordinate representation $\widehat{F} : \widehat{U} \to \widehat{V}$, hence it is the solution space of a regular system as in Example 2.26.   ◆

Example 2.37 (Spheres are Submanifolds). Example 2.27 shows that the n-dimensional sphere is an n-dimensional submanifold of $\mathbb{R}^{n+1}$.   ◆

Example 2.38. In the 3-dimensional projective space $\mathbb{R}P^3$, consider the subset

$$S := \left\{ [P^0 : P^1 : P^2 : P^3] \in \mathbb{R}P^3 : P^0 P^3 - P^1 P^2 = 0 \right\}.$$

We want to show that $S \subseteq \mathbb{R}P^3$ is a 2-dimensional submanifold. To do this, consider the affine charts (U_i, φ_i) on $\mathbb{R}P^3$, $i = 0, \ldots, 3$, and compute $\varphi_i(S \cap U_i) \subseteq \mathbb{R}^3$. For instance, if $i = 0$,

$$S \cap U_0 = \left\{ [1 : Q^1 : Q^2 : Q^3] : Q^3 - Q^1 Q^2 = 0 \right\}$$

and $\varphi_0(S \cap U_0)$ is the solution space of the regular system on $\mathbb{R}^3$

$$S : \left\{ t^3 - t^1 t^2 = 0. \right.$$

Similarly for $i = 1, 2, 3$.   ◆

Example 2.39 ("Trivial" Submanifolds). Let M be an n-dimensional manifold. It is easy to see that

- 0-dimensional submanifolds of M are exactly discrete subspaces and
- n-dimensional submanifolds are exactly open submanifolds

(do you see it?). ◆

There are two useful characterizations of submanifolds. We summarize them in the following

Proposition 2.40. *Let M be an n-dimensional manifold and let $S \subseteq M$ be a subset. The following three conditions are equivalent:*

(1) *S is a d-dimensional submanifold.*
(2) *For every $P_0 \in S$, there is a chart (W, ψ) around P_0 such that $\widehat{W} = U \times V$, with $U \subseteq \mathbb{R}^d$ and $V \subseteq \mathbb{R}^{n-d}$ open subsets, and $\psi(S \cap W)$ is the graph of a smooth map $f : U \to V$.*
(3) *For every $P_0 \in S$, there is a chart $(W_0, \chi = (x^1, \ldots, x^d, y^1, \ldots, y^{n-d}))$ around P_0 such that $\widehat{W}_0 = U_0 \times V_0$, with $U_0 \subseteq \mathbb{R}^d$ and $V_0 \subseteq \mathbb{R}^{n-d}$ open subsets, and $S \cap W_0$ is the zero locus of the y^a, i.e., $S \cap W_0 = \{p \in U_0 : y^1(p) = \cdots = y^{n-d}(p) = 0\}$ (see Figure 2.10).*

Proof.

(1) $\Rightarrow$ (2). It follows immediately from the Implicit Function Theorem, restricting to suitable subcharts.

(2) $\Rightarrow$ (3). First of all, let $U \subseteq \mathbb{R}^d$ be an open subset, and let $f : U \to \mathbb{R}^{n-d}$ be a smooth map. The map

$$\Phi_f : U \times \mathbb{R}^{n-d} \to U \times \mathbb{R}^{n-d}, \quad (P, Q) \mapsto (P, Q - f(P))$$

is a diffeomorphism with inverse Φ_{-f}. Additionally, Φ_f maps the graph of f to the graph of the zero map $0 : U \to \mathbb{R}^{n-d}$. Now, let S, P_0, and (W, ψ) be as in the statement. Then, composing with Φ_f, we get a new chart $(W, \chi := \Phi_f \circ \psi)$ such that $\chi(W \cap S)$ is the graph of the zero section. We can always restrict to a subchart (W_0, χ) around P_0 whose coordinate domain is of the form $\widehat{W}_0 = U_0 \times V_0$ (see Figure 2.11), and this concludes the proof.

(3) $\Rightarrow$ (1). The image $\chi(S \cap W_0)$ is given by the vanishing of the last $n - d$ coordinates, and this condition is a regular system. □

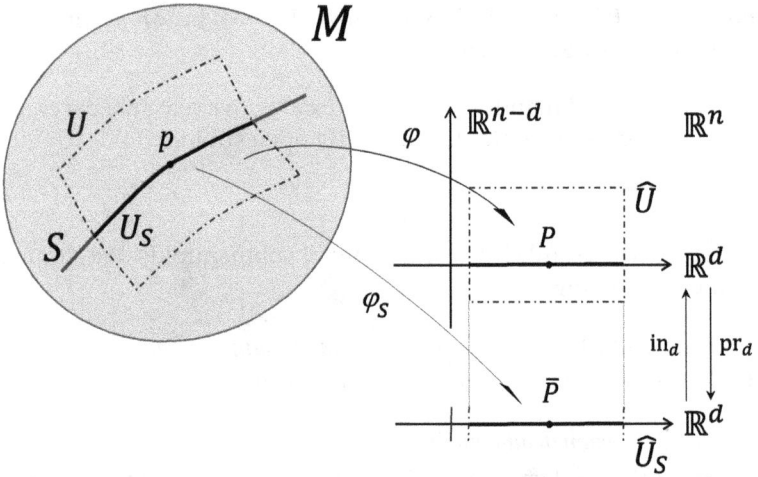

Figure 2.10.   The construction of a submanifold chart.

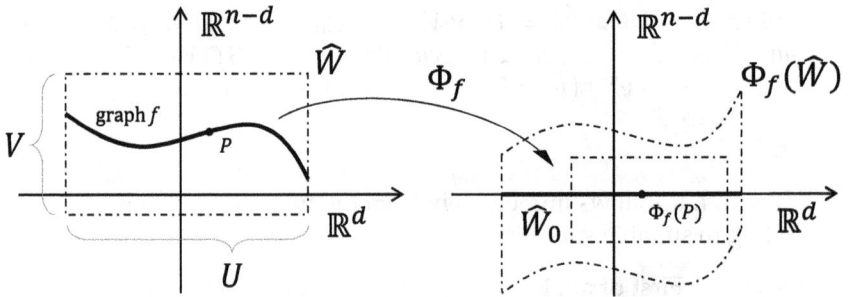

Figure 2.11.   The diffeomorphism Φ_f.

Let M be an n-dimensional manifold and let $S \subseteq M$ be a d-dimensional submanifold in M.

Definition 2.41 (Adapted Chart). A chart (W_0, χ) on M around a point $p \in S$ like in item (3) of Proposition 2.40 is said to be *adapted* to S.

Exercise 2.13. Let M, N be manifolds, and let $\Phi : M \to N$ be a diffeomorphism. Show that a subset $S \subseteq M$ is a d-dimensional submanifold if and only if so is $\Phi(S) \subseteq N$ (**Hint:** *Use Proposition 2.29*).

Our next aim is to show that submanifolds in a manifold M inherit from M a manifold structure. This is explained in the following theorem and its proof.

Theorem 2.42 (Submanifold Chart Theorem). *Let M be an n-dimensional manifold and let $S \subseteq M$ be a d-dimensional submanifold in M. There exists a unique smooth structure $\mathcal{A}_S$ on S such that*

(1) *$(S, \mathcal{A}_S)$ is a d-dimensional manifold,*
(2) *the inclusion $i_S : S \hookrightarrow M$ is a smooth map, and*
(3) *given a map between manifolds $F : N \to M$ taking values in S, i.e., $F(N) \subseteq S$, we have that $F : N \to M$ is smooth if and only if $F : N \to S$ is smooth.*

Proof. We begin with the existence. First of all, we consider two auxiliary smooth maps:

$$\mathrm{pr}_d : \mathbb{R}^n \to \mathbb{R}^d, \quad P = (P^1, \ldots, P^n) \mapsto \mathrm{pr}_d(P) := (P^1, \ldots, P^d),$$

and

$$\mathrm{in}_d : \mathbb{R}^d \to \mathbb{R}^n, \quad Q = (Q^1, \ldots, Q^d) \mapsto \mathrm{in}_d(Q) := (Q^1, \ldots, Q^d, 0, \ldots, 0).$$

In the following, it will be sometimes useful to keep in mind that

- pr_d maps open subsets to open subsets and
- in_d is a homeomorphism onto its image (equipped with the subspace topology)

(do you see it?). Now, let $p \in S$ and let (U, φ) be a chart on M around p adapted to S. According to the definition of adapted chart, we have, in particular, $\widehat{U} = \widehat{U}_S \times V \subseteq \mathbb{R}^d \times \mathbb{R}^{n-d} = \mathbb{R}^n$, for some open subsets $\widehat{U}_S \subseteq \mathbb{R}^d$ and $V \subseteq \mathbb{R}^{n-d}$. Put $U_S := S \cap U$. The composition

$$\varphi_S := \mathrm{pr}_d \circ \varphi : U_S \to \widehat{U}_S$$

is inverted by

$$\varphi^{-1} \circ \mathrm{in}_d : \widehat{U}_S \to U_S.$$

Hence (U_S, φ_S) is a chart on S around p. We will refer to any such chart as a *submanifold chart*.

Submanifold charts cover S. The next step is proving that they actually form an atlas. We have to prove that any two submanifold charts are compatible. So let $(U, \varphi), (V, \psi)$ be adapted charts on M around p, and let $(U_S, \varphi_S), (V_S, \psi_S)$ be the induced submanifold charts. First of all, $\varphi_S(U_S \cap V_S) \subseteq \widehat{U}_S$ is an open subset, indeed in_d maps it to

$$\varphi(U \cap V) \cap \left(\widehat{U}_S \times \{0\}\right)$$

which is open in the subspace $\widehat{U}_S \times \{0\} \subseteq \widehat{U}$. Similarly, $\psi_S(U_S \cap V_S) \subseteq \widehat{V}_S$ is an open subset. The transition map

$$\psi_S \circ \varphi_S^{-1} : \varphi_S(U_S \cap V_S) \to \psi_S(U_S \cap V_S)$$

can be written as the following composition of smooth maps:

$$\psi_S \circ \varphi_S^{-1} = \text{pr}_d \circ (\psi \circ \varphi^{-1}) \circ \text{in}_d$$

(see Figure 2.12), hence it is smooth. Exchanging the role of (U, φ) and (V, ψ) shows that the inverse $\varphi_S \circ \psi_S^{-1}$ is also smooth so that the transition map is a diffeomorphism and $(U_S, \varphi_S), (V_S, \psi_S)$ are compatible charts.

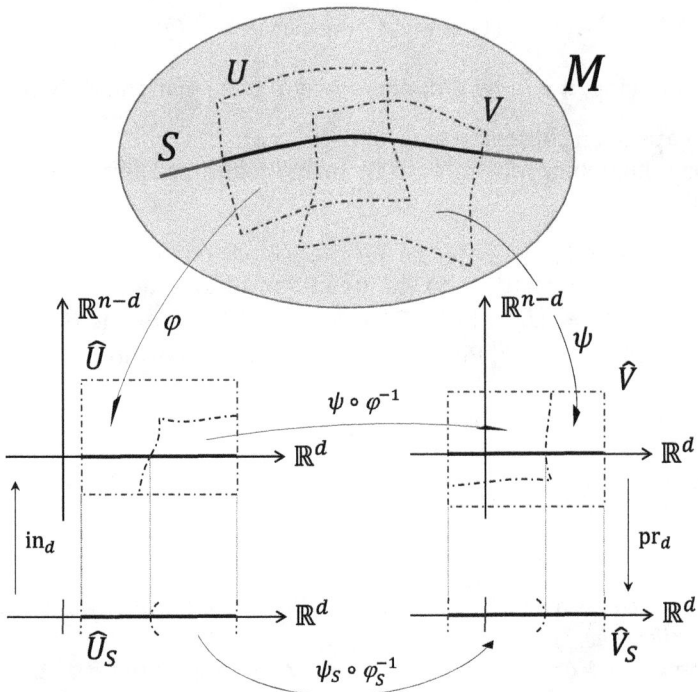

Figure 2.12. Submanifold charts are compatible.

The atlas $\{(U_S, \varphi_S)\}$ on S consisting of submanifold charts will be referred to as the *submanifold atlas*, and it is contained in a unique smooth structure denoted $\mathcal{A}_S$. In order to prove that $(S, \mathcal{A}_S)$ is a smooth manifold, we have to show that the atlas topology $\tau_{\mathcal{A}_S}$ induced by the maximal atlas $\mathcal{A}_S$, or, equivalently, the submanifold atlas, is Hausdorff and II-countable. Note that it is enough to prove that $\tau_{\mathcal{A}_S}$ coincides with the subspace topology τ_S. To do this, we use Proposition 1.35. We have to show that for every submanifold chart (U_S, φ_S), $U_S \subseteq S$ is an open subset with respect to the subspace topology, which is obvious because $U_S = S \cap U$ and $U \subseteq M$ is an open subset, and, additionally, that $\varphi_S : U_S \to \widehat{U}_S$ is a homeomorphism when $U_S \subseteq M$ is equipped with the subspace topology. For the continuity of φ_S, write φ_S as the following continuous composition of continuous maps:

$$\varphi_S = \mathrm{pr}_d \circ \varphi \circ i_S : U_S \to \widehat{U}_S.$$

For the continuity of the inverse φ_S^{-1}, note that $\varphi_S^{-1} : \widehat{U}_S \to U_S$ is the restriction to the subspace $U_S \subseteq U$ in the codomain of the continuous composition of continuous maps

$$\varphi^{-1} \circ \mathrm{in}_d : \widehat{U}_S \to U,$$

hence φ_S^{-1} is continuous with respect to the subspace topology (see Example 1.32). We conclude that $\tau_{\mathcal{A}_S} = \tau_S$, in particular $(S, \tau_{\mathcal{A}_S})$ is Hausdorff and II-countable, hence it is a smooth (d-dimensional) manifold.

Now we have to show that the manifold $(S, \mathcal{A}_S)$ enjoys Properties (2) and (3) in the statement. For the smoothness of $i_S : S \hookrightarrow M$, note that, if (U_S, φ_S) is a submanifold chart induced by an adapted chart (U, φ), then $i_S(U_S) \subseteq U$ showing that i_S has coordinate representations. Additionally, the coordinate representation $\widehat{i}_S : \widehat{U}_S \to \widehat{U}$ with respect to the charts (U_S, φ_S) and (U, φ) is simply in_d which is smooth. So i_S is smooth. Finally, take a map between manifolds $F : N \to M$ such that $F(N) \subseteq S$. For the sake of clarity, we denote by $F_S : N \to S$ the restriction of F to S in the codomain. We have to show that F is smooth if and only if so is F_S. If F_S is smooth, then $F = i_S \circ F_S$ is the smooth composition of smooth maps, hence it is smooth. Conversely, let F be smooth. Take a point $q \in N$, let (U, φ) be an adapted chart around $F(q)$, and let (U_S, φ_S) be the induced submanifold chart. As F is a continuous map, $F^{-1}(U) \subseteq N$ is an open subset, hence there is a chart (V, ψ) on N around q such that $V \subseteq F^{-1}(U)$, i.e., $F(V) \subseteq U$. But F takes values in S, hence $F_S(V) = F(V) \subseteq U \cap S = U_S$, showing that F_S has coordinate representations. It remains to show that the coordinate representation $\widehat{F}_S : \widehat{V} \to \widehat{U}_S$ with respect to the charts $(V, \psi), (U_S, \varphi_S)$ is

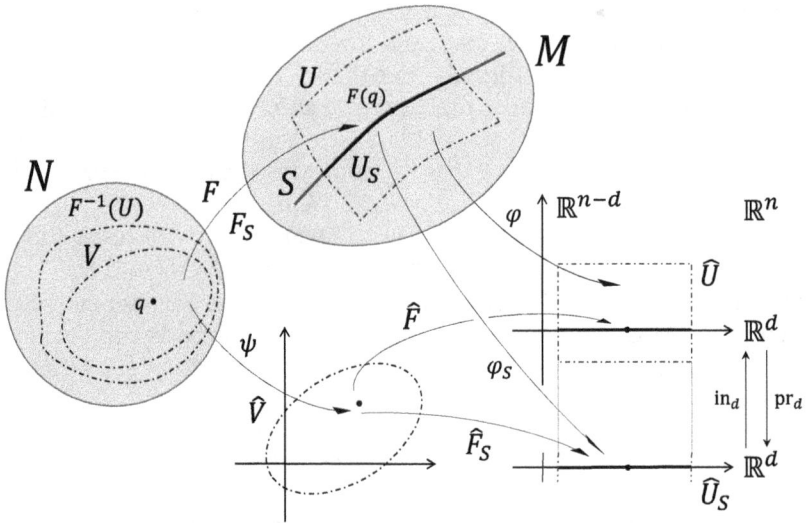

Figure 2.13.   Smoothness of a map taking values in a submanifold.

smooth. It is easy to see that, actually, $\widehat{F}_S = \text{pr}_d \circ \widehat{F}$; in particular, it is a smooth composition of smooth maps, see Figure 2.13. This concludes the proof of Property (3) for the smooth structure $\mathcal{A}_S$.

We conclude the proof of the theorem showing that $\mathcal{A}_S$ is the only smooth structure with Properties (1), (2), and (3) in the statement. So, suppose $\mathcal{A}'$, $\mathcal{A}''$ are smooth structures on S with those properties. We denote by S' the manifold $(S, \mathcal{A}')$ and by S'' the manifold $(S, \mathcal{A}'')$. We want to show that $S' = S''$, i.e., $\mathcal{A}' = \mathcal{A}''$. To do this, it is enough to prove that the identity map id : $S \to S$ induces a diffeomorphism id : $S' \to S''$ (do you see it?). So, denote by $i_{S'} : S' \hookrightarrow M$ the inclusion. As S' has Property (2), $i_{S'}$ is smooth. Additionally, it takes values in S. As S'' has Property (3), it follows that the restriction of $i_{S'}$ to S'' in the codomain is also smooth. But the latter restriction is exactly id : $S' \to S''$. Exchanging the role of S' and S'', we see that the inverse map id : $S'' \to S'$ is also smooth, hence id : $S' \to S''$ is a diffeomorphism as claimed, and this concludes the proof.   □

Definition 2.43 (Submanifold Atlas). A chart (U_S, φ_S) on S as in the proof of Theorem 2.42 is called a *submanifold chart*, and the atlas $\{(U_S, \varphi_S)\}$ is a *submanifold atlas*.

Example 2.44. Let M be a manifold and let $\mathcal{U} \subseteq M$ be an open submanifold. In particular, $\mathcal{U}$ is a submanifold. Actually, the submanifold atlas on $\mathcal{U}$ agrees with the atlas $\mathcal{A}_{\mathcal{U}}$ constructed at the end of the previous chapter.   ◆

Example 2.45. In $\mathbb{R}^{n+1}$, consider the n-dimensional sphere $S^n \subseteq \mathbb{R}^{n+1}$. We already explained that S^n is an n-dimensional submanifold. So, it inherits a smooth structure $\mathcal{A}_{S^n}$ from $\mathbb{R}^{n+1}$, and $(S^n, \mathcal{A}_{S^n})$ is an n-dimensional manifold. We want to show that $\mathcal{A}_{S^n}$ is not a *new* smooth structure on S^n but it coincides with the smooth structure $\mathcal{A}$ containing the stereographic atlas (or, equivalently, the atlas consisting of orthogonal projections). We do this via Theorem 2.42. We already know that $i_{S^n} : (S^n, \mathcal{A}) \to \mathbb{R}^{n+1}$ is smooth (Exercise 2.4). It remains to show that $(S^n, \mathcal{A})$ has Property (3) in the statement of Theorem 2.42. So, let $F : N \to \mathbb{R}^{n+1}$ be a smooth map taking values in S^n. We want to show that the restriction $F_{S^n} : N \to S^n$ of F to S^n in the codomain is smooth. We use a trick: we interpret F as a map with values in $\mathbb{R}^{n+1} \setminus \{0\}$:

$$F : N \to \mathbb{R}^{n+1} \setminus \{0\}.$$

As such it is still smooth. Now, there is a smooth projection

$$\pi : \mathbb{R}^{n+1} \setminus \{0\} \to (S^n, \mathcal{A}), \quad P \mapsto \pi(P) := \frac{P}{\|P\|}$$

(show that π is actually smooth as an exercise). Conclude noting that $F_{S^n} = \pi \circ F$, i.e., it is the smooth composition of smooth maps. ◆

Exercise 2.14. Show that a submanifold $T \subseteq S$ in a submanifold $S \subseteq M$ is a submanifold in M and that S and M induce the same smooth structure on T. (**Hint:** *Use adapted charts on S, and Proposition 2.34, to show that there are charts in M, where $T \subseteq M$ looks like the solution space of a regular system.*)

We conclude this chapter discussing briefly smooth functions on submanifolds. So let $S \subseteq M$ be a submanifold in a manifold, and, as usual, denote by $i_S : S \hookrightarrow M$ the (smooth) inclusion. For every smooth function $f \in C^\infty(M)$, its restriction $f|_S$ to S is a smooth function on S, indeed

$$f|_S = i_S^*(f) = f \circ i_S :$$

the pull-back of f along the smooth map i_S. However, in general, not all smooth functions $g \in C^\infty(S)$ are restrictions of smooth functions on M as the following counter-example shows.

Example 2.46. In $M = \mathbb{R}$ with coordinate t, consider the open submanifold $S = \mathbb{R} \setminus \{0\}$. The function

$$\frac{1}{t} : S \to \mathbb{R}$$

is a smooth function, but it is not the restriction of a smooth function on $\mathbb{R}$. ◆

However, every smooth function g on S is *locally* the restriction of a smooth function on M in the sense of the following.

Lemma 2.47 (Local Extension Lemma). *Let M be a manifold, and let $S \subseteq M$ be a submanifold. For every smooth function $g \in C^\infty(S)$ and every point $p_0 \in S$, there exist*

(1) *a smooth function $\widetilde{g} \in C^\infty(M)$ and*
(2) *an open neighborhood $V \subseteq M$ of p_0*

such that $\widetilde{g}$ and g agree on $S \cap V$, i.e., $g|_{S \cap V} = \widetilde{g}|_{S \cap V}$.

Proof.   The statement is a corollary of the Bump Function Theorem (Theorem 2.4). Choose an adapted chart (U, φ) around p_0, and let (U_S, φ_S) be the induced submanifold chart. Let $\widehat{g} : \widehat{U}_S \to \mathbb{R}$ be the coordinate representation of g with respect to the chart (U_S, φ_S). The composition

$$\widetilde{g}' := \widehat{g} \circ \mathrm{pr}_d \circ \varphi : U \to \mathbb{R}$$

is a smooth function on U, i.e., $\widetilde{g}' \in C^\infty(U)$. It is easy to see that $\widetilde{g}'|_{U_S} = g|_{U_S}$, however, in general, $\widetilde{g}'$ is not the restriction to U of a smooth function on M. We can cure this problem locally, using a bump function. So let $\beta \in C^\infty(M)$ be a smooth function such that $\mathrm{supp}\,\beta \subseteq U$ and $\beta|_V = 1$ for some open neighborhood $V \subseteq U$ of p_0. Consider the function $\widetilde{g} : M \to \mathbb{R}$ defined by

$$\widetilde{g}(p) := \begin{cases} \beta(p)\widetilde{g}'(p) & \text{if } p \in U \\ 0 & \text{if } p \in M \setminus \mathrm{supp}\,\beta \end{cases}.$$

Clearly, $\widetilde{g}$ is well defined, and it is smooth by the Gluing Lemma. Finally, for every p in $S \cap V$, we have

$$\widetilde{g}(p) = \beta(p)\widetilde{g}'(p) = g(p),$$

as claimed. $\square$

Chapter 3

Tangent Vectors

In this chapter, we begin developing a *differential calculus on manifolds*. The "smoothness" of manifolds allows us to "take derivatives". In particular, we can define tangent vectors and tangent spaces to manifolds. Tangent vectors at some point p of a manifold M can be informally understood as infinitesimal displacements from p. Similarly, the tangent space at p should be understood as the "first-order, linear approximation" to M. As such, tangent spaces carry natural structures of vector spaces. We can also define the tangent map at p to a smooth map F: it is the "first-order, linear approximation" to F.

3.1 Tangent Spaces

In order to motivate Definition 3.2 of tangent vector, we begin with a familiar situation: let M be a submanifold in the standard Euclidean space $\mathbb{R}^n$, and consider a *parameterized curve* in M, i.e., a smooth map $\gamma : I \to M$ from an open interval $I \subseteq \mathbb{R}$ to M. We denote by t the standard coordinate on I and interpret t as the "time". As γ takes values in $\mathbb{R}^n$, we can understand γ as a vector valued map: $\gamma = (\gamma^1, \ldots, \gamma^n)$. Its *velocity* at time $t_0 \in I$ is the vector

$$\frac{d}{dt}\Big|_{t_0} \gamma := \left(\frac{d\gamma^1}{dt}(t_0), \ldots, \frac{d\gamma^n}{dt}(t_0) \right).$$

More importantly, it can be well interpreted as a *tangent vector to M at the point* $p = \gamma(t_0)$ (see Figure 3.1). Our intuition confirms that every tangent vector to M at p is the velocity of some curve "passing through" p at

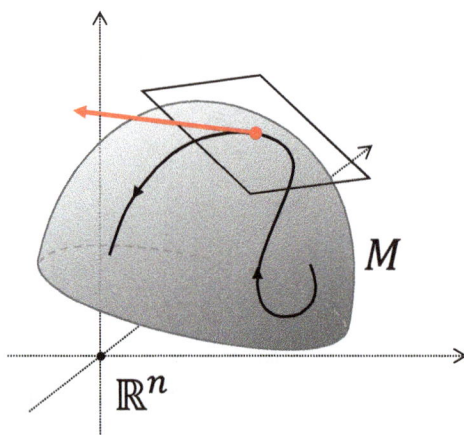

Figure 3.1.   Velocity of a curve in a submanifold of $\mathbb{R}^n$.

some time. We would like to extend this picture to a generic manifold M. First, we need to define curves in M. It is clear how to do this:

Definition 3.1 (Smooth Curve). A *smooth curve* (or, simply, a *curve*) in M is a smooth map $\gamma : I \to M$ from some open interval $I \subseteq \mathbb{R}$ to M.

Given a curve $\gamma : I \to M$, we will usually denote by t the standard coordinate on I and interpret it as the "time". The next step is defining the *velocity of γ at the time $t_0 \in I$*. However, this is not as easy as in the case when M is a submanifold in $\mathbb{R}^n$ because we don't know (yet) how to take derivatives of a map with values in a manifold. Note that, however defined, the velocity of γ should measure how fast the point $\gamma(t) \in M$ changes when t changes in I. In order to go back to a more familiar situation, we pick a "test function" $f \in C^\infty(M)$ and, instead of measuring directly how fast $\gamma(t)$ changes at the time t_0, we measure first how fast the real number $f(\gamma(t))$ changes at the time t_0. This can be easily done taking the derivative at t_0 of the real valued function $f \circ \gamma : I \to \mathbb{R}$:

$$\frac{d}{dt}|_{t_0} f \circ \gamma.$$

In order to get rid of the dependence on the test function f, we repeat this *for every function $f \in C^\infty(M)$*. In this way, we get a map

$$\dot{\gamma}(t_0) : C^\infty(M) \to \mathbb{R}, \quad f \mapsto \dot{\gamma}(t_0)(f) := \frac{d}{dt}|_{t_0} f \circ \gamma. \tag{3.1}$$

This map measures well enough how fast $\gamma(t)$ changes. Hence we call it the *velocity of γ at the time t_0* and denote it also by

$$\frac{d}{dt}\Big|_{t=t_0}\gamma(t) \quad \text{or} \quad \frac{d}{dt}\Big|_{t_0}\gamma \quad \text{or} \quad \frac{d\gamma}{dt}(t_0).$$

The main properties of $\dot{\gamma}(t_0)$ are summarized in the following

---

Exercise 3.1. Show that $\dot{\gamma}(t_0) : C^\infty(M) \to \mathbb{R}$ is $\mathbb{R}$-linear and satisfies the following identity: for all $f, g \in C^\infty(M)$,

$$\dot{\gamma}(t_0)(fg) = f(\gamma(t_0))\dot{\gamma}(t_0)(g) + g(\gamma(t_0))\dot{\gamma}(t_0)(f).$$

We interpret $\dot{\gamma}(t_0)$ as a *tangent vector* to M at the point $\gamma(t_0)$. Even more, given a manifold M, and a point $p \in M$, we axiomatize the properties of $\dot{\gamma}(t_0)$ stated in Exercise 3.1 and give the following

Definition 3.2 (Tangent Vector). A *tangent vector* to M at the point p is a linear map

$$v : C^\infty(M) \to \mathbb{R}$$

satisfying the following *Leibniz rule at the point p*: for all $f, g \in C^\infty(M)$,

$$v(fg) = f(p)v(g) + g(p)v(f). \tag{3.2}$$

The set of all tangent vectors to M at p is called the *tangent space* to M at p and denoted by T_pM.

Remark 3.3. Given a tangent vector v to M at the point p, we will always assume p as part of the *defining data* of v. From this point of view, a tangent vector should be better defined as a pair (p, v), where $p \in M$ is a point and v is as in Definition 3.2. Doing this we avoid anomalies like the zero map $0 : C^\infty(M) \to \mathbb{R}$ being a tangent vector at all points. We will explicitly understand tangent vectors as pairs (p, v) when defining the *tangent bundle* later on in this chapter. $\diamond$

Later on in this chapter, we will actually show that every tangent vector in the sense of Definition 3.2 is the velocity of some curve at some time. We now give another example of a tangent vector besides the velocity of a curve.

Example 3.4 (Coordinate Vectors). Let M be an n-dimensional manifold and let $p \in M$ be a point. Fix a chart $(U, \varphi = (x^1, \ldots, x^n))$ on M around p, and, for each $i = 1, \ldots, n$, define the map

$$\frac{\partial}{\partial x^i}\Big|_p : C^\infty(M) \to \mathbb{R}, \quad f \mapsto \frac{\partial}{\partial x^i}\Big|_p f := \frac{\partial \widehat{f}}{\partial t^i}(\varphi(p)),$$

where, as usual, $\widehat{f} = f \circ \varphi^{-1} : \widehat{U} \to \mathbb{R}$ is the coordinate representation of f. The map $\frac{\partial}{\partial x^i}\big|_p$ is called the i-th *coordinate vector* and it is indeed a tangent vector to M at p. The proof is left as Exercise 3.2. ◆

Exercise 3.2. Show that the i-th coordinate vector $\frac{\partial}{\partial x^i}\big|_p$ is a tangent vector to M at p.

In the following lemma, we present two properties of tangent vectors following from the Leibniz rule (and the linearity).

Lemma 3.5 (Elementary Properties of Tangent Vectors). *Let M be a manifold, let $p \in M$ be a point, and let v be a tangent vector to M at p, then*

- $v(c) = 0$ *for every constant function* $c = \text{const} \in C^\infty(M)$,
- v *is a local operator, i.e.,* $v(f) = v(g)$ *for every two functions* $f, g \in C^\infty(M)$ *coinciding on an open neighborhood of p. Yet in other words, $v(f)$ does only depend on the values of f around p.*

Proof. First of all,

$$v(1) = v(1 \cdot 1) = 2 \cdot v(1), \quad \text{hence } v(1) = 0,$$

where, in the second step, we used the Leibniz rule. Now, let $c \in C^\infty(M)$ be a constant function. Then c identifies with a real number, also denoted by $c \in \mathbb{R}$, and $c = c \cdot 1$, so that

$$v(c) = v(c \cdot 1) = c \cdot v(1) = 0,$$

as claimed, where, in the second step, we used the $\mathbb{R}$-linearity of v.

For the second part of the statement, let $f, g \in C^\infty(M)$ be functions that agree *around* p, i.e., $f|_{\mathcal{U}} = g|_{\mathcal{U}}$ for some open neighborhood $\mathcal{U} \subseteq M$ of p, then $h := f - g$ vanishes around p, i.e., $h|_{\mathcal{U}} = 0$. If we prove that v vanishes on h, we have done by linearity. Indeed, if $v(h) = 0$, then

$$v(f) - v(g) = v(f - g) = v(h) = 0.$$

So it remains to prove that v vanishes on every function h such that $h|_{\mathcal{U}} = 0$. To do this, consider a bump function $\beta \in C^\infty(M)$ relative to the data $(p, \mathcal{U})$, i.e., $\beta = 1$ around p, and $\text{supp}\,\beta \subseteq \mathcal{U}$, and denote $\eta = 1 - \beta$.

It follows that $h = \eta \cdot h$ (do you see it?) so that

$$v(h) = v(\eta \cdot h) = \eta(p)v(h) + h(p)v(\eta) = 0$$

because $h(p) = \eta(p) = 0$. This concludes the proof. $\qquad\square$

We conclude this section remarking that the tangent space T_pM to a manifold M at a point $p \in M$ is a vector space. First note that, by definition of tangent vectors, T_pM is a subset in the space $\mathrm{Hom}_{\mathbb{R}}(C^\infty(M), \mathbb{R})$ of $\mathbb{R}$-linear maps $C^\infty(M) \to \mathbb{R}$ (the dual space of the vector space $C^\infty(M)$):

$$T_pM \subseteq \mathrm{Hom}_{\mathbb{R}}(C^\infty(M), \mathbb{R}).$$

An easy check, that we leave as Exercise 3.3, reveals that T_pM is actually a vector subspace. In particular, it is a vector space.

> **Exercise 3.3.** Show that the tangent space to M at the point p is a vector subspace in the space $\mathrm{Hom}_{\mathbb{R}}(C^\infty(M), \mathbb{R})$ dual to the vector space $C^\infty(M)$.

Our intuition suggests that the tangent space T_pM should share the same dimension as M. This is indeed true, according to the following.

Theorem 3.6 (Tangent Vector Theorem). *Let M be an n-dimensional manifold, let $p \in M$ be a point, and let $(U, \varphi = (x^1, \ldots, x^n))$ be a chart on M around p. Then the coordinate vectors*

$$\left(\frac{\partial}{\partial x^1}\Big|_p, \ldots, \frac{\partial}{\partial x^n}\Big|_p \right)$$

form a frame in T_pM (called the coordinate frame*). In particular, T_pM is finitely generated and $\dim T_pM = \dim M = n$.*

We postpone a detailed proof of the Tangent Vector Theorem to the next section. For now, we only prove it in the case $M = \mathbb{R}^n$. The proof relies on the following

Lemma 3.7 (Hadamard Lemma). *Let $f \in C^\infty(\mathbb{R}^n)$ be a smooth function on $\mathbb{R}^n$, and let $P_0 = (P_0^1, \ldots, P_0^n) \in \mathbb{R}^n$ be a point. Then there exist smooth functions $g_1, \ldots, g_n \in C^\infty(\mathbb{R}^n)$ such that*

$$f = f(P_0) + \sum_{i=1}^n (t^i - P_0^i) \cdot g_i, \tag{3.3}$$

and, additionally,

$$g_i(P_0) = \frac{\partial f}{\partial t^i}(P_0), \quad i = 1, \ldots, n. \tag{3.4}$$

Proof. Let $P \in \mathbb{R}^n$ be an arbitrary point, and, for every $\epsilon \in \mathbb{R}$, denote

$$P(\epsilon) = P_0 + \epsilon(P - P_0) = (P^1(\epsilon), \dots, P^n(\epsilon)).$$

So, when $P = P_0$, $P(\epsilon) = P_0$ constantly, otherwise, it is the generic point on the line through P and P_0. In any case, $P(0) = P_0$ and $P(1) = P$. Compute

$$f(P) - f(P_0) = f(P(1)) - f(P(0))$$

$$= \int_0^1 \frac{d}{d\epsilon} f(P(\epsilon)) \, d\epsilon$$

$$= \sum_{i=1}^n \int_0^1 \frac{\partial f}{\partial t^i}(P(\epsilon)) \frac{dP^i(\epsilon)}{d\epsilon} \, d\epsilon$$

$$= \sum_{i=1}^n \left(P^i - P_0^i \right) \cdot \int_0^1 \frac{\partial f}{\partial t^i}(P(\epsilon)) \, d\epsilon.$$

Now, define $g_i \in C^\infty(\mathbb{R}^n)$ by putting

$$g_i(P) := \int_0^1 \frac{\partial f}{\partial t^i}(P(\epsilon)) \, d\epsilon.$$

So

$$f(P) = f(P_0) + \sum_{i=1}^n \left(P^i - P_0^i \right) \cdot g_i(P),$$

and (3.3) follows from the arbitrariness of P. Finally, we check (3.4). Remember that, when $P = P_0$, we have $P(\epsilon) = P_0$ constantly, so, for all $i \in \{1, \dots, n\}$,

$$g_i(P_0) = \int_0^1 \frac{\partial f}{\partial t^i}(P_0) \, d\epsilon = \frac{\partial f}{\partial t^i}(P_0) \int_0^1 d\epsilon = \frac{\partial f}{\partial t^i}(P_0). \qquad \square$$

We are now ready to prove the Tangent Vector Theorem in the case $M = \mathbb{R}^n$. In other words, we are proving the following

Lemma 3.8 (Tangent Vector Theorem for $\mathbb{R}^n$). *Consider the n-dimensional standard Euclidean space $\mathbb{R}^n$ with standard coordinates $(t^1, \dots, t^n)$, and let*

$P_0 \in \mathbb{R}^n$ *be a point. Then the coordinate vectors*

$$\left(\frac{\partial}{\partial t^1} \Big|_{P_0}, \ldots, \frac{\partial}{\partial t^n} \Big|_{P_0} \right) \tag{3.5}$$

form a frame of $T_{P_0}\mathbb{R}^n$.

Proof. Note, preliminarily, that the coordinate vectors (3.5) are exactly the partial derivatives (of smooth functions $\mathbb{R}^n \to \mathbb{R}$) at P_0. Now, let $a^1, \ldots, a^n \in \mathbb{R}$ be such that the linear combination

$$v := \sum_{i=1}^{n} a^i \frac{\partial}{\partial t^i} \Big|_{P_0}$$

vanishes. In particular, $v(t^j) = 0$ for all $j \in \{1, \ldots, n\}$, i.e.,

$$0 = \sum_{i=1}^{n} a^i \frac{\partial}{\partial t^i} \Big|_{P_0} t^j = \sum_{i=1}^{n} a^i \delta_i^j = a^j$$

for all j, where δ_i^j is the *Kronecker delta*. This shows that the $\frac{\partial}{\partial t^i} \Big|_{P_0}$ are independent. Finally, we use the Hadamard Lemma (Lemma 3.7) to show that they generate $T_{P_0}\mathbb{R}^n$. So let $v \in T_{P_0}\mathbb{R}^n$ be any tangent vector, and let $f \in C^\infty(\mathbb{R}^n)$ be an arbitrary smooth function. We compute $v(f)$ by writing f in the *Hadamard form* (3.3):

$v(f)$

$$= v \left(f(P_0) + \sum_{i=1}^{n} (t^i - P_0^i) \cdot g_i \right) \qquad \text{(Hadamard Lemma)}$$

$$= v \left(f(P_0) \right) + \sum_{i=1}^{n} v \left((t^i - P_0^i) \cdot g_i \right) \qquad \text{(v is $\mathbb{R}$-linear)}$$

$$= \sum_{i=1}^{n} v \left((t^i - P_0^i) \cdot g_i \right) \qquad \text{(v vanishes on constants)}$$

$$= \sum_{i=1}^{n} \left(v(t^i - P_0^i) \cdot g_i(P_0) + (P_0^i - P_0^i) \cdot v(g_i) \right) \quad \text{(Leibniz rule at the point P_0)}$$

$$= \sum_{i=1}^{n} v(t^i) \frac{\partial}{\partial t^i} \Big|_{P_0} f, \qquad \text{(Equation (3.4))}$$

and, from the arbitrariness of f,

$$v = \sum_{i=1}^{n} v(t^i) \frac{\partial}{\partial t^i}|_{P_0}$$

is indeed a linear combination of the coordinate vectors. $\square$

3.2   Tangent Maps

Before proving the Tangent Vector Theorem, we need to elaborate a little bit more on tangent vectors. In particular, we have to study how do tangent vectors interact with smooth maps and define the *tangent map* to a smooth map. So, let M, N be smooth manifolds, let $F : M \to N$ be a smooth map, and let $p \in M$ be a point.

Definition 3.9 (Tangent Map). The *tangent map* to $F : M \to N$ at the point $p \in M$ is the linear map

$$d_p F : T_p M \to T_{F(p)} N, \quad v \mapsto d_p F(v) := v \circ F^*. \tag{3.6}$$

We leave as Exercise 3.4 to prove that the composition of the pull-back $F^* : C^\infty(N) \to C^\infty(M)$ followed by a tangent vector $v : C^\infty(M) \to \mathbb{R}$ to M at the point p is indeed a tangent vector to N at the point $F(p)$ and that $d_p F$ is indeed a linear map.

> **Exercise 3.4.** Prove that the tangent map $d_p F$ in (3.6) is well defined (i.e., $d_p F(v) \in T_{F(p)} N$ for every $v \in T_p M$) and $\mathbb{R}$-linear.

Example 3.10. Let M be a manifold, let $\gamma : I \to M$ be a smooth curve, and let $t_0 \in I$. It immediately follows from the definition of velocity $\dot\gamma(t_0)$ of γ at the time t_0 (3.1) that $\dot\gamma(t_0)$ is the image of the coordinate tangent vector $\frac{d}{dt}|_{t_0} \in T_{t_0} I$ under the tangent map $d_{t_0} \gamma$:

$$\dot\gamma(t_0) = d_{t_0} \gamma \left(\frac{d}{dt}|_{t_0} \right)$$

(do you see it?).   ◆

Proposition 3.11 (Chain Rule). *Let* M, N, Q *be smooth manifolds, and let* $F : M \to N$ *and* $G : N \to Q$ *be smooth maps. For every point* $p \in M$,

(1) *the tangent map to the identity* $\mathrm{id}_M : M \to M$ *is the identity:*

$$d_p \mathrm{id}_M = \mathrm{id}_{T_p M},$$

(2) *the tangent map to the composition $G \circ F : M \to Q$ is the composition of the tangent maps:*

$$d_p(G \circ F) = d_{F(p)}G \circ d_pF, \tag{3.7}$$

(*Equation (3.7) is also known as the* chain rule),
(3) *the tangent map to a diffeomorphism $\Phi : M \to N$ is an isomorphism whose inverse is the tangent map to Φ^{-1}:*

$$(d_p\Phi)^{-1} = d_{\Phi(p)}\Phi^{-1}.$$

Proof.   Left as Exercise 3.5. $\square$

Exercise 3.5. Prove Proposition 3.11.

Remark 3.12. It follows from Proposition 3.11 and the Tangent Vector Theorem (Theorem 3.6, not proved yet in full generality) that two diffeomorphic manifolds M, N share the same dimension: $\dim M = \dim N$. In particular, there are no diffeomorphisms between open subsets in Euclidean spaces of different dimensions. $\Diamond$

We now look at the tangent spaces to a submanifold $S \subseteq M$ in a manifold M. As usual, denote by $i_S : S \hookrightarrow M$ the inclusion and recall that it is a smooth map. Let $p \in S$ be a point. Our intuition suggests that the tangent space to S at p is a subspace of the tangent space to M at the same point p and, additionally, if $S = \mathcal{U} \subseteq M$ is an *open* submanifold, that $\mathcal{U}$ and M share the same tangent space. Our intuition is correct, as rigorously stated in the following

Proposition 3.13. *Let $S \subseteq M$ be a submanifold in a manifold M, and let $p \in S$ be a point. Then the tangent map to the inclusion $i_S : S \hookrightarrow M$ at the point p is injective (hence a monomorphism of vector spaces). If, additionally, $S = \mathcal{U} \subseteq M$ is an open submanifold, then $d_p i_{\mathcal{U}}$ is also surjective, hence an isomorphism.*

Proof.   Consider the tangent map

$$d_p i_S : T_pS \to T_pM.$$

It is enough to show that $\ker d_p i_S = \{0\}$. So let $v \in T_pS$ be such that $d_p i_S(v) = 0$. We have to show that $v(g) = 0$ for all smooth functions g on S. So, take $g \in C^\infty(S)$ and use the Local Extension Lemma (Lemma 2.47) to pick a smooth function $\tilde{g} \in C^\infty(M)$ such that g and $\tilde{g}$ agree on an open neighborhood of p in S. As tangent vectors are local operators (Lemma 3.5, Point (2)), we have

$$v(g) = v(\tilde{g}|_S) = v(i_S^*(\tilde{g})) = d_p i_S(v)(\tilde{g}) = 0.$$

This concludes the proof of the first part of the statement. For the second part, let $S = \mathcal{U} \subseteq M$ be an open submanifold. We have to show that, for any $v \in T_p M$, there exists $w \in T_p \mathcal{U}$ such that $v = d_p i_\mathcal{U}(w)$. We define $w : C^\infty(\mathcal{U}) \to \mathbb{R}$ as follows. Let $g \in C^\infty(\mathcal{U})$, and let $\widetilde{g} \in C^\infty(M)$ be a smooth function on M that coincides with g around p (it exists in view of the Local Extension Lemma). Put

$$w(g) := v(\widetilde{g}).$$

First of all, as tangent vectors are local operators, $v(\widetilde{g})$ does only depend on the values of $\widetilde{g}$ around p, hence $w(g)$ is well defined. Second, if f, g are functions on S, and $\widetilde{f}, \widetilde{g} \in C^\infty(M)$ coincide with them around p, then $a\widetilde{f} + b\widetilde{g}, \widetilde{f}\widetilde{g}$ coincide with $af + bg, fg$ around p for all $a, b \in \mathbb{R}$. So

$$w(af + bg) = v(a\widetilde{f} + b\widetilde{g}) = av(\widetilde{f}) + bv(\widetilde{g}) = aw(f) + bw(g)$$

and

$$w(fg) = v(\widetilde{f}\widetilde{g}) = \widetilde{f}(p)v(\widetilde{g}) + \widetilde{g}(p)v(\widetilde{f}) = f(p)w(g) + g(p)w(f).$$

This shows that w is a tangent vector. Finally, let $h \in C^\infty(M)$ and compute

$$d_p i_\mathcal{U}(w)(h) = w(i_\mathcal{U}^*(h)) = w(h|_\mathcal{U}) = v(h),$$

and, from the arbitrariness of h, $d_p i_\mathcal{U}(w) = v$. This concludes the proof. $\square$

In the following, we will use $d_p i_S$ to identify $T_p S$ with its image (unless otherwise stated). In other words, we will interpret $T_p S$ as a subspace in $T_p M$, and write $T_p S \subseteq T_p M$. In particular, if $\mathcal{U} \subseteq M$ is an open submanifold, then $T_p \mathcal{U} = T_p M$. So a tangent vector $v \in T_p M$ can be safely applied to any smooth function f defined only *locally around* p, i.e., any smooth function $f \in C^\infty(\mathcal{U})$ on some open neighborhood $\mathcal{U} \subseteq M$ of p, and we will do this without further comments. We are finally ready to prove the Tangent Vector Theorem (Theorem 3.6).

Proof of the Tangent Vector Theorem 3.6. Let M be an n-dimensional manifold, let $p \in M$ be a point, and let $(U, \varphi = (x^1, \ldots, x^n))$ be a chart on M around p. The claim is that the coordinate vectors

$$\left(\frac{\partial}{\partial x^1}\Big|_p, \ldots, \frac{\partial}{\partial x^n}\Big|_p \right) \tag{3.8}$$

form a frame of T_pM. To prove this, denote by P the image $\varphi(p)$, and consider the chain of isomorphisms

$$T_pM \xleftarrow{\;d_pi_U\;} T_pU \xrightarrow{\;d_p\varphi\;} T_p\widehat{U} \xrightarrow{\;d_{p'}i_{\widehat{U}}\;} T_P\mathbb{R}^n. \tag{3.9}$$

It easily follows from the definition of coordinate vectors that (3.9) identify

$$\left(\frac{\partial}{\partial x^1}\Big|_p, \dots, \frac{\partial}{\partial x^n}\Big|_p\right) \quad \text{and} \quad \left(\frac{\partial}{\partial t^1}\Big|_P, \dots, \frac{\partial}{\partial t^n}\Big|_P\right)$$

(do you see it?). As the latter is a frame of $T_P\mathbb{R}^n$ (Lemma 3.8), the former must be a frame of T_pM as claimed. $\qquad\square$

Let M be an n-dimensional manifold, let $p \in M$ be a point, and let $(U, \varphi = (x^1, \dots, x^n))$ be a chart on M around p. In the following, the frame (3.8) of T_pM will be referred to as the *coordinate frame*. In view of the Tangent Vector Theorem, any tangent vector $v \in T_pM$ can be uniquely written in the form

$$v = \sum_{i=1}^{n} v^i \frac{\partial}{\partial x^i}\Big|_p \tag{3.10}$$

for some real numbers v^i. Note that, from the definition of coordinate vector,

$$\frac{\partial}{\partial x^i}\Big|_p x^j = \delta^j_i, \quad i, j = 1, \dots, n. \tag{3.11}$$

It follows that the components $v^1, \dots, v^n$ of a tangent vector $v \in T_pM$ in the coordinate frame are given by the action of v on the coordinates $x^1, \dots, x^n$, i.e.,

$$v^j = v(x^j), \quad j = 1, \dots, n. \tag{3.12}$$

Indeed

$$v(x^j) = \sum_{i=1}^{n} v^i \frac{\partial}{\partial x^i}\Big|_p x^j = \sum_{i=1}^{n} v^i \delta^j_i = v^j,$$

where we used (3.11).

From now on we adopt the *Einstein summation convention*: we will understand the *sum over pairs of upper-lower repeated indices* and simply write, e.g.,

$$v = v^i \frac{\partial}{\partial x^i}\Big|_p,$$

instead of (3.10).

Example 3.14 (Coordinate Expression for the Velocity of a Curve). Let $\gamma : I \to M$ be a curve, and let $t_0 \in I$. Denote $p_0 = \gamma(t_0)$. As γ is smooth, we can choose an open subinterval $J \subseteq I$ containing t_0 and a chart $(U, \varphi = (x^1, \ldots, x^n))$ on M around p_0, such that $\gamma(J) \subseteq U$. Denote by $\gamma^i = \gamma^*(x^i) : J \to \mathbb{R}$ the *components* of (the coordinate representation of) γ with respect to the chart (U, φ). Then, for every $t \in J$,

$$\dot{\gamma}(t) = v^i \frac{\partial}{\partial x^i}\Big|_{\gamma(t)},$$

for some $v^i \in \mathbb{R}$ given by

$$\dot{\gamma}(t)(x^i) = \frac{d}{dt} x^i(\gamma(t)) = \frac{d\gamma^i}{dt}(t).$$

Summarizing

$$\dot{\gamma}(t) = \frac{d\gamma^i}{dt}(t) \frac{\partial}{\partial x^i}\Big|_{\gamma(t)}. \tag{3.13}$$

♦

Clearly, if we change coordinates around p, the coordinate frame changes. We want to find out how. So let $(\tilde{U}, \tilde{\varphi} = (\tilde{x}^1, \ldots \tilde{x}^n))$ be another chart on M around p. We denote by $t^1, \ldots, t^n$ the standard coordinates on $\varphi(U)$ and, to avoid confusion, by $\tilde{t}^1, \ldots, \tilde{t}^n$ the standard coordinates on $\tilde{\varphi}(\tilde{U})$. We also put $P = \varphi(p)$. The transition matrix

$$M_{\tilde{\mathcal{R}}, \mathcal{R}} = (a_i^j)_{i=1,\ldots,n}^{j=1,\ldots,n} \in GL(n, \mathbb{R})$$

between the frames

$$\mathcal{R} = \left(\frac{\partial}{\partial x^1}\Big|_p, \ldots, \frac{\partial}{\partial x^n}\Big|_p \right) \quad \text{and} \quad \tilde{\mathcal{R}} = \left(\frac{\partial}{\partial \tilde{x}^1}\Big|_p, \ldots, \frac{\partial}{\partial \tilde{x}^n}\Big|_p \right)$$

is given by

$$a_i^j = \left(\text{the } j\text{-th component of } \frac{\partial}{\partial x^i}\Big|_p \text{ in the frame } \tilde{\mathcal{R}} \right) = \frac{\partial}{\partial x^i}\Big|_p \tilde{x}^j,$$

and, using the definition of coordinate vector, we find

$$\frac{\partial}{\partial x^i}\big|_p \tilde{x}^j = \frac{\partial}{\partial t^i}\big|_p \tilde{x}^j \circ \varphi^{-1}$$

$$= \frac{\partial}{\partial t^i}\big|_p \tilde{x}^j \circ \tilde{\varphi}^{-1} \circ \tilde{\varphi} \circ \varphi^{-1}$$

$$= \frac{\partial}{\partial t^i}\big|_p \tilde{t}^j \circ (\tilde{\varphi} \circ \varphi^{-1})$$

$$= \frac{\partial(\tilde{\varphi} \circ \varphi^{-1})^j}{\partial t^i}(P).$$

This shows that $M_{\mathcal{R},\tilde{\mathcal{R}}}$ is exactly the Jacobian matrix at the point $P = \varphi(p)$ of the transition map

$$\tilde{\varphi} \circ \varphi^{-1} : \varphi(U \cap \tilde{U}) \to \tilde{\varphi}(U \cap \tilde{U}).$$

Next we compute the representative matrix of the tangent map with respect to two coordinate frames. Namely, let M, N be manifolds with $\dim M = m, \dim N = n$, let $F : M \to N$ be a smooth map, and let $p \in M$ be a point. Choose a chart $(U, \varphi = (x^1,\ldots,x^m))$ on M around p and a chart $(V, \psi = (y^1,\ldots,y^n))$ on N around $F(p)$ such that $F(U) \subseteq V$. Put $P = \varphi(p)$. The restriction $F : U \to V$ is completely determined by the functions $F^a := y^a \circ F = F^*(y^a) \in C^\infty(U), a = 1,\ldots,n$. The representative matrix

$$M_{\mathcal{S},\mathcal{R}}^{d_pF} = (b_i^a)_{i=1,\ldots,m}^{a=1,\ldots,n} \in M(n, m; \mathbb{R})$$

of the tangent map $d_pF : T_pM \to T_{F(p)}N$ with respect to the coordinate frames

$$\mathcal{R} = \left(\frac{\partial}{\partial x^1}\big|_p,\ldots,\frac{\partial}{\partial x^m}\big|_p\right) \quad \text{and} \quad \mathcal{S} = \left(\frac{\partial}{\partial y^1}\big|_{F(p)},\ldots,\frac{\partial}{\partial y^n}\big|_{F(p)}\right)$$

is given by

$$b_i^a = \left(\text{the } a\text{-th component of } d_pF\left(\frac{\partial}{\partial x^i}\big|_p\right) \text{ in the frame } \mathcal{S}\right)$$

$$= d_pF\left(\frac{\partial}{\partial x^i}\big|_p\right)(y^a) = \frac{\partial}{\partial x^i}\big|_p F^*(y^a) = \frac{\partial}{\partial x^i}\big|_p F^a,$$

and using the definition of coordinate vector, we find

$$b_i^a = \frac{\partial}{\partial x^i}\Big|_p F^a = \frac{\partial}{\partial t^i}\Big|_p F^a \circ \varphi^{-1} = \frac{\partial \widehat{F}^a}{\partial t^i}(P),$$

where the $\widehat{F}^a$ are the components of the coordinate representation $\widehat{F} : \widehat{U} \to \widehat{V}$ of F. This shows that $M_{\mathcal{S},\mathcal{R}}^{d_p F}$ is exactly the Jacobian matrix at the point $P = \varphi(p)$ of $\widehat{F}$.

We conclude this section discussing two further applications of the Tangent Vector Theorem. We begin showing that every tangent vector is the velocity of a suitable (non-unique) curve.

Proposition 3.15. *Let M be a manifold, and let $p \in M$. For every tangent vector $v \in T_p M$, there exists a curve $\gamma : I \to M$, such that $0 \in I$, $\gamma(0) = p$, and $\dot{\gamma}(0) = v$.*

Proof. Let $n = \dim M$, choose a chart $(U, \varphi = (x^1, \ldots, x^n))$ around p, and put $P = \varphi(p)$. Then

$$v = v^i \frac{\partial}{\partial x^i}\Big|_p$$

for some n-tuple $V := (v^1, \ldots, v^n) \in \mathbb{R}^n$. Consider the following curve in $\mathbb{R}^n$:

$$\widehat{\gamma} : \mathbb{R} \to \mathbb{R}^n, \quad t \mapsto \widehat{\gamma}(t) := P + tV.$$

The preimage $\widehat{\gamma}^{-1}(\widehat{U})$ is an open neighborhood of 0 in $\mathbb{R}$. In particular, there is an open interval $I \subseteq \mathbb{R}$ containing 0, such that $\widehat{\gamma}(I) \subseteq \widehat{U}$. Denote again by $\widehat{\gamma} : I \to \widehat{U}$ the restriction, and let $(\gamma^1, \ldots, \gamma^n)$ be its components. The curve $\widehat{\gamma}$ is the coordinate representation of a unique curve $\gamma = \varphi^{-1} \circ \widehat{\gamma} : I \to M$ (taking values in U). The curve γ is the one we are looking for (see Figure 3.2). To show this, we compute the velocity $\dot{\gamma}(0)$. From (3.13),

$$\dot{\gamma}(0) = \frac{d\gamma^i}{dt}(0) \frac{\partial}{\partial x^i}\Big|_{\gamma(0)} = v^i \frac{\partial}{\partial x^i}\Big|_p = v.$$

This concludes the proof.   □

Exercise 3.6. Let $F : M \to N$ be a smooth map between manifolds, and let $\gamma : I \to M$ be a curve in M. Show that, for every $t_0 \in I$, the tangent map to F transforms the velocity of a curve γ into the velocity of the *transformed curve* $F \circ \gamma$, i.e.,

$$d_{\gamma(t_0)}F\left(\frac{d\gamma}{dt}(t_0)\right) = \frac{d(F \circ \gamma)}{dt}(t_0). \tag{3.14}$$

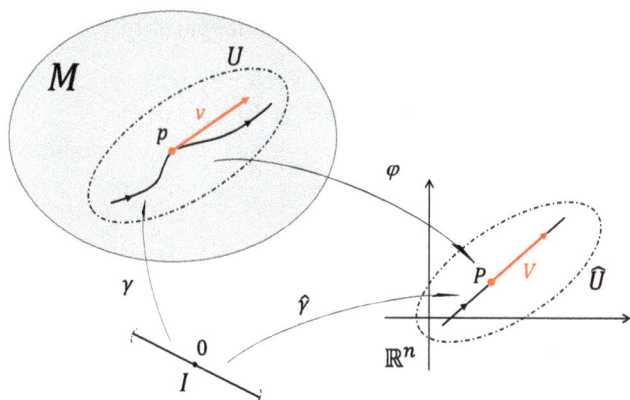

Figure 3.2.   Every tangent vector is the velocity of a curve.

Note that, in view of Proposition 3.15, Equation (3.14) can be understood as an alternative definition of the tangent map.

Our second application of the Tangent Vector Theorem is a description of the tangent spaces to a vector space. So, let V be an n-dimensional real vector space. In particular, V is an n-dimensional manifold. We begin noting that there are two (*a priori* different) ways to understand the velocity of a curve $\gamma : I \to V$ in V at time $t_0 \in I$: first of all, as the vector $\dot{\gamma}(t_0)$ in the tangent space $T_{\gamma(t_0)}V$ and second, as a vector in V, denoted $\gamma'(t_0)$, as follows. Choose a frame $\mathcal{R} = (e_1, \ldots, e_n)$ in V. Then, for all $t \in I$, $\gamma(t) = v^i(t)e_i$, where the $v^i : I \to \mathbb{R}$ are smooth functions. Put

$$\gamma'(t_0) = \frac{dv^i}{dt}(t_0)e_i \in V.$$

It is easy to see that $\gamma'(t_0)$ is actually independent of the choice of $\mathcal{R}$ (do you see it?).

Exercise 3.7. Let V be a finite dimensional real vector space, let $v_1, \ldots, v_k$ be (non-necessarily independent) vectors in V, and let $f_1, \ldots, f_k : I \to \mathbb{R}$ be smooth functions on some open interval $I \subseteq \mathbb{R}$. Show that the map

$$\gamma : I \to V, \quad t \mapsto \gamma(t) := f_1(t)v_1 + \cdots + f_k(t)v_k$$

is a smooth curve in V and that

$$\gamma'(t) = \frac{df_1}{dt}(t)v_1 + \cdots + \frac{df_k}{dt}(t)v_k$$

for all $t \in I$.

We are now ready to describe the tangent spaces to V.

Proposition 3.16 (Tangent Space to a Vector Space). *For any $v_0 \in V$, there is a canonical vector space isomorphism $V \cong T_{v_0}V$ given by*

$$V \to T_{v_0}V, \quad v \mapsto v^\uparrow := \frac{d}{dt}\Big|_{t=0}(v_0 + tv). \tag{3.15}$$

Additionally, for any curve $\gamma : I \to V$, and any $t_0 \in I$, we have

$$\gamma'(t_0)^\uparrow = \dot{\gamma}(t_0),$$

i.e., the isomorphism $V \cong T_{\gamma(t_0)}V$ identifies the "two velocities" $\gamma'(t_0)$ and $\dot{\gamma}(t_0)$ of γ.

Probably, the definition in (3.15) requires some explanations. For any $v_0, v \in V$, we are considering the smooth curve $\Gamma : \mathbb{R} \to V, t \mapsto \Gamma(t) := v_0 + tv$ in V (do you see that Γ is smooth?), and $v^\uparrow \in T_{v_0}V$ is the velocity of Γ at time 0 (see Figure 3.3).

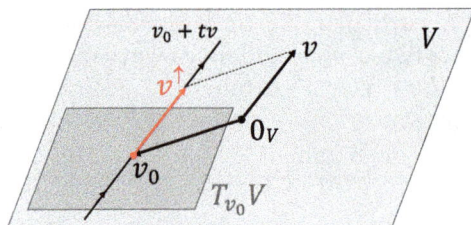

Figure 3.3.   The tangent space to a vector space V is canonically isomorphic to V.

Proof of Proposition 3.16. We have to show that the map $V \to T_{v_0}V$, $v \mapsto v^\uparrow$ is linear and bijective. We take care of both things at once. Choose a frame $\mathcal{R} = (e_1, \ldots, e_n)$ in V, and let

$$(V, \varphi_\mathcal{R} = (x^1, \ldots, x^n))$$

be the associated chart on V. Recall that, for any $v \in V$, $x^i(v)$ is, by definition, the i-th component v^i of v in the frame $\mathcal{R}$. Let

$$\mathcal{R}_T = \left(\frac{\partial}{\partial x^1}\Big|_{v_0}, \ldots, \frac{\partial}{\partial x^n}\Big|_{v_0} \right)$$

be the coordinate frame in $T_{v_0}V$ induced by the chart $(V, \varphi_\mathcal{R})$. Finally, let $\varphi_{\mathcal{R}_T} : T_{v_0}V \to \mathbb{R}^n$ be the coordinate isomorphism induced by $\mathcal{R}_T$. We want to compute $\varphi_{\mathcal{R}_T}(v^\uparrow)$. To do this, remember that the i-th component of a velocity vector in the coordinate frame is the time derivative of the i-th component of the coordinate representation of the curve (see (3.13)). As $v^\uparrow$ is the velocity at time 0 of the curve Γ, we have, in particular, that the coordinate vector of $v^\uparrow$ in the coordinate frame $\mathcal{R}_T$ is

$$\varphi_{\mathcal{R}_T}(v^\uparrow) = \left(\frac{d\Gamma^1}{dt}(0),\ldots,\frac{d\Gamma^n}{dt}(0)\right),$$

where $\varphi_{\mathcal{R}} \circ \Gamma = (\Gamma^1,\ldots,\Gamma^n) : \mathbb{R} \to \mathbb{R}^n$ is the coordinate representation of Γ. But

$$\Gamma^i = v_0^i + tv^i,$$

hence

$$\frac{d\Gamma^i}{dt}(0) = v^i$$

for all $i = 1,\ldots,n$. We conclude that

$$\varphi_{\mathcal{R}_T}(v^\uparrow) = \left(v^1,\ldots,v^n\right) = \varphi_{\mathcal{R}}(v).$$

This shows that the map $V \to T_{v_0}V$, $v \mapsto v^\uparrow$, is given by the composition of the vector space isomorphism $\varphi_{\mathcal{R}} : V \to \mathbb{R}^n$, followed by the vector space isomorphism $\varphi_{\mathcal{R}_T}^{-1} : \mathbb{R}^n \to T_{v_0}V$:

$$v^\uparrow = (\varphi_{\mathcal{R}_T}^{-1} \circ \varphi_{\mathcal{R}})(v), \quad \text{for all } v \in V.$$

Hence, it is a vector space isomorphism as well.

For the second part of the proposition, let $\gamma : I \to V$, $t \mapsto \gamma(t) = \gamma^i(t)e_i$ be a curve in V, and let $t_0 \in I$. By definition of $\gamma'(t_0)$, we have

$$\varphi_{\mathcal{R}}(\gamma'(t_0)) = \left(\frac{d\gamma^1}{dt}(t_0),\ldots,\frac{d\gamma^n}{dt}(t_0)\right).$$

On the other hand, the coordinate representation of γ with respect to the chart $(V, \varphi_{\mathcal{R}})$ is $\varphi_{\mathcal{R}} \circ \gamma = (\gamma^1,\ldots,\gamma^n) : I \to \mathbb{R}^n$, and, from (3.13) again,

$$\dot{\gamma}(t_0) = \frac{d\gamma^i}{dt}(t_0)\frac{\partial}{\partial x^i}\Big|_{\gamma(t_0)},$$

so that

$$\varphi_{\mathcal{R}_T}(\dot{\gamma}(t_0)) = \left(\frac{d\gamma^1}{dt}(t_0),\ldots,\frac{d\gamma^n}{dt}(t_0)\right) = \varphi_{\mathcal{R}}(\gamma'(t_0)),$$

showing that $\gamma'(t_0)^\uparrow = \dot{\gamma}(t_0)$ as claimed. $\qquad\square$

In the following, given a curve $\gamma : I \to V$ in a finite dimensional real vector space V, we will always interpret its velocity $\dot{\gamma}(t_0)$ at time $t_0 \in I$ as a vector in V, identifying $\dot{\gamma}(t_0)$ and $\gamma'(t_0)$ via the isomorphism $V \cong T_{\gamma(t_0)}V$, unless otherwise stated.

3.3 The Tangent Bundle

Let M be an n-dimensional manifold. Tangent vectors to M can be organized into a new manifold called the *tangent bundle to M* and denoted TM. In this section, we construct TM, including its smooth structure, and discuss its first properties. First of all, we put

$$TM := \coprod_{p \in M} T_p M = \{(p, v) : p \in M \text{ and } v \in T_p M\}.$$

When it is clear what is p, a point (p, v) in TM will be also denoted simply by v (see also Remark 3.3). The set TM comes with a natural surjection $\tau : TM \to M$, $(p, v) \mapsto p$, and the pair (TM, τ) is called the *tangent bundle to M*. Sometimes we understand τ and call TM itself the tangent bundle. The smooth structure on M induces an atlas on TM as we now show.

Begin with a chart $(U, \varphi = (x^1, \ldots, x^n))$ on M, and define a chart $(TU, T\varphi)$ on TM as follows. First of all, put

$$TU := \tau^{-1}(U) = \coprod_{p \in U} T_p M = \{(p, v) \in TM : p \in U\}.$$

As the tangent spaces to U identify canonically with the tangent spaces to M at points of U, TU is exactly the tangent bundle to U. Next define a map:

$$T\varphi : TU \to \widehat{U} \times \mathbb{R}^n, \quad (p, v) \mapsto T\varphi(p, v) := (\varphi(p); v(x^1), \ldots, v(x^n)).$$

Note that the last n entries of $T\varphi(p, v)$ are nothing but the components of v in the coordinate frame

$$\mathcal{R}_p = \left(\frac{\partial}{\partial x^1}\Big|_p, \ldots, \frac{\partial}{\partial x^n}\Big|_p \right)$$

of $T_p M$. It is clear that $(TU, T\varphi)$ is a $2n$-dimensional chart on TM. Every such chart on TM is called a *standard charts*. Let $(TU, T\varphi)$ be the standard chart on TM induced by a chart $(U, \varphi = (x^1, \ldots, x^n))$ on M. By an abuse of notation, the first n components of $T\varphi$ are denoted again by $(x^1, \ldots, x^n)$. The last n components are denoted $(\dot{x}^1, \ldots, \dot{x}^n)$, and we write

$$(TU, T\varphi = (x^1, \ldots, x^n; \dot{x}^1, \ldots, \dot{x}^n)).$$

Proposition 3.17 (Smooth Structure on the Tangent Bundle). *Standard charts on TM form an atlas. With the associated smooth structure, TM is a smooth $2n$-dimensional manifold, and $\tau : TM \to M$ is a smooth map.*

Proof. As the charts on M cover M, the standard charts on TM cover TM. Next we show that any two standard charts are compatible. So, take two charts $(U, \varphi = (x^1, \ldots, x^n)), (\tilde{U}, \tilde{\varphi} = (\tilde{x}^1, \ldots, \tilde{x}^n))$ on M and the associated standard charts $(TU, T\varphi), (T\tilde{U}, T\tilde{\varphi})$. If $U \cap \tilde{U} = \varnothing$, then $TU \cap T\tilde{U} = \varnothing$, and $(TU, T\varphi), (T\tilde{U}, T\tilde{\varphi})$ are compatible. If $U \cap \tilde{U} \neq \varnothing$, then $TU \cap T\tilde{U} = T(U \cap \tilde{U}) \neq \varnothing$ and we have to show that the transition map

$$T\tilde{\varphi} \circ (T\varphi)^{-1} : T\varphi(TU \cap T\tilde{U}) \to T\tilde{\varphi}(TU \cap T\tilde{U})$$

is a diffeomorphism between open subsets of $\mathbb{R}^{2n}$. To do this, first note that

$$T\varphi(TU \cap T\tilde{U}) = T\varphi(T(U \cap \tilde{U})) = \varphi(U \cap \tilde{U}) \times \mathbb{R}^n$$

is indeed an open subset and similarly for $T\tilde{\varphi}(TU \cap T\tilde{U})$. Now, check the smoothness of $T\tilde{\varphi} \circ (T\varphi)^{-1}$: take a point $(P; v^1, \ldots, v^n) \in \varphi(U \cap \tilde{U}) \times \mathbb{R}^n$ and compute

$$(T\tilde{\varphi} \circ (T\varphi)^{-1})(P; v^1, \ldots, v^n)$$

$$= T\tilde{\varphi}\left(\varphi^{-1}(P); v^i \frac{\partial}{\partial x^i}\Big|_{\varphi^{-1}(P)}\right)$$

$$= \left((\tilde{\varphi} \circ \varphi^{-1})(P); v^i \frac{\partial}{\partial x^i}\Big|_{\varphi^{-1}(P)} \tilde{x}^1, \ldots, v^i \frac{\partial}{\partial x^i}\Big|_{\varphi^{-1}(P)} \tilde{x}^n\right)$$

$$= \left((\tilde{\varphi} \circ \varphi^{-1})(P); v^i \frac{\partial(\tilde{\varphi} \circ \varphi^{-1})^1}{\partial t^i}(P), \ldots, v^i \frac{\partial(\tilde{\varphi} \circ \varphi^{-1})^n}{\partial t^i}(P)\right).$$

As $(U, \varphi), (\tilde{U}, \tilde{\varphi})$ are compatible, the first n entries depend smoothly on P. The last n entries depend smoothly on P and linearly, hence smoothly, on $v^1, \ldots, v^n$. This shows that $T\tilde{\varphi} \circ (T\varphi)^{-1}$ is smooth. Changing the roles of (U, φ) and $(\tilde{U}, \tilde{\varphi})$ reveals that $T\varphi \circ (T\tilde{\varphi})^{-1}$ is smooth as well. We conclude that both transition maps $T\tilde{\varphi} \circ (T\varphi)^{-1}$ and $T\varphi \circ (T\tilde{\varphi})^{-1}$ are diffeomorphisms, and this concludes the proof of the first part of the statement. The atlas on TM consisting of standard charts will be called the *standard atlas*.

For the second part of the statement, we have to show that the atlas topology induced by the standard atlas is Hausdorff and II-countable. We use the criterion provided by Proposition 1.41. For the Hausdorff property, take $(p, v), (q, w) \in TM$. If $p = q$, then p, q belong to the same coordinate domain U, hence $(p, v), (q, w)$ belong to the same *standard coordinate domain* TU. If $p \neq q$, we can find disjoint coordinate domains U, V such that $p \in U$ and $q \in V$. It follows that TU, TV are disjoint standard coordinate domains such that $(p, v) \in TU$ and $(q, w) \in TV$. For II-countability, we cover M by countably many charts (U, φ). Hence TM is covered by the standard charts $(TU, T\varphi)$, and they are countably many. We leave it to the reader to prove the very last part of the statement (about the smoothness of $\tau : TM \to M$) as Exercise 3.8. $\square$

Exercise 3.8. Complete the proof of Proposition 3.17 showing that the natural projection $\tau : TM \to M$ is smooth.

We now come to a special class of maps $M \to TM$ that will play an important role in the following chapter: *sections* of TM.

Definition 3.18 (Section of the Tangent Bundle). A *section* of TM is a map $s : M \to TM$ that inverts $\tau : TM \to M$ on the right, i.e., $\tau \circ s = \mathrm{id}_M$. In other words, a section is the assignment of a tangent vector $s(p)$ to M at the point p, for every point $p \in M$ (see Figure 3.4).

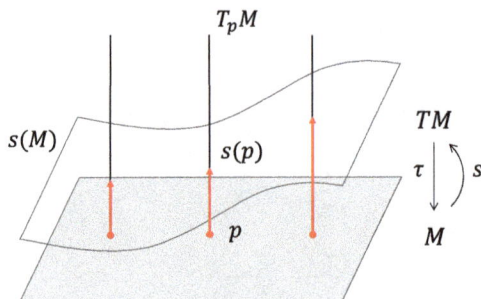

Figure 3.4.   A section of the tangent bundle.

There is a natural algebraic structure on the space of sections of TM: They form a *module over the algebra* $C^\infty(M)$. More generally, let A be a real algebra.

Definition 3.19 (Module). A *module over A*, or simply an *A-module*, is a set $\mathcal{M}$ equipped with two operations $+, \cdot$ such that

(1) $+ : \mathcal{M} \times \mathcal{M} \to \mathcal{M}$ is an interior operation and $(\mathcal{M}, +)$ is an abelian group,
(2) $\cdot : A \times \mathcal{M} \to \mathcal{M}$ is an exterior operation and
- $\alpha \cdot (\beta \cdot \mu) = (\alpha \cdot \beta) \cdot \mu$,
- $(\alpha + \beta) \cdot \mu = \alpha \cdot \mu + \beta \cdot \mu$,
- $\alpha \cdot (\mu + \nu) = \alpha \cdot \mu + \alpha \cdot \nu$,
- $1 \cdot \mu = \mu$,

for all $\alpha, \beta \in A$ and $\mu, \nu \in \mathcal{M}$.

A *submodule* of an A-module $\mathcal{M}$ is a non-empty subset $\mathcal{N} \subseteq \mathcal{M}$ which is stable under both operations in $\mathcal{M}$.

In other words, a module satisfies all the axioms of a vector space except that the ring of scalars is now a generic algebra, not necessarily a field. We usually write $\alpha\mu$, instead of $\alpha \cdot \mu$, for $\alpha \in A$ and $\mu \in \mathcal{M}$. As the first trivial example, note that any real vector space is an $\mathbb{R}$-module. Conversely, every module over a real algebra is also a real vector space. A submodule $\mathcal{N}$ of a module $\mathcal{M}$ is also a module with the restricted operations.

In many respects, the theory of modules is similar to the theory of vector spaces. For instance, linearly independent systems and systems of generators are defined exactly as for vector spaces, except that now linear combinations are taken with coefficients in the algebra rather than just in the field. A basis of a module is a system of independent generators. The main difference between modules and vector spaces is that a module may not possess a basis. Linear maps between modules over the same algebra A are defined exactly as for vector spaces, except that now they are compatible with the multiplication by a scalar in the algebra rather than just in the field. In order to distinguish explicitly this kind of linearity from $\mathbb{R}$-linearity, we call it A-*linearity*, and A-linear maps between A-modules are also called A-*module homomorphisms*. An invertible A-linear map is called an A-module *isomorphism*, and its inverse is also an A-module isomorphism.

The following is our main example of a module.

Example 3.20 (Module of Sections of TM). Let M be a manifold. The space of sections of the tangent bundle TM is a $C^\infty(M)$-module with the following *point-wise* operations:

$$+ : (s_1, s_2) \mapsto s_1 + s_2, \qquad (s_1 + s_2)(p) := s_1(p) + s_2(p),$$
$$\cdot : (f, s) \mapsto fs, \qquad\qquad (fs)(p) := f(p)s(p),$$

where $f \in C^\infty(M)$ and s, s_1, s_2 are sections. The *zero section* is the section mapping a point p to the zero tangent vector $0 \in T_pM$, for all p.   ◆

We are mainly interested in *smooth sections*. Discussing the smoothness of a section is easier than discussing the smoothness of a generic map. Indeed, let M be a manifold, and let s be a section of the tangent bundle TM. For every chart (U, φ) on M, the section s maps U to TU. This shows that all sections have coordinate representations. Now, look at the restriction

$$s : U \to TU.$$

As already remarked in the discussion following Definition 2.9, s is smooth provided only the pull-backs $s^*(x^i)$, $s^*(\dot{x}^i)$ of the coordinates on TU are smooth, $i = 1, \ldots, n$. Now, by definition of section, for every $p \in U$,

$$s^*(x^i)(p) = x^i(s(p)) = x^i(p),$$

which is automatically smooth in its argument p. We conclude that the section s is smooth provided only the functions $s^*(\dot{x}^i)$ on U are smooth for some family of charts (U, φ) covering M. In the following, we denote $s^i := s^*(\dot{x}^i)$ and call them the *components* of the section s in the chart (U, φ). They are implicitly given by

$$s(p) = s^i(p)\frac{\partial}{\partial x^i}\Big|_p, \quad p \in U.$$

We can now rephrase the preceding discussion saying that a section s of TM is smooth if and only if its components s^i are smooth in every chart of an atlas.

Proposition 3.21. *Smooth sections of TM form a submodule in the module of all sections.*

Proof.   Left as Exercise 3.9.   □

Exercise 3.9. Prove Proposition 3.21. (**Hint:** *Prove that, for every three sections s, s_1, s_2, any smooth function f, and any chart (U, φ) on M the following identities hold:*

$$(s_1 + s_2)^i = s_1^i + s_2^i,$$

$$(fs)^i = f|_U s^i$$

for all $i \in \{1, \ldots, n\}$. Conclude that, if s, s_1, s_2 are smooth, then so are $s_1 + s_2$ and fs.)

We denote by $\Gamma(TM)$ the $C^\infty(M)$-module of smooth sections of the tangent bundle TM. In the following, by "*section*" we will always mean "*smooth section*", unless otherwise stated.

We conclude this chapter discussing the tangent bundle to a submanifold. So let M be an n-dimensional manifold, and let $S \subseteq M$ be a d-dimensional submanifold. As $T_pS \subseteq T_pM$ for all $p \in S$, the tangent bundle to S can be seen as a subset in the tangent bundle to TM: $TS \subseteq TM$.

Proposition 3.22. *Let M be an n-dimensional manifold, and let $S \subseteq M$ be a d-dimensional submanifold. Then the tangent bundle to S is a 2-dimensional submanifold in the tangent bundle to M (and the submanifold atlas is compatible with the standard atlas on TS).*

Proof. Let $p_0 \in S$, and let $(U, \varphi = (y^1, \ldots, y^d; z^1, \ldots, z^{n-d}))$ be an adapted chart on M around p_0. This means that

$$S \cap U = \{p \in U : z^a(p) = 0, \text{ for all } a = 1, \ldots, n - d\}.$$

Consider also the submanifold chart $(U_S, \varphi_S = (y_S^1, \ldots, y_S^d))$. As we know, the coordinate representation of the inclusion $i_S : S \hookrightarrow M$ with respect to the charts (U_S, φ_S) and (U, φ) is

$$\widehat{i_S} = \text{in}_d : \widehat{U}_S \to \widehat{U}, \quad (t^1, \ldots, t^d) \mapsto \text{in}_d(t^1, \ldots, t^d) = (t^1, \ldots, t^d, 0, \ldots, 0).$$

It follows that the image of a tangent vector

$$v = v^i \frac{\partial}{\partial y_S^i}\Big|_p \in T_pS$$

to S at a point $p \in U_S$ under the tangent map $d_p i_S : T_pS \to T_pM$ is

$$d_p i_S(v) = v^i \frac{\partial}{\partial y^i}\Big|_p \in T_pM.$$

We conclude that a tangent vector

$$w = v^i \frac{\partial}{\partial y^i}\Big|_p + w^a \frac{\partial}{\partial z^a}\Big|_p \in T_p M, \quad p \in U_S,$$

is in $T_p S$ if and only if

$$w^a = \dot{z}^a(w) = 0, \quad \text{for all } a \in \{1, \dots, n-d\}.$$

In other words,

$$TS \cap TU = \{(p, w) \in TU : z^a(p, w) = \dot{z}^a(p, w) = 0, \text{ for all } a = 1, \dots, d\},$$

showing that $(TU, T\varphi)$ is an adapted chart to TS (up to a permutation of the coordinates). The last part of the statement follows from the easy remark that, for every chart (U, φ) on M adapted to S, the submanifold chart induced on TS by $(TU, T\varphi)$ agrees with $(TU_S, T\varphi_S)$: the standard chart induced by the submanifold chart (U_S, φ_S) (do you see it?). So TS can be covered by charts that are simultaneously in the submanifold atlas and the standard atlas. $\qquad\square$

Chapter 4

Full Rank Smooth Maps

By definition, the rank of a smooth map $F : M \to N$ between manifolds is the rank of the tangent map. So the rank is a non-negative integer valued function on M, bounded from the above by the minimum of the dimensions of M and N. Different smooth maps $F, G : M \to N$ may look very different, even locally, depending on how does their rank vary on M. However, all constant rank smooth maps, with fixed rank, say r, look locally the same. This is the main content of the *Rank Theorem*, which is a nice and important example of a *"local normal form theorem"* or *"local classification theorem"* in Differential Geometry. In this chapter, we begin defining the simplest possible examples of constant rank maps: *immersions*, *submersions*, and *local diffeomorphisms*. Thereafter, we state the Rank Theorem and discuss a couple of consequences.

4.1 Full Rank Maps

Let M, N be smooth manifolds, with $\dim M = m$ and $\dim N = n$, and let $F : M \to N$ be a smooth map.

Definition 4.1 (Rank of a Smooth Map). The *rank* of F is the non-negative integer valued map:

$$\operatorname{rank} F : M \to \mathbb{N}_0, \quad p \mapsto \operatorname{rank}_p F := \operatorname{rank} d_p F = \dim \left(\operatorname{im} d_p F \right).$$

- F is an *immersion* if $m \leq n$, and $\operatorname{rank}_p F = m$ for all $p \in M$ (in other words, the tangent map $d_p F$ is injective for all $p \in M$).
- F is a *submersion* if $m \geq n$, and $\operatorname{rank}_p F = n$ for all $p \in M$ (in other words, the tangent map $d_p F$ is surjective for all $p \in M$).

- F is a *local diffeomorphism* if $m = n$, and $\operatorname{rank}_p F = m = n$ for all $p \in M$ (in other words, the tangent map $d_p F$ is an isomorphism for all $p \in M$).

Let $F : M \to N$ be a smooth map between manifolds as above, and let $p \in M$. Choose $(U, \varphi = (x^1, \ldots, x^m))$, a chart on M around p, and $(V, \psi = (y^1, \ldots, y^n))$, a chart on N around $F(p)$, such that $F(U) \subseteq V$. In particular, we can consider the coordinate representation $\widehat{F} : \widehat{U} \to \widehat{V}$. Put $P = \varphi(p)$, and recall that the Jacobian matrix

$$J_{\widehat{F}}(P) = \left(\frac{\partial \widehat{F}^a}{\partial t^i}(P) \right)^{a=1,\ldots,n}_{i=1,\ldots,m}$$

is exactly the representative matrix of the tangent map $d_p F : T_p M \to T_{F(p)} N$ in the coordinate frames

$$\left(\frac{\partial}{\partial x^1}\big|_p, \ldots, \frac{\partial}{\partial x^m}\big|_p \right) \quad \text{and} \quad \left(\frac{\partial}{\partial y^1}\big|_{F(p)}, \ldots, \frac{\partial}{\partial y^n}\big|_{F(p)} \right).$$

It follows that

$$\operatorname{rank}_p F = \operatorname{rank} J_{\widehat{F}}(P).$$

We now provide a list of examples.

Example 4.2. Consider the standard Euclidean spaces $\mathbb{R}^2$ and $\mathbb{R}^3$ with standard coordinates (u, v) and (x, y, z), respectively, and the smooth map

$$F : \mathbb{R}^2 \to \mathbb{R}^3, \quad (u, v) \mapsto F(u, v) := (u^2, uv, v^2).$$

The Jacobian matrix of F is

$$J_F = \begin{pmatrix} \frac{\partial F^1}{\partial u} & \frac{\partial F^1}{\partial v} \\ \frac{\partial F^2}{\partial u} & \frac{\partial F^2}{\partial v} \\ \frac{\partial F^3}{\partial u} & \frac{\partial F^3}{\partial v} \end{pmatrix} = \begin{pmatrix} 2u & 0 \\ v & u \\ 0 & 2v \end{pmatrix}.$$

We conclude that

$$\operatorname{rank}_P F = \operatorname{rank} J_F(P) = \begin{cases} 0 & \text{if } P = (0,0) \\ 2 & \text{if } P \neq (0,0) \end{cases}.$$

In particular, F is neither an immersion nor a submersion (but $F : \mathbb{R}^2 \setminus \{(0,0)\} \to \mathbb{R}^3$ is an immersion — do you see it?). ◆

Example 4.3. Let m, n be non-negative integers with $m \leq n$. The smooth map

$$\mathrm{in}_m : \mathbb{R}^m \to \mathbb{R}^n, \ \ P = (P^1, \ldots, P^m) \mapsto \mathrm{in}_m(P) := (P^1, \ldots, P^m, 0, \ldots, 0)$$

is an immersion. Indeed, its Jacobian matrix is

$$J_{\mathrm{in}_m} = \begin{pmatrix} 1 & \cdots & 0 \\ \vdots & \ddots & \vdots \\ 0 & \cdots & 1 \\ 0 & \cdots & 0 \\ \vdots & \ddots & \vdots \\ 0 & \cdots & 0 \end{pmatrix},$$

whose rank is constant and equal to m. ◆

Example 4.4. Consider the standard Euclidean space $\mathbb{R}^2$ with coordinates (x, y). The curve

$$\gamma : \mathbb{R} \to \mathbb{R}^2, \ \ t \mapsto (\sin t, \cos t)$$

is an immersion. Indeed, its Jacobian matrix is

$$J_\gamma = (\cos t, -\sin t),$$

whose rank is constant and equal to 1. Alternatively, we could argue as follows. For any $t_0 \in \mathbb{R}$, the tangent space $T_{t_0}\mathbb{R}$ is spanned by $\frac{d}{dt}|_{t_0}$, whose image under the tangent map $d_{t_0}\gamma : T_{t_0}\mathbb{R} \to T_{\gamma(t_0)}\mathbb{R}^2$ is

$$d_{t_0}\gamma \left(\frac{d}{dt}|_{t_0} \right) = \dot{\gamma}(t_0) = \cos t_0 \frac{\partial}{\partial x}|_{\gamma(t_0)} - \sin t_0 \frac{\partial}{\partial y}|_{\gamma(t_0)} \neq 0.$$

It follows that $d_{t_0}\gamma$ is injective for all $t_0 \in \mathbb{R}$, hence γ is an immersion. More generally, a curve $\gamma : I \to M$ in a manifold M is an immersion if and only if the velocity $\dot{\gamma}(t_0)$ is non-zero for all $t_0 \in I$. ◆

Example 4.5. In view of Proposition 3.13, the inclusion $i_S : S \hookrightarrow M$ of a submanifold $S \subseteq M$ in a manifold M is always an immersion. ◆

Example 4.6. Let m, n be non-negative integers with $m \geq n$. The smooth map

$$\mathrm{pr}_n : \mathbb{R}^m \to \mathbb{R}^n, \quad P = (P^1, \dots, P^m) \mapsto \mathrm{pr}_n(P) := (P^1, \dots, P^n)$$

is a submersion. Indeed, its Jacobian matrix is

$$J_{\mathrm{pr}_n} = \begin{pmatrix} 1 & \cdots & 0 & 0 & \cdots & 0 \\ \vdots & \ddots & \vdots & \vdots & \ddots & \vdots \\ 0 & \cdots & 1 & 0 & \cdots & 0 \end{pmatrix},$$

whose rank is constant and equal to n. ◆

Example 4.7. Consider the standard Euclidean space $\mathbb{R}^{n+1}$ with coordinates $(t^0, \dots, t^n)$. We already know that the canonical projection

$$\pi : \mathbb{R}^{n+1} \setminus \{0\} \to \mathbb{R}P^n, \quad P \mapsto \pi(P) = [P],$$

is smooth (see Example 2.11). We want to show that it is a submersion. To do this, pick a point $P = (P^0, \dots, P^n) \in \mathbb{R}^{n+1} \setminus \{0\}$, and let $i \in \{0, \dots, n\}$ be such that $P^i \neq 0$. So $(V_i := \{t^i \neq 0\}, \mathrm{id})$ is a chart around P, and $\pi(V_i) \subseteq U_i$, where, as usual, we denote by (U_i, φ_i) the i-th affine chart on $\mathbb{R}P^n$. Recall from Example 2.11 that the coordinate representation of π with respect to these charts is

$$\widehat{\pi} : V_i \to \mathbb{R}^n, \quad (t^0, \dots, t^n) \mapsto \widehat{\pi}(t^0, \dots, t^n) = \left(\frac{t^0}{t^i}, \dots, \frac{\widehat{t^i}}{t^i}, \dots, \frac{t^n}{t^i} \right).$$

The Jacobian matrix of $\widehat{\pi}$ is

$$J_{\widehat{\pi}} = \begin{pmatrix} \frac{1}{t^i} & \cdots & 0 & -\frac{t^0}{(t^i)^2} & 0 & \cdots & 0 \\ \vdots & \ddots & \vdots & \vdots & \vdots & \ddots & \vdots \\ 0 & \cdots & \frac{1}{t^i} & -\frac{t^{i-1}}{(t^i)^2} & 0 & \cdots & 0 \\ 0 & \cdots & 0 & -\frac{t^{i+1}}{(t^i)^2} & \frac{1}{t^i} & \cdots & 0 \\ \vdots & \ddots & \vdots & \vdots & \vdots & \ddots & \vdots \\ 0 & \cdots & 0 & -\frac{t^n}{(t^i)^2} & 0 & \cdots & \frac{1}{t^i} \end{pmatrix},$$

whose rank is constant and equal to n. In particular, $\mathrm{rank}_P \, \pi = n$. So, from the arbitrariness of P, π is a submersion. ◆

Example 4.8. Let $\mathcal{U} \subseteq \mathbb{R}^m$ be an open subset, and let $F : \mathcal{U} \to \mathbb{R}^n$ be a smooth map, with $m \geq n$. Assume that 0 is a regular value of F. We already remarked that, in this case, rank $J_F = n$ around every point $P \in Z(F)$ (see the discussion following Example 2.28). Hence rank $F = n$ in an open neighborhood $\mathcal{V} \subseteq \mathcal{U}$ of $Z(F)$, and $F|_\mathcal{V} : \mathcal{V} \to \mathbb{R}^n$ is a submersion. Conversely, if there exists an open neighborhood $\mathcal{V} \subseteq \mathcal{U}$ of $Z(F)$ such that $F|_\mathcal{V} : \mathcal{V} \to \mathbb{R}^n$ is a submersion, then 0 is a regular value of F (do you see it?). ◆

Example 4.9. Every diffeomorphism is a local diffeomorphism (see Proposition 3.11). ◆

Example 4.10. Consider the curve $\gamma : \mathbb{R} \to \mathbb{R}^2$ from Example 4.4. Obviously, it takes values in the circle $S^1 \subseteq \mathbb{R}^2$ and can be restricted to S^1 in the codomain. Denote by $\gamma' : \mathbb{R} \to S^1$ the restriction. From the Submanifold Chart Theorem (Theorem 2.42), γ' is a smooth map. Additionally, γ is the composition of γ' followed by the inclusion $i_{S^1} : S^1 \hookrightarrow \mathbb{R}^2$. Hence, for any $t \in \mathbb{R}$,

$$d_t\gamma = d_t(i_{S^1} \circ \gamma') = d_{\gamma(t)}i_{S^1} \circ d_t\gamma',$$

where we used the chain rule (3.7). As $d_t\gamma$ is injective, $d_t\gamma'$ is injective as well, and, from the arbitrariness of t, γ' is also an immersion. Finally, $\dim \mathbb{R} = \dim S^1 = 1$, hence γ' is a local diffeomorphism. ◆

Example 4.11. Consider the canonical projection $\pi : \mathbb{R}^{n+1} \smallsetminus \{0\} \to \mathbb{R}P^n$, and restrict it to the sphere $S^n \subseteq \mathbb{R}^{n+1} \smallsetminus \{0\}$ to get a new smooth map $\pi' : S^n \to \mathbb{R}P^n$. We want to show that π' is a local diffeomorphism. As $\dim S^n = \dim \mathbb{R}P^n = n$, it is enough to check that π' is a submersion. To do this, consider the projection

$$\Pi : \mathbb{R}^{n+1} \smallsetminus \{0\} \to S^n, \quad P \mapsto \Pi(P) := \frac{P}{\|P\|}.$$

We already mentioned in Example 2.45 that Π is smooth. Additionally, π is the composition of Π followed by π' (do you see it?). Hence, for any $P \in \mathbb{R}^{n+1} \smallsetminus \{0\}$,

$$d_P\pi = d_{\Pi(P)}\pi' \circ d_P\Pi.$$

As $d_P\pi$ is surjective, $d_{\Pi(P)}\pi'$ is surjective as well. Finally, from the surjectivity of Π, every point in S^n is of the form $\Pi(P)$, and π' is a submersion (hence a local diffeomorphism). ◆

Exercise 4.1. Let $M_1, \ldots, M_k$ be manifolds, and consider the product manifold $M_1 \times \cdots \times M_k$. Show that the projection

$$\mathrm{pr}_i : M_1 \times \cdots \times M_k \to M_i, \quad (p_1, \ldots, p_k) \mapsto \mathrm{pr}_i(p_1, \ldots, p_k) := p_i,$$

onto the i-th factor is a submersion for all $i = 1, \ldots, k$.

4.2 Rank Theorem

Theorem 4.12 (Rank Theorem). *Let M, N be smooth manifolds, $\dim M = m$, $\dim N = n$, let $F : M \to N$ be a smooth map, and let $r \leq \min\{m, n\}$ be a non-negative integer. The following conditions are equivalent:*

(1) *The rank of F is constant and equal to r.*
(2) *For every point $p \in M$, there exist a chart (U, φ) in M, centered at p (recall that this means that $p \in U$, and $\varphi(p) = 0$), and a chart (V, ψ) in N, centered at $F(p)$, such that $F(U) \subseteq V$, and the coordinate representation $\widehat{F} : \widehat{U} \to \widehat{V}$ of F reads*

$$\widehat{F}(t^1, \ldots, t^m) = (t^1, \ldots, t^r, 0, \ldots, 0) = (\mathrm{in}_r \circ \mathrm{pr}_r)(t^1, \ldots, t^m).$$

In particular,

- *if $m \leq n$, then F is an immersion if and only if, around every point of M, it has a coordinate representation $\widehat{F}$ such that*

$$\widehat{F}(t^1, \ldots, t^m) = (t^1, \ldots, t^m, 0, \ldots, 0) = \mathrm{in}_m(t^1, \ldots, t^m),$$

- *if $m \geq n$, then F is a submersion if and only if, around every point of M, it has a coordinate representation $\widehat{F}$ such that*

$$\widehat{F}(t^1, \ldots, t^m) = (t^1, \ldots, t^n) = \mathrm{pr}_n(t^1, \ldots, t^m),$$

- *if $m = n$, then F is a local diffeomorphism if and only if, around every point of M, it has a coordinate representation $\widehat{F}$ such that*

$$\widehat{F}(t^1, \ldots, t^m) = (t^1, \ldots, t^m) = \mathrm{id}(t^1, \ldots, t^m).$$

Proof.

(1) $\Rightarrow$ (2) Omitted (for a detailed proof see, e.g., Lee (2013)).
(2) $\Rightarrow$ (1) Left as Exercise 4.2. $\qquad\qquad\qquad\qquad\qquad\qquad\square$

> **Exercise 4.2.** Prove the implication **(2)** $\Rightarrow$ **(1)** in the statement of the Rank Theorem (Theorem 4.12).

In practice, the Rank Theorem says that the local structure of a smooth map $F : M \to N$ of constant rank r does only depend on $\dim M, \dim N$, and r. Yet in other words, all smooth maps F of constant rank r between an m-dimensional and an n-dimensional manifold do locally look the same. For this reason, we also say that the Rank Theorem is a *local normal form theorem*. We now discuss some consequences of the rank theorem.

Proposition 4.13. *Let M, N be smooth manifolds, $\dim M = m$, $\dim N = n$, and let $F : M \to N$ be a smooth map such that $\operatorname{rank} F$ is constant and equal to $r \leq \min\{m, n\}$. Then*

- *for every point $p \in M$, there is an open neighborhood $U \subseteq M$ of p, such that $F(U) \subseteq N$ is an r-dimensional submanifold, and the restriction $F : U \to F(U)$ is a submersion,*
- *for every point $q \in F(M)$, the preimage $S_q := F^{-1}(q)$ is an $(m - r)$-dimensional submanifold, and, for every point $p \in S_q$, the tangent space $T_p S_q \subseteq T_p M$ is the kernel of the tangent map $d_p F : T_p M \to T_q N$.*

Proof. Let $p \in M$, and let (U, φ) and (V, ψ) be charts as in the statement of the Rank Theorem. As pr_r maps open subsets to open subsets, and in_r is a homeomorphism onto its image,

$$W := (\operatorname{pr}_r \circ \widehat{F})(\widehat{U}) \subseteq \mathbb{R}^r$$

is an open neighborhood of 0 in $\mathbb{R}^r$, and there exist open subsets $W \subseteq \mathcal{W} \subseteq \mathbb{R}^r$ and $W' \subseteq \mathbb{R}^{n-r}$, such that $0 \in W \times W' \subseteq \widehat{V}$. Consider the subcharts $(V_0 := \psi^{-1}(W \times W'), \psi)$, and $(U_0 := F^{-1}(V_0) \cap U, \varphi)$. We have

$$\psi(V_0 \cap F(U)) = \psi(F(U_0)) = W \times \{0\}$$

$$= \left\{ (Q^1, \ldots, Q^n) \in W \times W' : Q^{r+1} = \cdots = Q^n = 0 \right\}.$$

Hence (V_0, ψ) is an adapted chart to $F(U)$, around $F(p)$. We conclude that $F(U)$ is an r-dimensional submanifold. Finally, the adapted chart (V_0, ψ) induces a submanifold chart $(W_0, \psi_0 := \operatorname{pr}_r \circ \psi)$ with $W_0 = V_0 \cap F(U)$ and $\widehat{W}_0 = W$. The coordinate representation $\widehat{F} : \widehat{U}_0 \to \widehat{W}_0$ of $F : U \to F(U)$

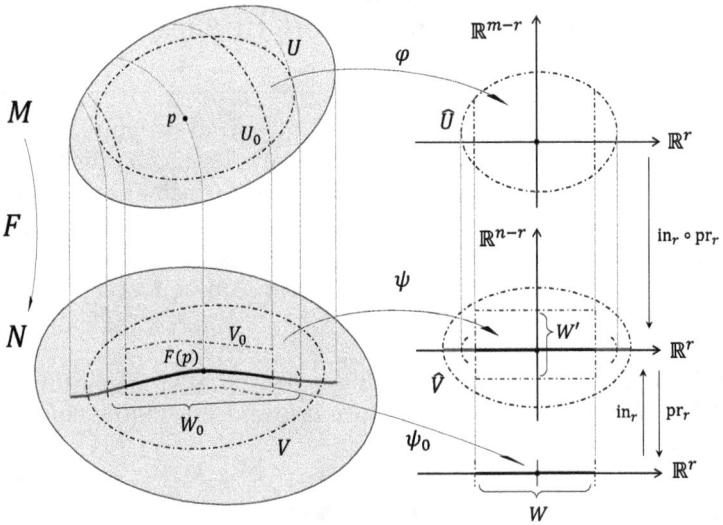

Figure 4.1.　Locally, the image of a constant rank map is a submanifold.

with respect to the charts (U_0, φ), (W_0, ψ_0) is given by

$$\widehat{F}(t^1, \ldots, t^m) = (t^1, \ldots, t^r) = \mathrm{pr}_r(t^1, \ldots, t^m),$$

showing that $F : U_0 \to W_0$ is a submersion (see Figure 4.1).

For the second item in the statement, let $q \in N$ be a point in the image of F, and let $p \in S_q$. This simply means that $F(p) = q$. Take charts (U, φ) and (V, ψ) as in the statement of the Rank Theorem. Then

$$\varphi(S_q \cap U) = \varphi(F^{-1}(q) \cap U) = \widehat{F}^{-1}(0) = Z(\widehat{F}).$$

But $Z(\widehat{F})$ is the solution space of the following regular system or r equations:

$$\begin{cases} t^1 = 0 \\ \quad \vdots \\ t^r = 0 \end{cases}.$$

It follows that S_q is an $(m - r)$-dimensional submanifold. It remains to compute the tangent space $T_p S_q$. First of all, note that $T_p S_q$ and $\ker d_p F$ are both $(m - r)$-dimensional subspaces in $T_p M$. Indeed, $T_p S_q$ is the tangent space to an $(m - r)$-dimensional submanifold, and $\ker d_p F$ is the kernel of a rank r linear map. So, in order to prove that $T_p S_q = \ker d_p F$, it is enough

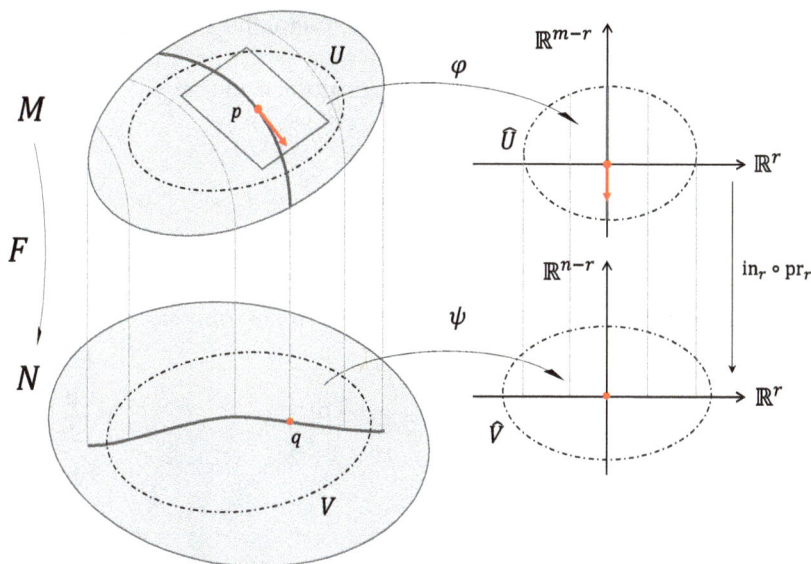

Figure 4.2. Fibers of a constant rank map are submanifolds.

to check that $T_p S_q \subseteq \ker d_p F$. To do this, pick a tangent vector $v \in T_p S_q$. In view of Proposition 3.15, v is the velocity of some curve $\gamma : I \to S_q$ in S_q at some time $t_0 \in I : v = \dot{\gamma}(t_0)$. Finally, use Exercise 3.6 to compute

$$d_p F(v) = d_p(F)\left(\frac{d\gamma}{dt}(t_0)\right) = \frac{d(F \circ \gamma)}{dt}(t_0).$$

But $(F \circ \gamma)(t) = q$ constantly on I, so $d_p F(v)$ is the velocity of a constant curve, which is clearly 0 (see Figure 4.2). $\qquad\square$

Example 4.14 (Tangent Spaces to the Sphere). Consider the standard Euclidean space $\mathbb{R}^{n+1}$ with standard coordinates $t^1, \ldots, t^{n+1}$ and the smooth map

$$F : \mathbb{R}^{n+1} \setminus \{0\} \to \mathbb{R},$$

$$P = (P^1, \ldots, P^{n+1}) \mapsto F(P) := (P^1)^2 + \cdots + (P^{n+1})^2.$$

The rank of F at the point $P = (P^1, \ldots, P^{n+1}) \in \mathbb{R}^{n+1} \setminus \{0\}$ is

$$\text{rank}_P\, F = \text{rank}\, J_F(P) = \text{rank}\left(2P^1, \ldots, 2P^{n+1}\right) = 1,$$

constantly. So F is a constant rank map (actually a submersion). The n-dimensional sphere $S^n \subseteq \mathbb{R}^{n+1}$ can be seen as the preimage of 1 under $F \colon S^n = F^{-1}(1)$. It now follows from Proposition 4.13 that $T_P S^n = \ker d_P F$ for all $P = (P^1, \dots, P^{n+1}) \in S^n$. So, a tangent vector

$$V = V^i \frac{\partial}{\partial t^i}\big|_P$$

to $\mathbb{R}^{n+1}$ at some point $P \in S^n$ is tangent to S^n if and only if

$$0 = d_P F(V) = V^i \frac{\partial F}{\partial t^i}(P) \frac{d}{dt}\big|_{F(P)} = 2(V^1 P^1 + \dots + V^{n+1} P^{n+1}) \frac{d}{dt}\big|_1$$

which is equivalent to

$$V^1 P^1 + \dots + V^{n+1} P^{n+1} = 0,$$

i.e., $(V^1, \dots, V^{n+1})$ is orthogonal to $(P^1, \dots, P^{n+1})$, in agreement with our usual understanding from elementary Euclidean Geometry (see Figure 4.3). ◆

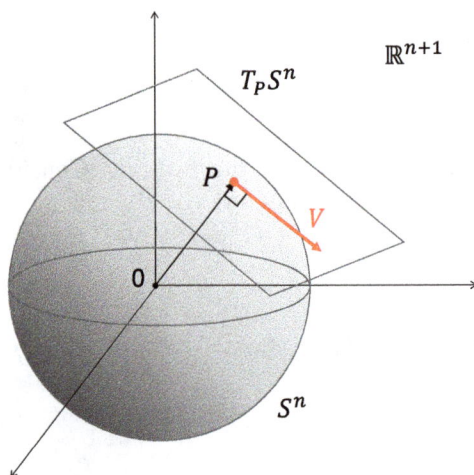

Figure 4.3.   The tangent space to the sphere at the point P is "orthogonal" to P.

Example 4.15 (The Orthogonal Group). The *orthogonal group* is the following subgroup of the general linear group:

$$O(n) := \left\{ X \in GL(n, \mathbb{R}) : X^t X = \mathbb{I}_n \right\},$$

where the "$(-)^t$" denotes transposition and $\mathbb{I}_n$ is the identity matrix. In other words, $O(n)$ is the *fiber over* $\mathbb{I}_n$ of the map:

$$F : GL(n, \mathbb{R}) \to M(n, \mathbb{R}), \quad X \mapsto X^t X,$$

i.e., $O(n) = F^{-1}(\mathbb{I}_n)$. It is clear that F is a smooth map (do you see it?). If we manage to prove that the rank of F is constant, it will follow from Corollary 4.13 that $O(n)$ is a submanifold in $GL(n, \mathbb{R})$. Now let $A \in GL(n, \mathbb{R})$ be any invertible matrix, and consider the maps:

$$\Phi_A : GL(n, \mathbb{R}) \to GL(n, \mathbb{R}), \quad X \mapsto XA$$

and

$$\Psi_A : M(n, \mathbb{R}) \to M(n, \mathbb{R}), \quad Y \mapsto A^t Y A.$$

They are smooth for every A (do you see it?). Additionally, they are inverted by the maps $\Phi_{A^{-1}}$ and $\Psi_{A^{-1}}$, respectively. Hence they are diffeomorphisms. A direct computation shows that the diagram

$$
\begin{array}{ccc}
GL(n, \mathbb{R}) & \xrightarrow{\Phi_A} & GL(n, \mathbb{R}) \\
{\scriptstyle F}\downarrow & & \downarrow{\scriptstyle F} \\
M(n, \mathbb{R}) & \xrightarrow{\Psi_A} & M(n, \mathbb{R})
\end{array}
$$

commutes, i.e., $F \circ \Phi_A = \Psi_A \circ F$. It then follows from the chain rule that

$$d_{XA} F \circ d_X \Phi_A = d_{X^t X} \Psi_A \circ d_X F, \quad \text{for all } X \in GL(n, \mathbb{R}).$$

As $d_X \Phi_A$ and $d_{X^t X} \Psi_A$ are both isomorphisms, we clearly have

$$\operatorname{rank}_X F = \operatorname{rank} d_X F = \operatorname{rank} d_{XA} F = \operatorname{rank}_{XA} F.$$

Finally, let $X' \in GL(n, \mathbb{R})$ be any other invertible matrix. Choosing, $A = X^{-1} X'$, we immediately get

$$\operatorname{rank}_X F = \operatorname{rank}_{XA} F = \operatorname{rank}_{X'} F.$$

Hence F has constant rank. One can actually show that

$$\operatorname{rank} F = \frac{n(n+1)}{2}.$$

We conclude that $O(n) \subseteq GL(n, \mathbb{R})$ is a submanifold of dimension $n^2 - n(n+1)/2 = n(n-1)/2$. ◆

Exercise 4.3 (The Special Linear and the Special Orthogonal Groups).
Prove that

(1) the special linear group

$$SL(n, \mathbb{R}) := \{X \in GL(n, \mathbb{R}) : \det X = 1\}$$

is an $(n^2 - 1)$-dimensional submanifold in $GL(n, \mathbb{R})$, and
(2) the special orthogonal group

$$SO(n) := O(n) \cap SL(n, \mathbb{R}) = \{X \in O(n) : \det X = 1\}$$

is an open submanifold in $O(n)$.

(Hint: *For point (1), proceed as in Example* 4.15. *For point (2), note that, since* $\det X = \pm 1$ *for all* $X \in O(n)$, *we have* $SO(n) = \{X \in O(n) : \det X > 0\}$.)

We conclude this section discussing one more corollary of the Rank Theorem.

Proposition 4.16 (Inverse Function Theorem). *Let M, N be manifolds, and let $F : M \to N$ be a smooth map. Then F is a local diffeomorphism if and only if, for every point $p \in M$, there exist an open neighborhood $U \subseteq M$ of p, and an open neighborhood $U' \subseteq N$ of $F(p)$, such that $F(U) \subseteq U'$ and the restriction $F : U \to U'$ is a diffeomorphism.*

Proof. The "if part" of the statement is obvious. Let's discuss the "only if" part. So let $F : M \to N$ be a local diffeomorphism, and let $p \in M$ be a point. Choose charts (U, φ), (V, ψ) as in the statement of the Rank Theorem. Then $\widehat{F} : \widehat{U} \to \widehat{V}$ acts as the identity. In particular, $\widehat{F}$ is injective, and $\widehat{U} = \widehat{F}(\widehat{U}) \subseteq \widehat{V}$. Put $U' := \psi^{-1}(\widehat{U})$. It is clear that $F(U) \subseteq U'$, and $F : U \to U'$ is a diffeomorphism as claimed (do you see it?). □

Note that the Inverse Function Theorem motivates the terminology "local diffeomorphism".

Corollary 4.17. *Let M, N be manifolds, and let F : M → N be a map. Then F is a diffeomorphism if and only if it is a bijective local diffeomorphism.*

Proof.   Left as Exercise 4.4. □

> **Exercise 4.4.** Prove Corollary 4.17.

4.3 Embeddings

If we specialize Proposition 4.13 to an immersion $i : N \to M$, we see that, restricting i to a sufficiently small open subset $U \subseteq N$, we can make $i(U)$ to be a submanifold. In general, we cannot choose $U = N$ as the following counter-example shows.

Example 4.18 (The Eight Figure). Consider $\mathbb{R}^2$ with standard coordinates x, y and the following curve:

$$i : \mathbb{R} \to \mathbb{R}^2, \quad t \mapsto i(t) := (\sin 2t, \sin t).$$

The image of i is called the *eight figure* (for obvious reasons, see Figure 4.4). The eight figure is not a submanifold. Indeed, around the origin, it is neither of the form $y = f(x)$ nor of the form $x = f(y)$ (for some smooth function f). Indeed, every vertical (or horizontal) line through a non-zero point in an open neighborhood $U \subseteq i(N)$ of 0 intersects U in at least two

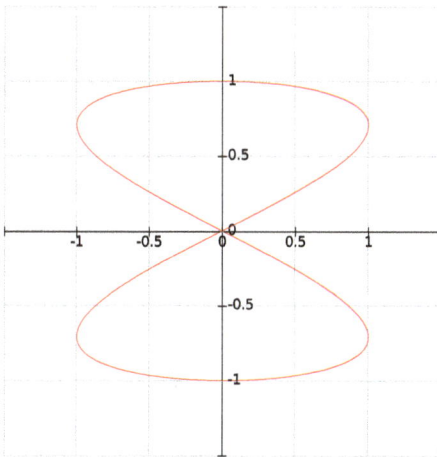

Figure 4.4.   The eight figure.

points (do you see it?). Note that the issue is still there if we restrict i to the interval $(-\pi, \pi)$ turning it in an *injective immersion*

$$i : (-\pi, \pi) \to \mathbb{R}^2.$$

Indeed $i(-\pi, \pi) = i(\mathbb{R})$ which is again the eight figure.

However, for every positive real number $\epsilon < \pi$, $i(-\pi + \epsilon, \pi - \epsilon)$ is a submanifold and

$$i : (-\pi + \epsilon, \pi - \epsilon) \to i(-\pi + \epsilon, \pi - \epsilon)$$

is a diffeomorphism. ◆

In the following proposition, given an injective immersion $i : N \to M$, we find conditions on i that guarantee that $i(N)$ is a submanifold. Before stating the result, we need some preliminary remarks. So, let $i : N \to M$ be an injective immersion. In particular, the restriction of i to its image $i(N)$ in the codomain is a bijection $i : N \to i(N)$. On $i(N)$, there are at least two topologies: the subspace topology τ_{sub} induced by the inclusion $i(N) \subseteq M$ and the topology τ_i induced from that of N via the bijection $i : N \to i(N)$, i.e.,

$$\tau_i := \{i(U) \subseteq i(N) : U \subseteq N \text{ is an open subset}\}.$$

In other words, τ_i is the unique topology on $i(N)$ such that $i : N \to (i(N), \tau_i)$ is a homeomorphism. In general, $\tau_{\text{sub}} \neq \tau_i$. For instance, when i is the injective immersion

$$i : (-\pi, \pi) \to \mathbb{R}^2$$

from Example 4.18, then, for every positive real number $\epsilon < \pi$, any open neighborhood of 0 in $(i(N), \tau_{\text{sub}})$ does necessarily intersect non-trivially $i(\pi - \epsilon, \pi)$ and $i(-\pi, -\pi + \epsilon)$. On the other hand, $i(-\pi/4, \pi/4) \in \tau_i$, but it does not intersect, for instance, $i(3\pi/4, \pi)$ nor $i(-\pi, -3\pi/4)$. The following proposition shows that this is precisely the reason why the eight figure is not a submanifold.

Proposition 4.19. *Let $i : N \to M$ be an injective immersion. Then the following two conditions are equivalent:*

(1) $i(N) \subseteq M$ *is a submanifold of the same dimension as N.*
(2) $\tau_{\text{sub}} = \tau_i$, *or, equivalently, $i : N \to (i(N), \tau_{\text{sub}})$ is a homeomorphism.*

Additionally, if (1), or, equivalently, (2), holds true, then $i : N \to i(N)$ is a diffeomorphism.

Proof.

(1) ⇒ (2) If $i(N)$ is a submanifold of the same dimension as N, then $i : N \to i(N)$ is a bijective local diffeomorphism (do you see it?), hence a diffeomorphism. In particular, it is a homeomorphism.

(2) ⇒ (1) Assume that $i : N \to (i(N), \tau_{\text{sub}})$ is a homeomorphism, pick a point $q \in i(N)$, and let $p \in N$ be the unique point such that $q = i(p)$. From Proposition 4.13, there is an open neighborhood $U \subseteq M$ of p such that $i(U) \subseteq M$ is a submanifold of the same dimension as N. On the other hand, we have $i(U) \in \tau_i = \tau_{\text{sub}}$. Hence there is an open subset $\mathcal{U} \subseteq M$ such that $i(U) = i(N) \cap \mathcal{U}$. Summarizing, we have showed that, for any $q \in i(N)$, there exists an open neighborhood $\mathcal{U} \subseteq M$ of q such that $i(N) \cap \mathcal{U}$ is a submanifold. This easily implies that $i(N)$ is itself a submanifold. We leave the details as Exercise 4.5. □

> **Exercise 4.5.** Conclude the proof of Proposition 4.19, showing that a subset $S \subseteq M$ in a manifold M is a submanifold if an only if it is locally so, i.e., for every point $q \in S$, there exists an open neighborhood $\mathcal{U} \subseteq M$ of q such that $S \cap \mathcal{U}$ is a submanifold.

Definition 4.20 (Embedding). An *embedding* is an injective immersion $i : N \to M$ satisfying one, hence both, of the equivalent conditions in Proposition 4.19.

Example 4.21 (Embeddings and Submanifolds). Let M be a manifold. The inclusion $i_S : S \hookrightarrow M$ of a submanifold $S \subseteq M$ is an embedding, and every embedding $i : N \to M$ can be written as the composition $i_S \circ \Phi$ of a diffeomorphism $\Phi : N \to S$ onto some submanifold $S \subseteq M$, followed by the inclusion $i_S : S \hookrightarrow M$. ♦

Example 4.22. Let d, n be non-negative integers such that $d \le n$. The smooth map

$$\text{in}_d : \mathbb{R}^d \to \mathbb{R}^n, \quad P = (P^1, \dots, P^d) \mapsto \text{in}_d(P) = (P^1, \dots, P^d, 0, \dots, 0)$$

is an embedding. Indeed it is an injective immersion, and its image is the d-dimensional submanifold $\mathbb{R}^d \times \{0\} \subseteq \mathbb{R}^d \times \mathbb{R}^{n-d} = \mathbb{R}^n$. ♦

An important class of embeddings is provided by *sections of surjective submersions*, as explained in Proposition 4.23. Before stating it, we give a definition. Let M, N be smooth manifolds, and let $\pi : M \to N$ and $i : N \to M$ be smooth maps such that $\pi \circ i = \text{id}_N$. In this situation, i is called a *(smooth) section* of π (cf. Section 3.3).

Proposition 4.23. *Let $\pi : M \to N$ be a smooth map of manifolds, and let $i : N \to M$ be a section of π. Then i is an embedding, and there is an open neighborhood $\mathcal{U}$ of $i(N)$ in M such that $\pi : \mathcal{U} \to N$ is a surjective submersion.*

Proof.   From the *section property* $\pi \circ i = \mathrm{id}_N$, it immediately follows that π is surjective and i is injective. Now, let $p \in N$. The chain rule at p reads

$$d_{i(p)}\pi \circ d_p i = d_p \mathrm{id}_N = \mathrm{id}_{T_p N}$$

so that $d_p i$ is injective and $d_{i(p)}\pi$ is surjective. From the arbitrariness of p, i is an immersion, hence an injective immersion, and $\mathrm{rank}_p \pi = \dim N$ for all $p \in i(N)$, hence for all p in an open neighborhood $\mathcal{U} \subseteq M$ of $i(N)$ in M. So, $\pi : \mathcal{U} \to N$ is a surjective submersion. It remains to prove that $i : N \to (i(N), \tau_{\mathrm{sub}})$ is a homeomorphism. It is enough to prove that the inverse map $i^{-1} : (i(N), \tau_{\mathrm{sub}}) \to N$ is continuous. But, from the section property $\pi \circ i = \mathrm{id}_N$ again, i^{-1} is the restriction of π to the subspace $i(N)$, hence it is indeed continuous. This concludes the proof. $\square$

Example 4.24. Let M, N be manifolds, and let $F : M \to N$ be a smooth map. Consider the map

$$\Gamma_F : M \to M \times N, \quad p \mapsto \Gamma_F(p) := (p, F(p)).$$

From Exercise 2.7, Γ_F is a smooth map. More precisely, it is a section of the projection

$$\mathrm{pr}_M : M \times N \to M, \quad (p, q) \mapsto \mathrm{pr}_M(p, q) := p$$

onto the first factor. So, from Proposition 4.23, Γ_F is an embedding. In particular, $\Gamma_F(M)$ is a submanifold of the same dimension as M in the product $M \times N$. As $\Gamma_F(M) = \mathrm{graph}\, F$, the latter fact is not new (see Example 2.36). The new fact is that $\Gamma_F(M)$ is, additionally, diffeomorphic to M.   ◆

Example 4.25 (The Segre Embedding). We want to show that the product $\mathbb{R}P^1 \times \mathbb{R}P^1$ of two copies of the *projective line* $\mathbb{R}P^1$ can be embedded into the 3-dimensional projective space $\mathbb{R}P^3$. More precisely, there is an embedding, called the *Segre embedding*,

$$\sigma : \mathbb{R}P^1 \times \mathbb{R}P^1 \to \mathbb{R}P^3,$$

given by

$$\sigma\left([s^0 : s^1], [t^0 : t^1]\right) := [s^0 t^0 : s^0 t^1 : s^1 t^0 : s^1 t^1], \quad [s^0 : s^1], [t^0 : t^1] \in \mathbb{R}P^1.$$
$$\tag{4.1}$$

To see this, first note that σ is well defined by (4.1) (do you see it?).

Second, we show that σ is a smooth map. Denote by $(U_0, \varphi_0), (U_1, \varphi_1)$ the usual affine charts on $\mathbb{R}P^1$ and by (V_i, ψ_i) the affine charts on $\mathbb{R}P^3$, $i = 0, \dots, 3$. It is clear that

$$\sigma(U_0 \times U_0) \subseteq V_0, \quad \sigma(U_0 \times U_1) \subseteq V_1,$$
$$\sigma(U_1 \times U_0) \subseteq V_2, \quad \sigma(U_1 \times U_1) \subseteq V_3,$$

showing that σ has coordinate representations around every point. Let's compute the first one:

$$\widehat{\sigma} : \widehat{U}_0 \times \widehat{U}_0 = \mathbb{R} \times \mathbb{R} \to \widehat{V}_0 = \mathbb{R}^3. \tag{4.2}$$

For any $(P, Q) \in \mathbb{R} \times \mathbb{R}$, we have

$$\begin{aligned} \widehat{\sigma}(P, Q) &= (\psi_0 \circ \sigma \circ (\varphi_0 \times \varphi_0)^{-1})(P, Q) \\ &= (\psi_0 \circ \sigma)\left([1 : P], [1 : Q]\right) \\ &= \psi_0\left([1 : Q : P : PQ]\right) \\ &= (Q, P, PQ), \end{aligned}$$

which depends smoothly on (P, Q). Similarly, all other coordinate representations are smooth (check it as an exercise).

Third, we show that σ is an immersion. Denote by x, y the standard coordinates on $\widehat{U}_0 \times \widehat{U}_0 = \mathbb{R} \times \mathbb{R}$. Then, the Jacobian matrix of the coordinate representation (4.2) is

$$\begin{pmatrix} \frac{\partial \widehat{\sigma}^1}{\partial x} & \frac{\partial \widehat{\sigma}^1}{\partial y} \\ \frac{\partial \widehat{\sigma}^2}{\partial x} & \frac{\partial \widehat{\sigma}^2}{\partial y} \\ \frac{\partial \widehat{\sigma}^3}{\partial x} & \frac{\partial \widehat{\sigma}^3}{\partial y} \end{pmatrix} = \begin{pmatrix} 0 & 1 \\ 1 & 0 \\ y & x \end{pmatrix},$$

whose rank is constant and equal to 2. This shows that $\operatorname{rank}_p \sigma = 2$ for all $p \in U_0 \times U_0$. Similarly, the rank of σ is 2 at any other point (check it as an exercise), and σ is an immersion.

Fourth, we show that σ is injective. So let

$$p = \left([s^0 : s^1], [t^0 : t^1]\right) \quad \text{and} \quad \tilde{p} = \left([\tilde{s}^0 : \tilde{s}^1], [\tilde{t}^0 : \tilde{t}^1]\right)$$

be two points in $\mathbb{R}P^1 \times \mathbb{R}P^1$ such that $\sigma(p) = \sigma(\tilde{p})$. This means that there exists a non-zero scalar $\lambda \in \mathbb{R}$ such that

$$s^i t^j = \lambda \tilde{s}^i \tilde{t}^j, \quad \text{for all } i, j \in \{0, 1\}.$$

Now, let j be such that $t^j \neq 0$. Then

$$s^i = \frac{\lambda \tilde{t}^j}{t^j} \tilde{s}^i, \quad \text{for all } i \in \{0, 1\}.$$

Hence, $[s^0 : s^1] = [\tilde{s}^0 : \tilde{s}^1]$. Similarly, $[t^0 : t^1] = [\tilde{t}^0 : \tilde{t}^1]$, so that $p = \tilde{p}$.

We conclude proving that $\sigma(\mathbb{R}P^1 \times \mathbb{R}P^1) \subseteq \mathbb{R}P^3$ is a 2-dimensional submanifold. To do this, we show that it coincides with the submanifold S from Example 2.38. So let $q = [P^0 : \cdots : P^3]$ be a point in $\mathbb{R}P^3$. We claim that q is in the image of σ if and only if

$$P^0 P^3 - P^1 P^2 = 0. \tag{4.3}$$

The "only if part" of the claim follows from an easy computation that we leave as an exercise. For the "if part", note that condition (4.3) can be re-written in the form

$$\det \begin{pmatrix} P^0 & P^1 \\ P^2 & P^3 \end{pmatrix} = 0. \tag{4.4}$$

Let $(P^0, P^1) \neq (0, 0)$. Then, from (4.4), there exists a scalar $a \in \mathbb{R}$ such that

$$(P^2, P^3) = (aP^0, aP^1)$$

so that

$$q = [P^0 : \cdots : P^3] = [P^0 : P^1 : aP^0 : aP^1] = \sigma\left([1 : a], [P^0 : P^1]\right).$$

The case $(P^2, P^3) \neq (0, 0)$ can be discussed in a similar way and we leave the obvious details to the reader.

Summarizing, we have proved that σ is an embedding. More generally, for any two non-negative integers n, m, there is an embedding, also called the *Segre embedding*, $\sigma : \mathbb{R}P^n \times \mathbb{R}P^m \to \mathbb{R}P^{nm+n+m}$. We will not discuss the details of the latter construction. For more details on the Segre embedding in Algebraic Geometry see, e.g., Harris (1992). ◆

Chapter 5

Vector Fields

In this chapter, we continue developing a differential calculus on manifolds, defining *vector fields*. There are two alternative (but equivalent) definitions of a vector field on a manifold. Geometrically, a vector field on a manifold M is the assignment of a tangent vector X_p to M at the point p, for every point $p \in M$, what we also call a *field of vectors*. Additionally, we require that X_p *depends smoothly* on p. This can be formalized rigorously via sections of the tangent bundle. There is another possible algebraic definition of a vector field, which is particularly efficient, and that we adopt as the primary definition: vector fields on M are *derivations of the algebra $C^\infty(M)$ of smooth functions on M*. This latter definition can be easily transported to other areas of Geometry like Algebraic Geometry or Supergeometry. We will provide both definitions and discuss their equivalence. We also discuss how do vector fields interact with smooth maps. In the following chapter, we will discuss two important (and interrelated) interpretations of vector fields: vector fields as *ordinary differential equations* and vector fields as *infinitesimal diffeomorphisms*.

5.1 Vector Fields: An Algebraic Definition

The most intuitive definition of a vector field on a manifold M is the geometric one: a *vector field* is a *field of vectors*, i.e., the assignment of a tangent vector $X_p \in T_pM$ for each $p \in M$ (see Figure 5.1). We will always require that X_p *depends smoothly* on p. This can be formalized noting that a field of vectors is the same as a section $s : M \to TM$ of the tangent bundle (see Section 3.3), for which the *smoothness* is a well-defined property. However, in these lecture notes, we will adopt a different, more algebraic, definition, similar in spirit to the very definition of a tangent vector. This choice

$$M$$

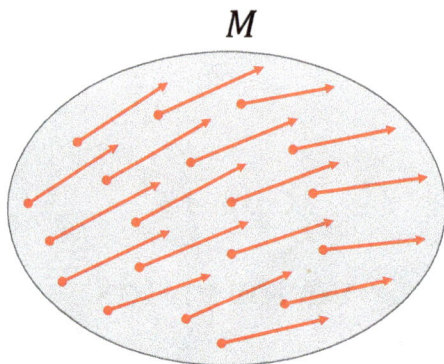

Figure 5.1.   A field of vectors on a manifold M.

is, partly, a matter of personal taste but, for the remaining part, is motivated by the efficiency of Algebra in several different situations in Differential Geometry. We begin with a purely algebraic definition: let A be a real algebra.

Definition 5.1 (Derivation). A *derivation* of A is an $\mathbb{R}$-linear map

$$X : A \to A$$

satisfying the following *Leibniz rule*: for all $\alpha, \beta \in A$,

$$X(\alpha\beta) = \alpha X(\beta) + \beta X(\alpha).$$

Now, let M be a manifold.

Definition 5.2 (Vector Field). A *vector field* on M is a derivation of the algebra $C^\infty(M)$ of smooth functions on M. The set of all vector fields on M is denoted by $\mathfrak{X}(M)$.

In this section, we elaborate on Definition 5.2, discussing its algebraic consequences for the space of vector fields. In the following section, we show that vector fields on M are equivalent to (smooth) fields of vectors, i.e., (smooth) sections of the tangent bundle TM.

Example 5.3 (Coordinate Vector Fields). Let M be an n-dimensional manifold and let $(U, \varphi = (x^1, \ldots, x^n))$ be a chart on M. For each $i = 1, \ldots, n$, define the map

$$\frac{\partial}{\partial x^i} : C^\infty(U) \to C^\infty(U), \quad f \mapsto \frac{\partial}{\partial x^i} f := \frac{\partial \widehat{f}}{\partial t^i} \circ \varphi,$$

where, as usual, $\widehat{f} = f \circ \varphi^{-1}$ is the coordinate representation of f. Note that, for each $f \in C^\infty(U)$, the coordinate representation of $\frac{\partial}{\partial x^i} f$ is exactly $\frac{\partial \widehat{f}}{\partial t^i}$:

$$\widehat{\frac{\partial}{\partial x^i}} f = \frac{\partial \widehat{f}}{\partial t^i}. \tag{5.1}$$

The map $\frac{\partial}{\partial x^i}$ is called the i-th *coordinate vector field* and it is indeed a vector field on U (beware, not on M!). The proof of this claim is left as Exercise 5.1. ◆

Exercise 5.1. Show that the i-th coordinate vector field $\frac{\partial}{\partial x^i}$ is a vector field on U.

We now show that the space $\mathfrak{X}(M)$ of vector fields on M does actually carry some algebraic structures. We begin defining the *sum of two vector fields* and the *product of a vector field by a smooth function*. So let $X, Y \in \mathfrak{X}(M)$ be two vector fields, and let $f \in C^\infty(M)$ be a smooth function. We define new maps as follows:

$$X + Y : C^\infty(M) \to C^\infty(M), \quad g \mapsto (X + Y)(g) := X(g) + Y(g),$$

and

$$f \cdot X : C^\infty(M) \to C^\infty(M), \quad g \mapsto (f \cdot X)(g) := f \cdot X(g).$$

Exercise 5.2. Show that

(1) for every $X, Y \in \mathfrak{X}(M)$, $X + Y$ is a vector field,
(2) for every $X \in \mathfrak{X}(M)$, and every $f \in C^\infty(M)$, $f \cdot X$ is a vector field,
(3) $(\mathfrak{X}(M), +, \cdot)$ is a $C^\infty(M)$-module.

We conclude this section discussing one more algebraic structure on $\mathfrak{X}(M)$: vector fields form a *Lie algebra*. We begin with the definition of Lie algebra.

Definition 5.4 (Lie Algebra). A *real Lie algebra*, or simply a *Lie algebra*, is a real vector space $\mathfrak{g}$ equipped with one more interior operation

$$[-, -] : \mathfrak{g} \times \mathfrak{g} \to \mathfrak{g}, \quad (v, w) \mapsto [v, w],$$

called the *Lie bracket*, with the following properties:

- $[-,-]$ is $\mathbb{R}$-bilinear,
- $[-,-]$ is skew-symmetric, i.e., $[v,w] = -[w,v]$ for all $v,w \in \mathfrak{g}$,
- $[-,-]$ satisfies the following *Jacobi identity*:

$$[u,[v,w]] + [v,[w,u]] + [w,[u,v]] = 0 \quad \text{for all } u,v,w \in \mathfrak{g}.$$

A *Lie subalgebra* of a Lie algebra $\mathfrak{g}$ is a vector subspace $\mathfrak{h} \subseteq \mathfrak{g}$ which is preserved by the Lie bracket, i.e., $[v,w] \in \mathfrak{h}$, for all $v,w \in \mathfrak{h}$.

Note that a Lie subalgebra $\mathfrak{h}$ of a Lie algebra $\mathfrak{g}$ is also a Lie algebra with the restricted Lie bracket. The following example is our main example of a Lie algebra (besides vector fields).

Example 5.5 (Lie Algebra of Endomorphisms). Let V be a real vector space. The real vector space $\mathrm{End}_{\mathbb{R}}(V)$ of ($\mathbb{R}$-linear) endomorphisms of V is a Lie algebra with the following Lie bracket:

$$[-,-] : (h,l) \mapsto [h,l] := h \circ l - l \circ h.$$

Prove this claim as an exercise! The Lie bracket in $\mathrm{End}_{\mathbb{R}}(V)$ is called the *commutator*, and two endomorphisms $h,l \in \mathrm{End}_{\mathbb{R}}(V)$ are said to *commute* if their commutator vanishes: $[h,l] = 0$. ♦

Proposition 5.6 (Lie Algebra of Vector Fields). *Let M be a manifold. Vector fields on M form a Lie subalgebra of the Lie algebra $\mathrm{End}_{\mathbb{R}}(C^{\infty}(M))$ of $\mathbb{R}$-linear endomorphisms of the real vector space $C^{\infty}(M)$. In other words, the commutator of two vector fields is a vector field.*

Proof. Left as Exercise 5.3. □

> **Exercise 5.3.** Prove Proposition 5.6.

In particular, the space $\mathfrak{X}(M)$ of vector fields, with the commutator, is a Lie algebra. Note that the commutator of vector fields is $\mathbb{R}$-bilinear, but it is not $C^{\infty}(M)$-bilinear (in general). Actually, the $C^{\infty}(M)$-module structure and the Lie algebra structure on $\mathfrak{X}(M)$ interact in a more complicated (but still nice) way, as explained in the following

Proposition 5.7. *Let $X,Y \in \mathfrak{X}(M)$ be vector fields and let $f \in C^{\infty}(M)$ be a smooth function. Then*

$$[X,fY] = X(f)Y + f[X,Y].$$

Proof. A simple computation left as Exercise 5.4. □

> **Exercise 5.4.** Prove Proposition 5.7.

5.2 Vector Fields and Fields of Vectors

Let M be a manifold. In this section, we prove that vector fields on M are equivalent to fields of vectors. For a precise statement, see Proposition 5.8. We begin noting that a vector field $X \in \mathfrak{X}(M)$ determines a field of vectors. Indeed, for each point $p \in M$, we can consider the map

$$X_p : C^\infty(M) \to \mathbb{R}, \quad f \mapsto X_p(f) := X(f)(p). \tag{5.2}$$

It can be proved that X_p is a tangent vector to M at p (see Exercise 5.5), called the *value at p* of the vector field X.

Exercise 5.5. Prove that, for every $p \in M$, the map X_p defined in (5.2) is a tangent vector to M at the point p. Prove also that, for every $X, Y \in \mathfrak{X}(M)$, and every $f \in C^\infty(M)$,

$$(X + Y)_p = X_p + Y_p,$$

$$(fX)_p = f(p)X_p.$$

Finally, let $(U, \varphi = (x^1, \ldots, x^n))$ be a chart around p. Prove that the value of the coordinate vector field $\frac{\partial}{\partial x^i}$ at p is exactly the coordinate tangent vector $\frac{\partial}{\partial x^i}\big|_p$.

The assignment $p \mapsto X_p$ is a field of vectors that can be interpreted as a(n *a priori* non-necessarily smooth) section s_X of the tangent bundle TM:

$$s_X : M \to TM, \quad p \mapsto s_X(p) := X_p.$$

We are now ready to state the main result of this section.

Proposition 5.8. *The map*

$$\mathfrak{X}(M) \to \Gamma(TM), \quad X \mapsto s_X \tag{5.3}$$

is a well-defined $C^\infty(M)$-module isomorphism.

Proof. We have to check several things. First of all, we have to check that, for every $X \in \mathfrak{X}(M)$, the section s_X is actually smooth. It is enough to check that, for every chart $(U, \varphi = (x^1, \ldots, x^n))$ on M, the components s_X^i of the section s_X in the chart (U, φ) (as defined at the end of Section 3.3) are smooth functions on U. To do this, we fix an arbitrary point p_0 in U

and show that s_X^i is smooth around p_0 (the smoothness of s_X^i then follows from the arbitrariness of p_0). So, let $p \in U$ and compute

$$s_X^i(p) = \dot{x}^i(s_X(p)) = \dot{x}^i(X_p)$$

$$= i\text{-th component of } X_p \text{ in the coordinate frame}$$

$$= X_p(x^i).$$

In order to continue our computation, we need to choose a function $\tilde{x}^i \in C^\infty(M)$ such that $\tilde{x}^i = x^i$ in an open neighborhood $V \subseteq U$ of p_0. Such function exists in view of the Local Extension Lemma (Lemma 2.47), and, from the locality of tangent vectors, we have that

$$s_X^i(p) = X_p(x^i) = X_p(\tilde{x}^i) = X(\tilde{x}^i)(p)$$

for all $p \in V$. As X maps smooth functions to smooth functions, we conclude that $s_X^i|_V$ is smooth, and (5.3) is well defined.

We leave it as Exercise 5.6 (Point (1)) to check that the map (5.3) is $C^\infty(M)$-linear. It remains to check that (5.3) is invertible (the $C^\infty(M)$-linearity of the inverse then follows from general properties of module isomorphisms). To do this, we define its inverse. Namely, given a section $s \in \Gamma(TM)$ of the tangent bundle, we construct a vector field $X_s \in \mathfrak{X}(M)$ in such a way that

$$s_{X_s} = s \quad \text{and} \quad X_{s_X} = X, \tag{5.4}$$

for all $s \in \Gamma(TM)$, and all $X \in \mathfrak{X}(M)$, i.e., in such a way that the map $\Gamma(TM) \to \mathfrak{X}(M), s \mapsto X_s$ inverts (5.3). For every $f \in C^\infty(M)$, we define

$$X_s(f) : M \to \mathbb{R}, \quad p \mapsto X_s(f)(p) := s(p)(f).$$

We have to show that

- $X_s(f) \in C^\infty(M)$ for all $s \in \Gamma(TM)$ and all $f \in C^\infty(M)$,
- $X_s : C^\infty(M) \to C^\infty(M)$ is a derivation of $C^\infty(M)$, hence a vector field, and
- the identities (5.4) hold for all $s \in \Gamma(TM)$ and all $X \in \mathfrak{X}(M)$.

We only discuss the first point and leave it to the reader to prove the remaining two points as Exercise 5.6 (Points (2) and (3)). Let $s \in \Gamma(TM)$ and $f \in C^\infty(M)$. In order to prove that $X_s(f)$ is smooth, we prove that the restriction $X_s(f)|_U$ is smooth for every chart (U, φ). So, let $p \in U$, and compute

$$X_s(f)(p) = s(p)(f) = s(p)(f|_U)$$
$$= \left(s^i(p)\frac{\partial}{\partial x^i}\Big|_p\right) f|_U = \left(s^i\frac{\partial}{\partial x^i}\right)_p f|_U = \left(s^i\frac{\partial}{\partial x^i}\right)(f|_U)(p),$$

where, in the first step of the second line, we used Exercise 5.5. As the s^i are smooth, $s^i\frac{\partial}{\partial x^i}$ is a vector field on U, and the last expression depends smoothly on p. This concludes the proof. □

Exercise 5.6. Complete the proof of Proposition 5.8 showing that

(1) the map $X \mapsto s_X$ is $C^\infty(M)$-linear,
(2) for every $s \in \Gamma(TM)$, the map $X_s : C^\infty(M) \to C^\infty(M)$ is a derivation,
(3) the map $s \mapsto X_s$ inverts $X \mapsto s_X$.

Proposition 5.8 has some important consequences. First of all, a vector field X on M can be "restricted" to an open submanifold $U \subseteq M$. Namely, X corresponds to a section $s : M \to TM$ of TM. Restricting s to U in the domain and to TU in the codomain, we get a section $s|_U : U \to TU$ of TU. Finally, $s|_U$ corresponds to a vector field on U that we denote $X|_U$, and call the *restriction of X to U*. Note that, by construction, for every point $p \in U$, the value of $X|_U$ at p agrees with the value of X at the same point p:

$$(X|_U)_p = X_p, \quad \text{for all } p \in U.$$

Hence, we also have $X|_U(f|_U) = X(f)|_U$ for all $X \in \mathfrak{X}(M)$, and all $f \in C^\infty(M)$. Now, let $X \in \mathfrak{X}(M)$, and $f \in C^\infty(U)$. Sometimes, abusing the notation, we will write $X(f)$ instead of $X|_U(f)$. In any case, $X(f) = X|_U(f)$ is a smooth function on U (not on M).

The second consequence of Proposition 5.8 (closely related to the previous one) is the following.

Lemma 5.9 (Gluing Lemma for Vector Fields). *Let M be a manifold, and let $C = \{U\}$ be an open cover of M. For every $U \in C$, let X_U be a vector field on U. If*

$$X_U|_{U \cap V} = X_V|_{U \cap V}$$

for every $\mathcal{U}, \mathcal{V} \in \mathcal{C}$, then there exists a unique vector field X on M such that

$$X|_{\mathcal{U}} = X_{\mathcal{U}}$$

for all $\mathcal{U} \in \mathcal{C}$.

Proof.   Left as Exercise 5.7.   □

Exercise 5.7. Prove the Gluing Lemma for Vector Fields (Lemma 5.9).

The third consequence of Proposition 5.8 is that, for every chart (U, φ), the $C^\infty(U)$-module of vector fields on U possesses a finite frame consisting of the coordinate vector fields, according to the following

Proposition 5.10. *Let M be an n-dimensional manifold, and let $(U, \varphi = (x^1, \ldots, x^n))$ be a chart on M. Then the coordinate vector fields*

$$\left(\frac{\partial}{\partial x^1}, \ldots, \frac{\partial}{\partial x^n} \right)$$

form a frame in $\mathfrak{X}(U)$ (called the coordinate frame), i.e., they are independent and generate $\mathfrak{X}(U)$. In particular, every vector field $X \in \mathfrak{X}(U)$ can be uniquely written in the form

$$X = X^i \frac{\partial}{\partial x^i}$$

for some smooth functions $X^i \in C^\infty(U)$.

Proof.   We first prove that the coordinate vector fields are independent. So let $f^1, \ldots, f^n$ be smooth functions on U such that

$$f^i \frac{\partial}{\partial x^i} = 0.$$

This means that $f^i \frac{\partial}{\partial x^i} g = 0$ for all $g \in C^\infty(U)$. In particular, we can choose $g = x^j, j \in \{1, \ldots, n\}$. From (5.1), $\frac{\partial}{\partial x^i} x^j = \delta_i^j$, hence

$$0 = f^i \frac{\partial}{\partial x^i} x^j = f^i \delta_i^j = f^j.$$

It remains to prove that every vector field $X \in \mathfrak{X}(U)$ can be written in the form

$$X = X^i \frac{\partial}{\partial x^i}$$

for some smooth functions $X^i \in C^\infty(U)$. To do this, consider the section $s_X \in \Gamma(TU)$ corresponding to U, and put $X^i = s_X^i$. This means that, for every $p \in U$

$$X_p = X^i(p)\frac{\partial}{\partial x^i}|_p = \left(X^i \frac{\partial}{\partial x^i} \right)_p,$$

where we used Exercise 5.5. As X and $X^i\frac{\partial}{\partial x^i}$ share the same values (in points of U), they must coincide. $\qquad\square$

We can combine the Gluing Lemma 5.9 with Proposition 5.10 to find that, for every vector field X on M,

(1) X is completely determined by its restrictions $X|_U$ to some coordinate domains U covering M,
(2) for every chart $(U, \varphi = (x^1, \ldots, x^n))$, the restriction $X|_U$ can be uniquely written in the form

$$X|_U = X^i \frac{\partial}{\partial x^i}$$

for some smooth functions $X^i \in C^\infty(U)$, called the *components* of X in the chart (U, φ).

The proof of Proposition 5.10 then shows that the components X^i of a vector field $X \in \mathfrak{X}(M)$ in the chart $(U, \varphi = (x^1, \ldots, x^n))$ coincide with the components of the corresponding section, and they are given by

$$X^i = X(x^i).$$

Now, let X be a vector field on M. How do the components X^i of X in a chart change under a change of coordinates? To answer this question, take two charts $(U, \varphi = (x^1, \ldots, x^n))$ and $(U, \tilde\varphi = (\tilde x^1, \ldots, \tilde x^n))$ with the same coordinate domain U. Then $X|_U$ can be written either as

$$X|_U = X^i \frac{\partial}{\partial x^i} \quad \text{or as} \quad X|_U = \tilde X^i \frac{\partial}{\partial \tilde x^i}$$

for some smooth functions $X^i, \tilde X^i \in C^\infty(U)$, and we want to find the relationship between those. We have

$$\tilde X^i = X(\tilde x^i) = X^j \frac{\partial}{\partial x^j} \tilde x^i.$$

In its turn, the matrix

$$\left(\frac{\partial}{\partial x^j}\tilde{x}^i\right)^{i=1,\dots,n}_{j=1,\dots,n}$$

is essentially the Jacobian matrix of the transition map $\tilde{\varphi}\circ\varphi^{-1}:\widehat{U}\to\widehat{\tilde{U}}$ (up to a composition with φ). Indeed, from (5.1),

$$\frac{\partial}{\partial x^j}\tilde{x}^i = \widehat{\frac{\partial}{\partial x^j}\tilde{x}^i}\circ\varphi = \frac{\partial\widehat{\tilde{x}}^i}{\partial t^j}\circ\varphi = \frac{\partial(\tilde{\varphi}\circ\varphi^{-1})^i}{\partial t^j}\circ\varphi,$$

where a hat "$\widehat{(-)}$" denotes the coordinate representation with respect to the chart (U,φ) (not $(U,\tilde{\varphi})$). Yet in other words, the coordinate representation of $\frac{\partial}{\partial x^j}\tilde{x}^i$ with respect to the chart (U,φ) is $\frac{\partial(\tilde{\varphi}\circ\varphi^{-1})^i}{\partial t^j}$:

$$\widehat{\frac{\partial}{\partial x^j}\tilde{x}^i} = \frac{\partial(\tilde{\varphi}\circ\varphi^{-1})^i}{\partial t^j}.$$

Exercise 5.8. Consider the 2-dimensional standard Euclidean space $\mathbb{R}^2$ with standard coordinates x,y and the *polar chart* $(U,\psi=(r,\phi))$ on it. Here $U=\mathbb{R}^2\smallsetminus\{y=0,\ x\geq 0\}$, $\widehat{U}=(0,+\infty)\times(0,2\pi)$ and

$$\psi:U\to\widehat{U},\quad (x,y)\mapsto\psi(x,y)=(r,\phi):=\left(\sqrt{x^2+y^2},\phi(x,y)\right),$$

where

$$\phi(x,y):=\begin{cases}\arccos\dfrac{x}{\sqrt{x^2+y^2}} & \text{if } y\geq 0 \\[3mm] -\arccos\dfrac{x}{\sqrt{x^2+y^2}} & \text{if } y<0\end{cases}.$$

The inverse map is

$$\psi^{-1}:\widehat{U}\to U,\quad (r,\phi)\mapsto\psi^{-1}(r,\phi)=(r\cos\phi,r\sin\phi).$$

Consider the following vector fields on $\mathbb{R}^2$:

$$X=x\frac{\partial}{\partial x}+y\frac{\partial}{\partial y}\quad\text{and}\quad Y=x\frac{\partial}{\partial y}-y\frac{\partial}{\partial x}.$$

Prove that

$$X|_U=r\frac{\partial}{\partial r}\quad\text{and}\quad Y|_U=\frac{\partial}{\partial\phi}.$$

Exercise 5.9. Consider the real *projective plane* $\mathbb{R}P^2$, and denote $(U_i, \varphi_i = (x_i, y_i))$ the standard affine charts on it, $i = 0, 1, 2$. Prove that there exists a unique vector field $X \in \mathfrak{X}(\mathbb{R}P^2)$ such that

$$X|_{U_0} = \left(1 + x_0^2\right) \frac{\partial}{\partial x_0} + x_0 y_0 \frac{\partial}{\partial y_0},$$

$$X|_{U_1} = -\left(1 + x_1^2\right) \frac{\partial}{\partial x_1} - x_1 y_1 \frac{\partial}{\partial y_1},$$

$$X|_{U_2} = x_2 \frac{\partial}{\partial y_2} - y_2 \frac{\partial}{\partial x_2}.$$

5.3   Vector Fields and Smooth Maps

In this section, we discuss how vector fields interact with smooth maps. Unlike smooth functions, there is no way to transport a vector field from a manifold to another manifold along a smooth map, i.e., there is no analog of the pull-back construction for vector fields, unless we restrict to diffeomorphisms. However, there is a way to define the compatibility of two vector fields, one on a manifold M and the other on a manifold N, with respect to a smooth map $F : M \to N$. So let M, N be manifolds, and let $F : M \to N$ be a smooth map.

Definition 5.11 (F-related Vector Fields). A vector field X on M and a vector field Y on N are *F-related* if

$$X \circ F^* = F^* \circ Y.$$

Before providing examples, we remark that the *F-relatedness* of two vector fields can be equivalently defined interpreting vector fields as fields of vectors, according to the following

Proposition 5.12. *A vector field X on M and a vector field Y on N are F-related if and only if, for every $p \in M$,*

$$(d_p F)(X_p) = Y_{F(p)}.$$

Proof.   Let $X \in \mathfrak{X}(M)$ and $Y \in \mathfrak{X}(N)$. Then X and Y are F-related if and only if $X(F^*(f)) = F^*(Y(f))$ for all $f \in C^\infty(N)$. In turn, this is equivalent to

$$X(F^*(f))(p) = F^*(Y(f))(p), \quad \text{for all } f \in C^\infty(N), \text{ and all } p \in M.$$

But

$$X(F^*(f))(p) = X_p(F^*(f)) = d_p F(X_p)(f)$$

and

$$F^*(Y(f))(p) = Y(f)(F(p)) = Y_{F(p)}(f).$$

This concludes the proof. $\qquad \square$

Example 5.13. In the standard Euclidean plane $\mathbb{R}^2$ with standard coordinates x, y, consider the following curve:

$$\gamma : \mathbb{R} \to \mathbb{R}^2, \quad t \mapsto \gamma(t) := (\cos t, \sin t).$$

Consider also the vector fields

$$X = \frac{d}{dt} \in \mathfrak{X}(\mathbb{R}) \quad \text{and} \quad Y = x\frac{\partial}{\partial y} - y\frac{\partial}{\partial x} \in \mathfrak{X}(\mathbb{R}^2).$$

Then X and Y are γ-related. Indeed, for every smooth function $f = f(x, y)$ on $\mathbb{R}^2$,

$$\begin{aligned}
X(\gamma^*(f)) &= \frac{d}{dt} f \circ \gamma = \frac{d}{dt} f(\cos t, \sin t) \\
&= \frac{\partial f}{\partial x}(\cos t, \sin t)\frac{d \cos t}{dt} + \frac{\partial f}{\partial y}(\cos t, \sin t)\frac{d \sin t}{dt} \\
&= -\sin t\frac{\partial f}{\partial x}(\cos t, \sin t) + \cos t\frac{\partial f}{\partial y}(\cos t, \sin t) \\
&= \left(\left(x\frac{\partial}{\partial y} - y\frac{\partial}{\partial x} \right) f \right) \circ \gamma = \gamma^*(Y(f)).
\end{aligned}$$

$\blacklozenge$

Example 5.14. Let d, n be non-negative integers, with $d \leq n$, and consider the standard Euclidean spaces $\mathbb{R}^d$ and $\mathbb{R}^n$, with coordinates $(t^1, \ldots, t^d)$ and

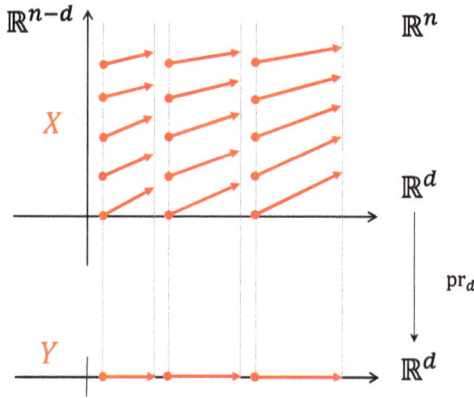

Figure 5.2. The vector fields X and Y from Example 5.14 are pr_d-related.

$(s^1, \ldots, s^d, r^1, \ldots, r^{n-d})$, respectively. Consider also the usual projection

$$\mathrm{pr}_d : \mathbb{R}^n \to \mathbb{R}^d, \quad P = (P^1, \ldots, P^n) \mapsto \mathrm{pr}_d(P) = (P^1, \ldots, P^d).$$

The vector fields

$$X = \mathrm{pr}_d^*(X^i)\frac{\partial}{\partial s^i} + Y^a \frac{\partial}{\partial r^a}$$

on $\mathbb{R}^n$ and

$$Y = X^i \frac{\partial}{\partial t^i}$$

on $\mathbb{R}^d$ are pr_d-related, for every choice of the $X^i = X^i(t^1, \ldots, t^d) \in C^\infty(\mathbb{R}^d)$, and the $Y^a = Y^a(s^1, \ldots, s^d, r^1, \ldots, r^{n-d}) \in C^\infty(\mathbb{R}^n)$ (Exercise 5.10, see Figure 5.2).

$\blacklozenge$

Exercise 5.10. Show that the vector fields X and Y from Example 5.14 are pr_d-related.

The case when F is the inclusion of a submanifold is particularly relevant. So let M be a manifold, and let $S \subseteq M$ be a submanifold. As usual, denote by $i_S : S \hookrightarrow M$ the inclusion.

Definition 5.15 (Vector Field Tangent to a Submanifold). A vector field X on M is *tangent* to S if there exists a vector field Y on S such that Y and X are

i_S-related. In this case, Y is denoted by $X|_S$ and called the *restriction* of X to S.

The following proposition shows that Definition 5.15 agrees with our intuition on a vector field $X \in \mathfrak{X}(M)$ being tangent to a submanifold $S \subseteq M$.

Proposition 5.16. *A vector field X on M is tangent to a submanifold $S \subseteq M$ if and only if, for every $p \in S$, $X_p \in T_pS$, and, in this case, $(X|_S)_p = X_p$.*

Proof. First assume that X is tangent to S, and let $X|_S$ be its restriction. Then, from Proposition 5.12,

$$X_p = X_{i_S(p)} = d_p i_S(X|_S)_p,$$

for all $p \in S$. This shows that $X_p \in T_pS$. Conversely, assume that $X_p \in T_pS$ for all $p \in S$. Think of X as a section s_X of TM. Then the hypothesis says that $s_X|_S$ takes values in TS. In particular, s_X can be restricted to S in the domain and to TS in the codomain, to get a section $s_X|_S : S \to TS$ of TS. As $TS \subseteq TM$ is a submanifold (Proposition 3.22), $s_X|_S$ is a smooth section. Denote by $Y \in \mathfrak{X}(S)$ the associated vector field. By construction $Y_p = d_p i_S(Y_p) = X_p$ for all $p \in S$, so, from Proposition 5.12 again, Y and X are i_S-related. $\qquad\square$

It immediately follows from the last part of Proposition 5.16 that, if the vector field $X \in \mathfrak{X}(M)$ is tangent to a submanifold $S \subseteq M$, then there is a *unique* restriction of X to S, i.e., a unique vector field $X|_S$ on S such that $X|_S$ and X are i_S-related (see Figure 5.3).

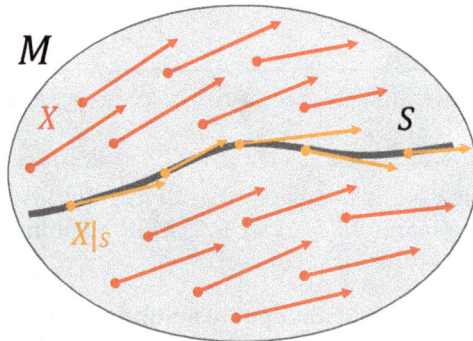

Figure 5.3.   A vector field tangent to a submanifold.

Example 5.17. Consider the standard 3-dimensional Euclidean space $\mathbb{R}^3$ with coordinates (x, y, z) and the vector field

$$X = y\frac{\partial}{\partial z} - z\frac{\partial}{\partial y} \in \mathfrak{X}(M).$$

Use Proposition 5.16 to show that X is tangent to the 2-dimensional sphere $S^2 \subseteq \mathbb{R}^3$. So let $P = (x_0, y_0, z_0) \in S^2$, i.e., $x_0^2 + y_0^2 + z_0^2 = 1$, and check that $X_P \in T_P S^2$. We have

$$X_P = \left(y\frac{\partial}{\partial z} - z\frac{\partial}{\partial y}\right)_P = y_0\frac{\partial}{\partial z}|_P - z_0\frac{\partial}{\partial y}|_P,$$

which is in $T_P S^2$ if and only if its component vector $(0, -z_0, y_0)$ is orthogonal to $P = (x_0, y_0, z_0)$ (Example 4.14). This is indeed the case as

$$0 \cdot x_0 - z_0 \cdot y_0 + y_0 \cdot z_0 = 0.$$

♦

Exercise 5.11. Consider the vector field X from Example 5.17. As X is tangent to the sphere S^2, it can be restricted to S^2 and then further restricted to $U_+ = S^2 \setminus \{P_+\}$. Denote by (A, B) the stereographic coordinates on U_+. Prove that

$$X|_{U_+} = AB\frac{\partial}{\partial A} + \frac{B^2 - A^2 + 1}{2}\frac{\partial}{\partial B}$$

(see Figure 5.4).

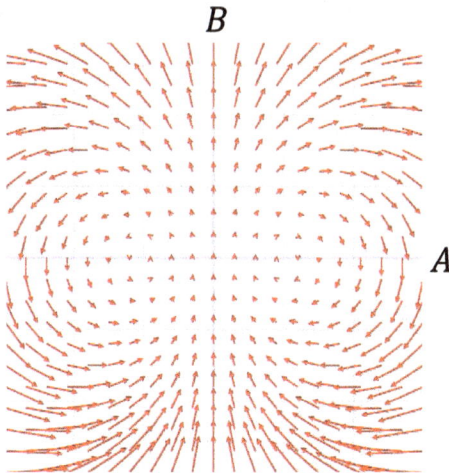

B

A

Figure 5.4. The vector field $AB\frac{\partial}{\partial A} + \frac{B^2-A^2+1}{2}\frac{\partial}{\partial B}$ from Exercise 5.11.

The following proposition explores the relationship between F-relatedness and the algebraic structures on vector fields.

Proposition 5.18. *Let M, N be manifolds, let $F : M \to N$ be a smooth map, let X, X_1, X_2 be vector fields on M, and let Y, Y_1, Y_2 be vector fields on N. If X, X_1, X_2 are F-related to Y, Y_1, Y_2 respectively, then*

- $X_1 + X_2$ *and* $Y_1 + Y_2$ *are F-related,*
- *for every* $f \in C^\infty(N)$, $F^*(f)X$ *and* fY *are F-related, and*
- $[X_1, X_2]$ *and* $[Y_1, Y_2]$ *are F-related.*

Proof.   Left as Exercise 5.12 $\hspace{4cm}$ $\square$

> **Exercise 5.12.** Prove Proposition 5.18.

Example 5.19. Let M be a manifold, let $S \subseteq M$ be a submanifold, and let X_1, X_2, X be vector fields on M that are tangent to S. According to our notation, denote by $X_1|_S, X_2|_S, X|_S$ the restrictions to S of X_1, X_2, X, respectively. It immediately follows from Proposition 5.18 that

- $X_1 + X_2$ is tangent to S, and $(X_1 + X_2)|_S = X_1|_S + X_2|_S$,
- fX is tangent to S for every $f \in C^\infty(M)$, and $(fX)|_S = f|_S X|_S$,
- $[X_1, X_2]$ tangent to S, and $[X_1, X_2]|_S = [X_1|_S, X_2|_S]$

(do you see it?). $\hspace{8cm}$ ◆

Example 5.20. Let M be an n-dimensional manifold, and let $\mathcal{U} \subseteq M$ be an open submanifold. As "vector fields can be restricted to open submanifolds", every vector field $X \in \mathfrak{X}(M)$ is tangent to $\mathcal{U}$, and, given two vector fields $X, Y \in \mathfrak{X}(M)$, the restriction $[X, Y]|_{\mathcal{U}}$ of their commutator is the commutator $[X|_{\mathcal{U}}, Y|_{\mathcal{U}}]$ of the restrictions. In particular, if $(U, \varphi = (x^1, \ldots, x^n))$ is a chart on M, and

$$X|_{\mathcal{U}} = X^i \frac{\partial}{\partial x^i} \quad \text{and} \quad Y|_{\mathcal{U}} = Y^i \frac{\partial}{\partial x^i},$$

then

$$[X, Y]|_{\mathcal{U}} = [X|_{\mathcal{U}}, Y|_{\mathcal{U}}] = Z^i \frac{\partial}{\partial x^i},$$

where the components Z^i are given by

$$
\begin{aligned}
Z^i &= [X|_{\mathcal{U}}, Y|_{\mathcal{U}}](x^i) \\
&= X|_{\mathcal{U}}(Y|_{\mathcal{U}}(x^i)) - Y|_{\mathcal{U}}(X|_{\mathcal{U}}(x^i)) \\
&= X|_{\mathcal{U}}(Y^i) - Y|_{\mathcal{U}}(X^i) \\
&= X^j \frac{\partial}{\partial x^j} Y^i - Y^j \frac{\partial}{\partial x^j} X^i.
\end{aligned}
$$

Summarizing

$$[X,Y]|_u = \left(X|_u(Y^i) - Y|_u(X^i) \right) \frac{\partial}{\partial x^i}$$
$$= \left(X^j \frac{\partial}{\partial x^j} Y^i - Y^j \frac{\partial}{\partial x^j} X^i \right) \frac{\partial}{\partial x^i}.$$

In particular,

$$\left[\frac{\partial}{\partial x^i}, \frac{\partial}{\partial x^j} \right] = 0, \quad \text{for all } i,j \in \{1,\dots,n\}.$$

♦

Finally, we show that vector fields can be pulled-back along diffeomorphisms.

Proposition 5.21. *Let M, N be manifolds, let $\Phi : M \to N$ be a diffeomorphism, and let Y be a vector field on N. Then there exists a unique vector field $\Phi^*(Y)$ on M, called the* pull-back *of Y along Φ, such that $\Phi^*(Y)$ and Y are Φ-related. The pull-back $\Phi^*(Y)$ is given by*

$$\Phi^*(Y) = \Phi^* \circ Y \circ (\Phi^{-1})^*$$

or, equivalently,

$$\Phi^*(Y)_p = (d_{\Phi(p)}\Phi^{-1})(Y_{\Phi(p)}), \quad \text{for all } p \in M. \tag{5.5}$$

Additionally, for any three vector fields Y, Y_1, Y_2 on N and any function $f \in C^\infty(N)$, we have

- $\Phi^*(Y_1 + Y_2) = \Phi^*(Y_1) + \Phi^*(Y_2)$,
- $\Phi^*(fY) = \Phi^*(f)\Phi^*(Y)$,
- $\Phi^*([Y_1, Y_2]) = [\Phi^*(Y_1), \Phi^*(Y_2)]$.

Finally, if $\Psi : N \to Q$ is another diffeomorphism, and $Z \in \mathfrak{X}(Q)$, then

$$\mathrm{id}_Q^*(Z) = Z \quad \text{and} \quad (\Psi \circ \Phi)^*(Z) = \Phi^*(\Psi^*(Z)).$$

Proof. Left as Exercise 5.13. □

> **Exercise 5.13.** Prove Proposition 5.21.

Example 5.22. Let M be an n-dimensional manifold, and let $(U, \varphi = (x^1, \dots, x^n))$ be a chart on M. Then, every function $f \in C^\infty(U)$ can be seen as the pull-back $\varphi^*(\widehat{f})$ of its coordinate representation $\widehat{f}$ along the diffeomorphism $\varphi : U \to \widehat{U}$. It follows that Identity (5.1) can be rephrased as follows:

$$\frac{\partial}{\partial x^i} = \varphi^* \left(\frac{\partial}{\partial t^i} \right) \tag{5.6}$$

(do you see it?). ◆

Remark 5.23. Let M be a manifold, and let $(U, \varphi = (x^1, \ldots, x^n))$ be a chart on it. Identity (5.6) basically says that, if we identify U and $\widehat{U}$ via φ, then the coordinate vector fields $\frac{\partial}{\partial x^i}$ identify with the standard partial derivatives $\frac{\partial}{\partial t^i}$. In particular, the $\frac{\partial}{\partial x^i}$ inherits all standard properties of standard derivatives, and, to stress this, in the following, for every function g on U, we simply write

$$\frac{\partial g}{\partial x^i} \quad \text{for} \quad \frac{\partial}{\partial x^i} g, \quad \text{or} \quad \frac{\partial^2 g}{\partial x^i \partial x^j} \quad \text{for} \quad \frac{\partial}{\partial x^i} \frac{\partial}{\partial x^j} g, \quad \text{etc.}$$

Let's make one example, take another manifold N, a chart $(V, \psi = (y^1, \ldots, y^n))$ on N, and a smooth map $F : M \to N$ such that $F(U) \subseteq V$. We already know that the restriction $F : U \to V$ is completely determined by its *components* $F^a = F^*(y^a)$. Take a function $f \in C^\infty(N)$, and its pull-back $F^*(f) = f \circ F$. It is easy to see using (5.6) that

$$\frac{\partial (f \circ F)}{\partial x^i} = \frac{\partial F^a}{\partial x^i} \cdot \left(\frac{\partial f}{\partial y^a} \circ F \right) \tag{5.7}$$

(check it as an exercise) which is the *chain rule* for coordinate vector fields. In the following, we will apply this and similar formulas without further comments. ◇

Chapter 6

Flows and Symmetries

In this chapter, we provide two more interpretations of vector fields. First of all, a vector field can be interpreted as an ordinary differential equation (ODE), more precisely, a system of explicit and autonomous ordinary differential equations (see, e.g., Arnold, 1983, 1992 for a treatment of vector fields as dynamical systems). The solutions of such system are the so-called *integral curves* of the vector field, and the *Theorem of Existence, Uniqueness, and Smooth Dependence on Initial Data of Solutions of a System of ODEs* from Calculus guarantees that integral curves are particularly well behaved. Second, a vector field can be seen as an *infinitesimal diffeomorphism*. Points of a manifold move along the integral curves of a vector field, and the vector field says what is the velocity of this displacement. Hence a vector field can be seen as an *infinitesimal displacement* of every point of the manifold. This displacement is smooth and (morally) one-to-one, hence it is a(n infinitesimal) diffeomorphism. This informal discussion will be made more rigorous along the following sections.

6.1 Integral Curves of a Vector Field

Let M be a manifold, and let X be a vector field on M. In this chapter, we will always denote by t the canonical coordinate on $\mathbb{R}$ and on any open interval $I \subseteq \mathbb{R}$.

Definition 6.1 (Integral Curve). An *integral curve* of X is a curve $\gamma : I \to M$ on M, defined on some open interval $I \subseteq \mathbb{R}$, such that the canonical coordinate vector field $\frac{d}{dt}$ on I and the vector field X are γ-related, i.e.,

$$\frac{d}{dt} \circ \gamma^* = \gamma^* \circ X,$$

or, equivalently,

$$\dot{\gamma}(t_0) = d_{t_0}\gamma\left(\frac{d}{dt}\big|_{t_0}\right) = X_{\gamma(t_0)}, \quad \text{for every } t_0 \in I.$$

If $t_0 \in I$ and $\gamma(t_0) = p_0 \in M$, we say that γ *passes through* p_0 *at time* t_0. In this case, if $t_0 = 0$ (so $\gamma(0) = p_0$), we also say that γ *starts from* p_0.

Now, we show that, morally, an integral curve is the solution of a system of ODEs. So, let $\gamma : I \to M$ be a curve, and let $t_0 \in I$. Denote $p_0 = \gamma(t_0)$. As γ is smooth, we can choose an open subinterval $J \subseteq I$ containing t_0 and a chart $(U, \varphi = (x^1, \ldots, x^n))$ on M around p_0 such that $\gamma(J) \subseteq U$. Denote by $\gamma^i = \gamma^*(x^i) : J \to \mathbb{R}$ the components of γ in the chart (U, φ). Then, for every $t \in J$,

$$\dot{\gamma}(t) = \frac{d\gamma^i}{dt}(t)\frac{\partial}{\partial x^i}\big|_{\gamma(t)}$$

(see (3.13)). On the other hand,

$$X|_U = X^i\frac{\partial}{\partial x^i},$$

for some $X^i \in C^\infty(U)$, and

$$X_{\gamma(t)} = X^i(\gamma(t))\frac{\partial}{\partial x^i}\big|_{\gamma(t)}.$$

We conclude that $\gamma : J \to M$ is an integral curve of X if and only if

$$\frac{d\gamma^i}{dt}(t) = X^i(\gamma(t)) = \widehat{X}^i(\gamma^1(t), \ldots, \gamma^n(t)), \quad \text{for all } t \in J,$$

which is an ODE for the coordinate representation

$$\widehat{\gamma} : J \to \widehat{U}, \quad t \mapsto \widehat{\gamma}(t) = (\gamma^1(t), \ldots, \gamma^n(t))$$

of γ. Most of the properties of integral curves are corollaries of the following standard result in Calculus, which we state without a proof.

Theorem 6.2 (Existence, Uniqueness, and Dependence on Initial Data).
Let $U \subseteq \mathbb{R}^n$ be an open subset, and let

$$\widehat{X} = \left(\widehat{X}^1, \ldots, \widehat{X}^n\right) : U \to \mathbb{R}^n$$

be a smooth (vector-valued) map. Consider the following system of ODEs

$$\frac{d\gamma^i}{dt}(t) = \widehat{X}^i(\gamma^1(t), \ldots, \gamma^n(t)), \quad i \in \{1, \ldots, n\}, \tag{6.1}$$

imposed on a curve $\widehat{\gamma} = (\gamma^1, \ldots, \gamma^n) : I \to U$ *in* U. *We have the following:*

(1) **Uniqueness:** *For every* $t_0 \in \mathbb{R}$, *every two solutions* $\gamma_1 : I_1 \to U$ *and* $\gamma_2 : I_2 \to U$ *of* (6.1) *such that* $t_0 \in I_1 \cap I_2$, *and* $\gamma_1(t_0) = \gamma_2(t_0)$, *agree on the (non-empty) intersection* $I_1 \cap I_2$ *of their domains, i.e.,* $\gamma_1|_{I_1 \cap I_2} = \gamma_2|_{I_1 \cap I_2}$.
(2) **Existence:** *For every* $t_0 \in \mathbb{R}$, *and every* $P_0 \in U$, *there is an open interval* I *containing* t_0, *and an open neighborhood* $V \subseteq U$ *of* P_0, *such that, for every* $P \in V$, *there exists a (necessarily unique) solution* $\gamma_P : I \to U$ *of* (6.1) *such that* $\gamma_P(t_0) = P$.
(3) **Smooth Dependence on Initial Data:** *For every* $t_0 \in \mathbb{R}$, *and every* $P_0 \in U$, *let* I, V, P, *and* γ_P *be as in Point* (2). *Then the map* $\Phi : I \times V \to U$, $(t, P) \mapsto \Phi(t, P) := \gamma_P(t)$ *depends smoothly on (both t and) P.*

Proof. Omitted (for a detailed proof, see, e.g., Hartman, 2002; Arnold, 1983, 1992; Lee, 2013). $\square$

The first corollary of Theorem 6.2 is the *existence and uniqueness* of *maximal integral curves* of a vector field (starting from a prescribed point). Before stating this result, we have to define maximal integral curves. So let M be a manifold, let X be a vector field on M, and let $p \in M$ be a point. Denote by $\mathcal{I}_p$ the set of integral curves of X starting from p.

Lemma 6.3. *The set* $\mathcal{I}_p$ *is non-empty, i.e., for every* $p \in M$, *there exists an integral curve starting from* p. *Moreover, any two integral curves* $\gamma_1 : I_1 \to M$, $\gamma_2 : I_2 \to M$ *in* $\mathcal{I}_p$, *agree on the intersection* $I_1 \cap I_2$ *of their domains.*

Proof. For the first part of the statement, choose a chart (U, φ) around p and look for an integral curve $\gamma : I \to M$ taking values in U. From Theorem 6.2, and the discussion preceding it, such γ exists. For the second part, let γ_1, γ_2 be as in the statement, and note that $I_1 \cap I_2$ is an open interval containing 0. Denote by $J \subseteq I_1 \cap I_2$ the subset consisting of times where γ_1, γ_2 agree:

$$J := \{t \in I_1 \cap I_2 : \gamma_1(t) = \gamma_2(t)\}.$$

The subset J is simultaneously open and closed in the interval $I_1 \cap I_2$. Indeed, for any $\bar{t}_0 \in J$, not only γ_1, γ_2 agree at $\bar{t}_0$, they also agree in a whole open interval $K \subseteq I_1 \cap I_2$ containing $\bar{t}_0$. The latter claim easily follows from the uniqueness part of Theorem 6.2 (do you see it? If not, check what happens in a chart around $\gamma_1(\bar{t}_0) = \gamma_2(\bar{t}_0)$). This shows that $J \subseteq I_1 \cap I_2$ is an open subset. To see that it is also closed, consider the map

$$(\gamma_1, \gamma_2) : I_1 \cap I_2 \to M \times M, \quad t \mapsto (\gamma_1(t), \gamma_2(t)).$$

As γ_1, γ_2 are smooth, hence continuous maps, it follows that (γ_1, γ_2) is continuous as well with respect to the product topology on $M \times M$ (Exercise 1.11). Moreover, J is the preimage under (γ_1, γ_2) of the diagonal

$$\Delta := \{(p, p) \in M \times M : p \in M\} \subseteq M \times M.$$

But M is a Hausdorff space, hence $\Delta \subseteq M \times M$ is a closed subspace (Exercise 1.12) and J is a closed subset of $I_1 \cap I_2$ as claimed (the preimage of a closed subset under a continuous map). As J is non-empty (it contains at least 0), and $I_1 \cap I_2$ is a connected space, J must coincide with the whole $I_1 \cap I_2$. This concludes the proof. $\qquad \square$

The set $\mathcal{I}_p$ is equipped with a *partial order* $\leq$ defined as follows: let $\gamma_1 : I_1 \to M$ and $\gamma_2 : I_2 \to M$ be in $\mathcal{I}_p$, i.e., γ_1, γ_2 are integral curves starting from p, then

$$\gamma_1 \leq \gamma_2 \quad \text{if, by definition,}$$

$$I_1 \subseteq I_2 \quad \text{(and, in this case, necessarily } \gamma_1 = \gamma_2|_{I_1}).$$

A *maximal integral curve* starting from p is, by definition, a *maximal element* of $\mathcal{I}_p$ with respect to this partial order, i.e., an integral curve γ of X starting from p such that, whenever γ' is another integral curve starting from p such that $\gamma \leq \gamma'$, in fact, we have $\gamma = \gamma'$. Yet in other words, γ cannot be extended to an integral curve starting from p and defined on a larger time interval.

Proposition 6.4 (Existence and Uniqueness of Maximal Integral Curves). *Let M be a manifold, and let X be a vector field on M. Then, for every point $p \in M$, there exists a unique maximal integral curve $\gamma_p : I_p \to M$ of X starting from p, and, for every other integral curve $\gamma : I \to M$ that starts from p, we have $\gamma \leq \gamma_p$, i.e., $I \subseteq I_p$, and $\gamma = \gamma_p|_I$.*

Proof. Define $I_p \subseteq \mathbb{R}$ as the union of the domains of all integral curves starting from p:

$$I_p := \bigcup_{(\gamma: I \to M) \in \mathcal{I}_p} I.$$

Being the union of open intervals containing 0, I_p is itself an open interval containing 0. Finally, we define a curve

$$\gamma_p : I_p \to M$$

as follows. For every $t \in I_p$, let $\gamma : I \to M$ be any integral curve starting from p such that $t \in I$, and put

$$\gamma_p(t) := \gamma(t).$$

From the above discussion, $\gamma_p(t)$ is independent of the choice of γ. Additionally, γ_p agrees with γ in a whole open neighborhood of t in I_p. Hence, from the Gluing Lemma for Smooth Maps, γ_p is a well-defined curve. As γ_p agrees locally, around every time in I_p, with an integral curve, it is an integral curve itself, and, by construction, its domain is the largest possible, so γ_p is maximal. The rest is clear. $\square$

Example 6.5. On $\mathbb{R}^2$ with standard coordinates (x, y) consider the coordinate vector field

$$X = \frac{\partial}{\partial x}.$$

A curve $\gamma = (x(t), y(t)) : I \to \mathbb{R}^2$ is an integral curve of X if it is a solution of the system of ODEs:

$$\begin{cases} \dfrac{dx}{dt}(t) = 1 \\[2mm] \dfrac{dy}{dt}(t) = 0 \end{cases}.$$

This means that

$$\begin{cases} x(t) = x(0) + t \\ y(t) = y(0) \end{cases}.$$

Hence, the maximal integral curve of X starting from $P = (x_0, y_0)$ is

$$\gamma_P : \mathbb{R} \to \mathbb{R}^2, \quad t \mapsto \gamma_P(t) = (x_0 + t, y_0)$$

(see Figure 6.1).

$\blacklozenge$

$$y$$

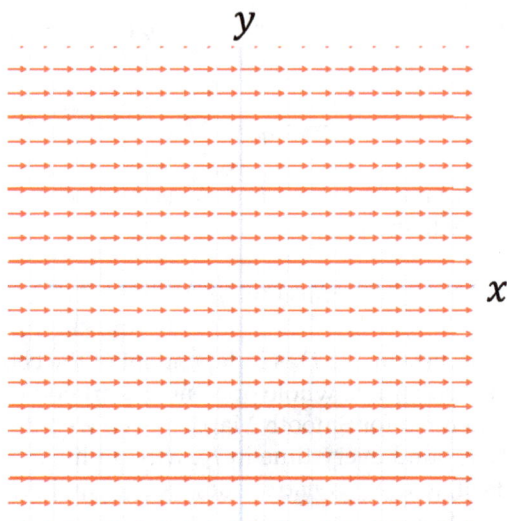

Figure 6.1.   The coordinate vector field $\frac{\partial}{\partial x}$ and its integral curves.

Example 6.6. On $\mathbb{R}^2$ with standard coordinates (x, y), consider the vector field

$$X = x\frac{\partial}{\partial x} + y\frac{\partial}{\partial y}.$$

A curve $\gamma = (x(t), y(t)) : I \to \mathbb{R}^2$ is an integral curve of X if it is a solution of the system of ODEs:

$$\begin{cases} \dfrac{dx}{dt}(t) = x(t) \\[2mm] \dfrac{dy}{dt}(t) = y(t) \end{cases}.$$

This means that

$$\begin{cases} x(t) = x(0)e^t \\ y(t) = y(0)e^t \end{cases}.$$

Hence, the maximal integral curve of X starting from $P = (x_0, y_0)$ is

$$\gamma_P : \mathbb{R} \to \mathbb{R}^2, \quad t \mapsto \gamma_P(t) = (x_0 e^t, y_0 e^t)$$

(see Figure 6.2).

$\blacklozenge$

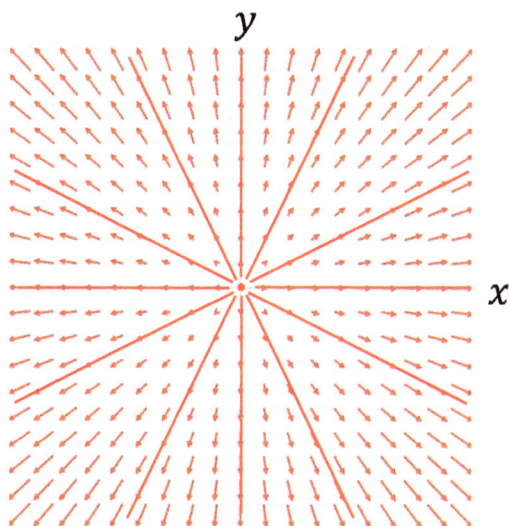

Figure 6.2. The vector field $x\frac{\partial}{\partial x} + y\frac{\partial}{\partial y}$ and its integral curves.

Example 6.7. On $\mathbb{R}^2$ with standard coordinates (x, y), consider the vector field

$$X = x\frac{\partial}{\partial y} - y\frac{\partial}{\partial x}.$$

A curve $\gamma = (x(t), y(t)) : I \to \mathbb{R}^2$ is an integral curve of X if it is a solution of the system of ODEs:

$$\begin{cases} \dfrac{dx}{dt}(t) = -y(t) \\[2mm] \dfrac{dy}{dt}(t) = x(t) \end{cases}.$$

This means that

$$\begin{cases} x(t) = x(0)\cos t - y(0)\sin t \\ y(t) = x(0)\sin t + y(0)\cos t \end{cases}.$$

Hence, the maximal integral curve of X starting from $P = (x_0, y_0)$ is

$$\gamma_P : \mathbb{R} \to \mathbb{R}^2, \quad t \mapsto \gamma_P(t) = (x_0\cos t - y_0\sin t, x_0\sin t + y_0\cos t)$$

(see Figure 6.3).

♦

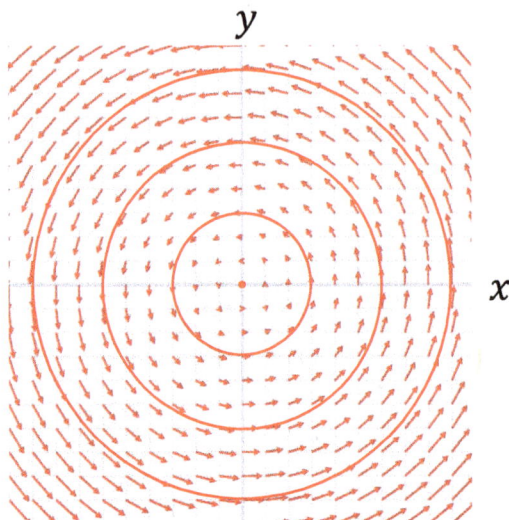

Figure 6.3.   The vector field $x\frac{\partial}{\partial y} - y\frac{\partial}{\partial x}$ and its integral curves.

Exercise 6.1. On $\mathbb{R}^2$ with standard coordinates (x, y), find the maximal integral curves of the vector field

$$X = x\frac{\partial}{\partial x} - y\frac{\partial}{\partial y}$$

(see Figure 6.4).

Note that maximal integral curves need not be defined on the whole $\mathbb{R}$ as the following counter-example shows.

Example 6.8. Consider the standard line $\mathbb{R}$ with standard coordinate x and the open submanifold $(0, +\infty) \subseteq \mathbb{R}$ in it. Additionally, consider the coordinate vector field $X = \frac{d}{dx}\big|_{(0,+\infty)}$ on $(0, +\infty)$. Let $x_0 \in \mathbb{R}$. Any integral curve $\gamma : I \to \mathbb{R}$ of X starting from x_0 is of the type

$$\gamma : I \to (0, +\infty), \quad t \mapsto \gamma(t) = x_0 + t.$$

In particular, $-x_0$ cannot belong to I, not even when $I = I_{x_0}$, and γ is the maximal integral curve starting from x_0.   ◆

We can change the "initial point" of an integral curve by composing with a *translation*. Namely, let $\bar{t} \in \mathbb{R}$. Then the translation

$$\tau_{\bar{t}} : \mathbb{R} \to \mathbb{R}, \quad t \mapsto \tau_{\bar{t}}(t) := t + \bar{t}$$

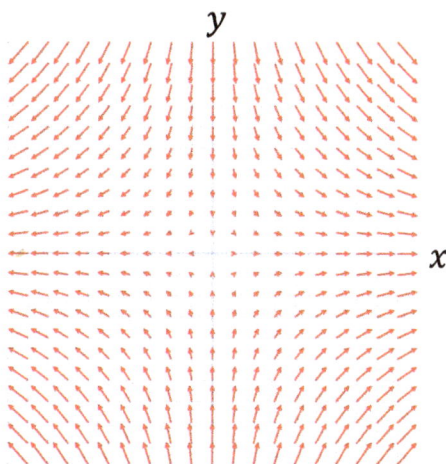

Figure 6.4. The vector field $x\frac{\partial}{\partial x} - y\frac{\partial}{\partial y}$.

is a diffeomorphism whose inverse is $\tau_{-\bar{t}}$. For any open interval $I \subseteq \mathbb{R}$, we denote by $I + \bar{t} := \tau_{\bar{t}}(I)$ its image under the translation $\tau_{\bar{t}}$.

Lemma 6.9 (Translation Lemma). *Let M be a manifold, let X be a vector field on M, and let $\gamma : I \to M$ be an integral curve of X. Then, for every $\bar{t} \in \mathbb{R}$, the composition*

$$\gamma \circ \tau_{\bar{t}} : I - \bar{t} \to M$$

is an integral curve. If, additionally, $I = I_p$, for some point $p \in M$, $\gamma = \gamma_p : I_p \to M$ is the maximal integral curve of X that starts from p, and $\bar{t} \in I_p$, then $I_p - \bar{t} = I_{\gamma_p(\bar{t})}$, and $\gamma \circ \tau_{\bar{t}} : I_p - \bar{t} \to M$ is the maximal integral curve of X that starts from $\gamma_p(\bar{t})$.

Proof. For the first part of the statement, we compute the velocity

$$\frac{d(\gamma \circ \tau_{\bar{t}})}{dt}(t)$$

of the curve $\gamma \circ \tau_{\bar{t}}$ at time $t \in I - \bar{t}$. So, let $f \in C^{\infty}(M)$ and compute

$$\frac{d(\gamma \circ \tau_{\bar{t}})}{dt}(t)(f) = \frac{d}{dt}f(\gamma(\tau_{\bar{t}}(t)))$$

$$= \frac{d}{ds}\Big|_{s=t}f(\gamma(s + \bar{t}))$$

$$= \frac{d}{ds'}\Big|_{s'=t+\bar{t}}f(\gamma(s')) \cdot \frac{d}{ds}\Big|_{s=t}(s + \bar{t})$$

$$= \frac{d\gamma}{dt}(t + \bar{t})(f)$$

$$= X_{\gamma(t+\bar{t})}(f)$$

$$= X_{(\gamma \circ \tau_{\bar{t}})(t)}(f).$$

This shows that $\gamma \circ \tau_{\bar{t}}$ is an integral curve. For the second part of the statement, let $p \in M$ be a point and consider the maximal integral curve $\gamma_p : I_p \to M$ that starts from p. For every $\bar{t} \in I_p$, $\gamma \circ \tau_{\bar{t}} : I_p - \bar{t} \to M$ is an integral curve that starts from $\gamma_p(\bar{t})$. Indeed, first of all, as $\bar{t} \in I_p$, then $0 = \bar{t} - \bar{t} \in I_p - \bar{t}$. Now,

$$(\gamma_p \circ \tau_{\bar{t}})(0) = \gamma_p(0 + \bar{t}) = \gamma_p(\bar{t}).$$

It remains to prove that $\gamma \circ \tau_{\bar{t}}$ is a maximal integral curve (it will then follow that $I_p - \bar{t} = I_{\gamma_p(\bar{t})}$). So, let $\gamma' : I' \to M$ be another integral curve starting from $\gamma_p(\bar{t})$ and such that

$$\gamma_p \circ \tau_{\bar{t}} \leq \gamma', \quad \text{i.e.,} \quad I_p - \bar{t} \subseteq I'.$$

In particular,

$$\gamma'|_{I_p - \bar{t}} = \gamma_p \circ \tau_{\bar{t}}.$$

We also have that $\gamma' \circ \tau_{-\bar{t}} : I' + \bar{t} \to M$ is an integral curve. Additionally,

$$I' + \bar{t} \supseteq I_p - \bar{t} + \bar{t} = I_p \ni 0,$$

so that $0 \in I' + \bar{t}$, and

$$(\gamma' \circ \tau_{-\bar{t}})(0) = \gamma'(0 - \bar{t}) = \gamma'(-\bar{t}) = \gamma_p(\tau_{\bar{t}}(-\bar{t})) = \gamma_p(\bar{t} - \bar{t}) = \gamma_p(0) = p,$$

where we used that $-\bar{t} = 0 - \bar{t} \in I_p - \bar{t}$. Summarizing, $\gamma' \circ \tau_{-\bar{t}} : I' + \bar{t} \to M$ is an integral curve that starts from p. As γ_p is the maximal integral curve that starts from p, then $\gamma' \circ \tau_{-\bar{t}} \leq \gamma_p$, i.e.,

$$I' + \bar{t} \subseteq I_p \Rightarrow I' \subseteq I_p - \bar{t}.$$

We conclude that $I' = I_p - \bar{t}$, and $\gamma' = \gamma_p \circ \tau_{\bar{t}}$. This shows that $\gamma_p \circ \tau_{\bar{t}}$ is a maximal integral curve and concludes the proof. $\qquad \square$

We conclude this section discussing the relationship between integral curves and smooth maps.

Proposition 6.10. *Let M, N be smooth manifolds, let $F : M \to N$ be a smooth map, and let X, Y be vector fields on M, N, respectively. Then X and Y are F-related if and only if F maps integral curves of X to integral curves of Y, i.e., for every integral curve $\gamma : I \to M$ of X, the curve $F \circ \gamma : I \to N$ is an integral curve of Y.*

Proof. Assume that X and Y are F-related, and let $\gamma : I \to M$ be an integral curve of X. Then,

$$
\frac{d}{dt} \circ (F \circ \gamma)^* = \frac{d}{dt} \circ \gamma^* \circ F^*
$$

$$
= \gamma^* \circ X \circ F^* \qquad (\gamma \text{ is an integral curve of } X)
$$

$$
= \gamma^* \circ F^* \circ Y \qquad (X \text{ is } F\text{-related to } Y)
$$

$$
= (F \circ \gamma)^* \circ Y,
$$

showing that $F \circ \gamma$ is an integral curve of Y.

Conversely, assume that F maps integral curves of X to integral curves of Y, and let $p \in M$ be a point. There exists an integral curve of X, say $\gamma : I \to M$, starting from p (e.g., the maximal one), and by hypothesis, $F \circ \gamma : I \to N$ is an integral curve of Y, hence,

$$
d_p F(X_p) = d_{\gamma(0)} F \left(\frac{d\gamma}{dt}(0) \right) = \frac{d(F \circ \gamma)}{dt}(0) = Y_{F(\gamma(0))} = Y_{F(p)},
$$

where, in the second step, we used Identity (3.14). This concludes the proof. □

6.2 Flow of a Vector Field

Let M be a manifold, and let X be a vector field on M. A point $p \in M$ "moves" along the maximal integral curve $\gamma_p : I_p \to M$. After a time t, it has moved to $\gamma_p(t)$. Fix $t \in \mathbb{R}$ and move all points p of M in this way. We get a map $\Phi_t : M \to M$, $p \mapsto \Phi_t(p) := \gamma_p(t)$ (possibly well defined only on a suitable subset of M). Roughly, the *flow* of X is, by definition, the 1-parameter family of maps $\{\Phi_t\}_t$ (see Figure 6.5 for a pictorial representation). The flow of X has several noteworthy properties that we discuss in this section. But we need a precise definition of the flow first. Note, preliminarily, that a 1-parameter family of maps $\Phi_t : M \to M$, $t \in \mathbb{R}$, can be encoded in a single map $\Phi : \mathbb{R} \times M \to M$ by putting $\Phi(t, p) = \Phi_t(p)$.

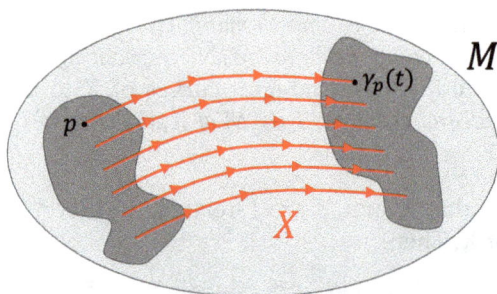

Figure 6.5.   The flow of a vector field X on a manifold M.

Given a vector field X on a manifold M, we will define the *flow* of X as a map $\Phi : D \to M$ from a suitable subset $D \subseteq \mathbb{R} \times M$ that we now identify. In practice, D will be the subset of $\mathbb{R} \times M$ consisting of pairs (t, p) such that the expression $\gamma_p(t)$ is well defined. We put

$$D := \big\{(t, p) \in \mathbb{R} \times M : t \in I_p\big\} \subseteq \mathbb{R} \times M. \tag{6.2}$$

Definition 6.11 (Flow of a Vector Field). The *flow* of X is the map

$$\Phi : D \to M, \quad (t, p) \mapsto \Phi(t, p) := \gamma_p(t).$$

The domain $D \subseteq \mathbb{R} \times M$ of Φ will be referred to as the *flow domain* of X.

We have some remarks on Definition 6.11. First of all, (6.2) guarantees that Φ is well defined. As $0 \in I_p$ for all $p \in M$, it is clear that D contains $\{0\} \times M \subseteq \mathbb{R} \times M$. More precisely, for all $p \in M$,

$$D \cap (\mathbb{R} \times \{p\}) = I_p \times \{p\}.$$

In particular, if $I_p = \mathbb{R}$ for all $p \in M$, then $D = \mathbb{R} \times M$. However, D is usually a proper subset in $\mathbb{R} \times M$ (see Example 6.12 and Exercise 6.2).

Example 6.12 (A Non-Complete Vector Field). Consider the standard line $\mathbb{R}$, with standard coordinate x, the open submanifold $(0, +\infty) \subseteq \mathbb{R}$, and the coordinate vector field $\frac{d}{dx}\big|_{(0,+\infty)}$. For any point $x_0 \in (0, +\infty)$, the maximal integral curve γ_{x_0} of $\frac{d}{dx}\big|_{(0,+\infty)}$ starting from x_0 is

$$\gamma_{x_0} : I_{x_0} = (-x_0, +\infty) \to (0, +\infty), \quad t \mapsto \gamma_{x_0}(t) := x_0 + t$$

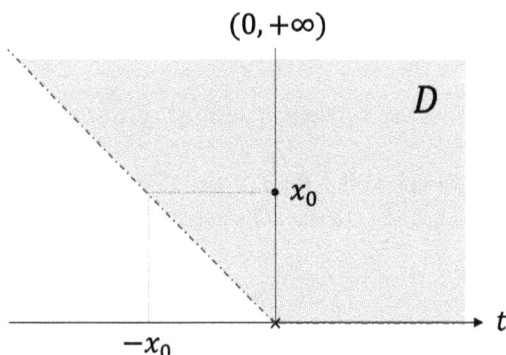

Figure 6.6.   The flow domain of $\frac{d}{dx}|_{(0,+\infty)}$.

(see also Example 6.8). It follows that the flow domain of $\frac{d}{dx}|_{(0,+\infty)}$ is

$$D = \{(t, x_0) \in \mathbb{R} \times (0, +\infty) : -x_0 < t\},$$

which is a proper subset in $\mathbb{R} \times (0, +\infty)$ (see Figure 6.6).

$\blacklozenge$

Exercise 6.2. Consider the standard line $\mathbb{R}$, with standard coordinate x, and the open submanifold $(-1, +1) \subseteq \mathbb{R}$. Compute the flow domain of the vector field $\frac{d}{dx}|_{(-1,+1)}$.

The flow Φ can be encoded in a 1-parameter family of maps Φ_t as follows. For any $t \in \mathbb{R}$, define

$$M_t := \{p \in M : (t, p) \in D\} = \{p \in M : t \in I_p\} \subseteq M.$$

Note that M_t might be the empty subset (can you provide an example?). In any case, there is a well-defined map

$$\Phi_t : M_t \to M, \quad p \mapsto \Phi_t(p) := \Phi(t, p) = \gamma_p(t),$$

and Φ is equivalently encoded in the family of the Φ_t that we denote $\{\Phi_t\}_t$ and we also call the flow of X. The main properties of the flow are stated in the following

Theorem 6.13 (Fundamental Theorem of Flows). *Let M be a manifold, let X be a vector field on M, and let $\Phi : D \to M$ be the flow of X. Then*

(1) *the flow domain D is an open neighborhood of $\{0\} \times M$ and the flow $\Phi :$ $D \to M$ is a smooth map,*

(2) *the flow* Φ *enjoys the following* group properties: $M_0 = M$, *and* $\Phi_0 = \mathrm{id}_M$, *and, additionally,*

$$(\Phi_s \circ \Phi_t)(p) = \Phi_{t+s}(p) \tag{6.3}$$

whenever both sides are well defined ($s, t \in \mathbb{R}$, $p \in M$),

(3) *for every* $t \in \mathbb{R}$, $M_t \subseteq M$ *is a (possibly empty) open subset,* $\Phi_t(M_t) \subseteq M_{-t}$, *and, if* $M_t \neq \varnothing$, $\Phi_t : M_t \to M_{-t}$ *is a diffeomorphism with inverse* $\Phi_{-t} : M_{-t} \to M_t$.

Proof (A Sketch). Point (1) is a consequence of Theorem 6.2 and we omit the (technical) proof (for all the details see, e.g., Lee, 2013).

Point (2) is a consequence of the Translation Lemma (Lemma 6.9), as we now show. First note that the left-hand side of Equation (6.3) is well defined when

$$t \in I_p \quad \text{and} \quad s \in I_{\Phi_t(p)} = I_{\gamma_p(t)}, \tag{6.4}$$

while the right-hand side is well defined when

$$t + s \in I_p. \tag{6.5}$$

Actually, from the second part of the Translation Lemma, we have that (6.4) implies (6.5). So, let s, t, p be such that (6.4), and note that

$$\begin{aligned}
(\Phi_s \circ \Phi_t)(p) &= \Phi_s(\Phi_t(p)) \\
&= \gamma_{\Phi_t(p)}(s) \\
&= \gamma_{\gamma_p(t)}(s) \\
&= \gamma_p(\tau_t(s)) \\
&= \gamma_p(s + t) \\
&= \Phi_{s+t}(p),
\end{aligned}$$

where, in the fourth step, we used the Translation Lemma again (the maximal integral curve starting from $\gamma_p(t)$ is obtained from the maximal integral curve starting from p by composition with the translation τ_t).

For Point (3), first note that M_t is the preimage of the flow domain D under the map

$$\Gamma_t : M \to \mathbb{R} \times M, \quad p \mapsto \Gamma_t(p) := (t, p).$$

From Exercise 2.7, Γ_t is a smooth, hence continuous, map. As $D \subseteq \mathbb{R} \times M$ is an open subset, it follows that $M_t \subseteq M$ is an open subset. Second,

$$p \in M_t \quad \Rightarrow \quad t \in I_p \quad \Rightarrow \quad I_{\Phi_t(p)} = I_{\gamma_p(t)} = I_p - t,$$

where we used the Translation Lemma again. As $0 \in I_p$, we get

$$-t \in I_{\Phi_t(p)} \quad \Rightarrow \quad \Phi_t(p) \in M_{-t},$$

as claimed. Finally, assume that $M_t \neq \varnothing$. The map $\Phi_t : M_t \to M$ can be seen as the composition of $\Gamma_t : M_t \to D$ followed by the flow Φ. As both Φ and Γ_t are smooth, so is $\Phi_t : M_t \to M$, and its restriction to M_{-t} in the codomain. From the arbitrariness of t, $\Phi_{-t} : M_{-t} \to M$ is also smooth, and, from Point (2),

$$\Phi_t \circ \Phi_{-t} = \Phi_{-t} \circ \Phi_t = \mathrm{id}_M,$$

i.e., $\Phi_t : M_t \to M_{-t}$ and $\Phi_{-t} : M_{-t} \to M_t$ are mutually inverse diffeomorphisms. $\qquad\square$

Remark 6.14. The flow $\{\Phi_t\}_t$ of a vector field X is sometimes called the *local 1-parameter group of local diffeomorphisms* generated by X, and X is called the *infinitesimal generator* of $\{\Phi_t\}_t$. This terminology is motivated by the Fundamental Theorem of Flows. $\qquad\diamond$

As a vector field X is completely determined by its maximal integral curves γ_p, via $X_p = \dot{\gamma}_p(0)$, and the flow of X determines the maximal integral curves via $\Phi_t(p) = \gamma_p(t)$, it follows that a vector field is completely determined by its flow, in the sense that two vector fields with the same flow necessarily coincide.

Definition 6.15 (Complete Vector Field). A vector field X on a manifold M is *complete* if every maximal integral curve of X is defined on the whole $\mathbb{R}$, i.e., $I_p = \mathbb{R}$ for all $p \in M$. In other words, the flow of X is defined on the whole $\mathbb{R} \times M$.

Example 6.16 (The Infinitesimal Generator of Translations). On $\mathbb{R}^2$ with standard coordinates (x, y) consider the coordinate vector field

$$X = \frac{\partial}{\partial x}.$$

It follows from Example 6.5 that the flow of X is

$$\Phi : \mathbb{R} \times \mathbb{R}^2 \to \mathbb{R}^2, \quad (t, (x, y)) \mapsto \Phi(t, (x, y)) = (x + t, y).$$

As, for every t, Φ_t is a translation, we also say that X *generates the translations along the x-axis.* ◆

Example 6.17 (The Infinitesimal Generator of Dilations). On $\mathbb{R}^2$ with standard coordinates (x, y), consider the vector field

$$X = x\frac{\partial}{\partial x} + y\frac{\partial}{\partial y}.$$

It follows from Example 6.6 that the flow of X is

$$\Phi : \mathbb{R} \times \mathbb{R}^2 \to \mathbb{R}^2, \quad (t, (x, y)) \mapsto \Phi(t, (x, y)) = (xe^t, ye^t).$$

As, for every t, Φ_t is a dilation, we also say that X *generates the dilations.* ◆

Example 6.18 (The Infinitesimal Generator of Rotations). On $\mathbb{R}^2$ with standard coordinates (x, y), consider the coordinate vector field

$$X = x\frac{\partial}{\partial y} - y\frac{\partial}{\partial x}.$$

It follows from Example 6.7 that the flow of X is

$$\Phi : \mathbb{R} \times \mathbb{R}^2 \to \mathbb{R}^2,$$
$$(t, (x, y)) \mapsto \Phi(t, (x, y)) = (x\cos t - y\sin t, x\sin t + y\cos t).$$

As, for every t, Φ_t is a rotation, we also say that X *generates the rotations around the origin.* ◆

The vector fields from Examples 6.16, 6.17, and 6.18 are all complete. The vector field $\frac{d}{dx}\big|_{(0,+\infty)}$ from Example 6.12 is not complete.

Exercise 6.3. On $\mathbb{R}^2$ with standard coordinates (x, y), find the flow of the vector field

$$X = x\frac{\partial}{\partial x} - y\frac{\partial}{\partial y}$$

(see also Exercise 6.1). Is X complete?

We conclude this section describing a new construction involving the flow: the *Lie derivative along a vector field*. We begin with some informal remarks on vector fields and their flows. The flow $\{\Phi_t\}_t$ of a vector field X moves the points of a manifold. At the time 0, this displacement is trivial: it is just the identity map that does not move any point. At an arbitrary time, this displacement is a diffeomorphism. The vector field is the velocity of this displacement at the initial time: informally,

$$\Phi_t = \mathrm{id}_M + t \cdot X + O(t^2).$$

In this sense, X should be interpreted as an *infinitesimal diffeomorphism*, and the Fundamental Theorem of Flows then says that an infinitesimal diffeomorphism generates a whole 1-parameter family of diffeomorphisms. Note that, moving all points of M, the flow $\{\Phi_t\}_t$ of a vector field can often *move a structure* on M. By a *structure* on M we mean an additional datum, e.g., a *submanifold*, a *smooth function*, a *vector field*, etc. In particular, a function f moves via the pull-back $\Phi_t^*(f)$ (and similarly for vector fields). The Lie derivative is then the initial velocity of this displacement.

To be precise, let M be a manifold, let X be a vector field on M, with flow $\{\Phi_t\}_t$, and let $f \in C^\infty(M)$ be a smooth function. The *Lie derivative* of f along X is the function

$$\mathcal{L}_X f : M \to \mathbb{R}, \quad p \mapsto (\mathcal{L}_X f)(p) := \frac{d}{dt}\Big|_{t=0}\Phi_t^*(f)(p). \tag{6.6}$$

This definition requires some explanations. First of all, note that the expression $\Phi_t^*(f)(p)$ under the t-derivative does only make sense when $p \in M_t$, or, equivalently, $t \in I_p$. Indeed,

$$\Phi_t^*(f)(p) = f(\Phi_t(p)) = f(\gamma_p(t)). \tag{6.7}$$

Additionally, (6.7) shows that $\Phi_t^*(f)(p)$ depends smoothly on t in I_p which is an open neighborhood of 0. Hence, we can take the t-derivative of (6.7) at $t = 0$ and the Lie derivative $\mathcal{L}_X f$ is well defined.

Proposition 6.19 (Lie Derivative of a Function). *Let M be a manifold, let $X \in \mathfrak{X}(M)$ be a vector field, and let $f \in C^\infty(M)$ be a smooth function. Then the Lie derivative of f along X is simply given by*

$$\mathcal{L}_X f = X(f). \tag{6.8}$$

In particular, it is a smooth function.

Proof.   We have

$$(\mathcal{L}_X f)(p) = \frac{d}{dt}\big|_{t=0}\Phi_t^*(f)(p)$$

$$= \frac{d}{dt}\big|_{t=0} f(\gamma_p(t))$$

$$= \dot{\gamma}_p(0)(f)$$

$$= X_p(f)$$

$$= X(f)(p).$$

This concludes the proof. $\square$

Sometimes Definition (6.6) and Identity (6.8) are also written

$$X(f) = \mathcal{L}_X f = \frac{d}{dt}\big|_{t=0}\Phi_t^*(f).$$

It is also useful to compute the following *derivative at an arbitrary time t*:

$$\frac{d}{dt}\Phi_t^*(f)(p) = \frac{d}{ds}\big|_{s=t}\Phi_s^*(f)(p). \tag{6.9}$$

This is slightly more complicated. First of all, (6.9) does only make sense if the expression $\Phi_s^*(f)(p) = f(\gamma_p(s))$ depends smoothly on s in an open neighborhood of t. This is the case when $t \in I_p$ or, equivalently, $p \in M_t$ (do you see it?). In other words, the correspondence

$$\frac{d}{dt}\Phi_t^*(f) : M_t \to \mathbb{R}, \quad p \mapsto \frac{d}{dt}\Phi_t^*(f)(p) = \frac{d}{ds}\big|_{s=t}\Phi_s^*(f)(p)$$

is a well-defined function.

Proposition 6.20. *Let M be a manifold, let $X \in \mathfrak{X}(M)$ be a vector field with flow $\{\Phi_t\}_t$, and let $f \in C^\infty(M)$ be a smooth function. Then*

$$\frac{d}{dt}\Phi_t^*(f) = \Phi_t^*(X(f)), \tag{6.10}$$

for all t such that $M_t \neq \varnothing$. In particular, $\frac{d}{dt}\Phi_t^*(f)$ is a smooth function (on M_t).

Proof.   Let $p \in M_t$, and compute

$$\frac{d}{ds}\big|_{s=t}\Phi_s^*(f)(p) = \frac{d}{ds'}\big|_{s'=0} f(\Phi_{s'+t}(p)) \cdot \frac{d}{ds}\big|_{s=t}(s-t)$$

$$= \frac{d}{ds'}\big|_{s'=0} f(\Phi_{s'+t}(p)).$$

Now, we want to use the group property of the flow:

$$\Phi_{s'+t}(p) = \Phi_{s'}(\Phi_t(p)). \tag{6.11}$$

To do this, we have to make sure that $t \in I_p$ (exactly what we are already assuming) and $s' \in I_{\Phi_t(p)}$. The latter condition can be safely assumed as we are computing the s'-derivative at 0. So we can go on with the computation:

$$\frac{d}{ds}|_{s=t}\Phi_s^*(f)(p) = \frac{d}{ds'}|_{s'=0}f(\Phi_{s'+t}(p))$$

$$= \frac{d}{ds'}|_{s'=0}f(\Phi_{s'}(\Phi_t(p)))$$

$$= \frac{d}{ds'}|_{s'=0}(\Phi_{s'}^*f)(\Phi_t(p))$$

$$= (\mathcal{L}_X f)(\Phi_t(p))$$

$$= X(f)(\Phi_t(p))$$

$$= \Phi_t^*(X(f))(p),$$

which, as claimed, depends smoothly on $p \in M_t$. $\qquad\square$

Remark 6.21. We can also compute $\frac{d}{dt}\Phi_t^*(f)$ following a different path. Namely, we can use the group property of the flow in the form:

$$\Phi_{s'+t}(p) = \Phi_t(\Phi_{s'}(p)).$$

To do this, we have to make sure that $s' \in I_p$, which can be safely assumed as we are computing the s'-derivative at 0, and $t \in I_{\Phi_{s'}(p)}$, i.e., $\Phi_{s'}(p) = \gamma_p(s') \in M_t$. The latter condition can be guaranteed by taking s' sufficiently close to 0, e.g., in $I_p \cap \gamma_p^{-1}(M_t)$ (which is an open neighborhood of 0). So we can go on with the computation:

$$\frac{d}{ds}|_{s=t}\Phi_s^*(f)(p) = \frac{d}{ds'}|_{s'=0}f(\Phi_{s'+t}(p))$$

$$= \frac{d}{ds'}|_{s'=0}f(\Phi_t(\Phi_{s'}(p)))$$

$$= \frac{d}{ds'}|_{s'=0}\Phi_{s'}^*(\Phi_t^*(f))(p).$$

Exactly the same computation as for the Lie derivative (of a function) shows that

$$\frac{d}{ds'}|_{s'=0}\Phi_{s'}^*(\Phi_t^*(f))(p) = X|_{M_t}(\Phi_t^*(f))(p),$$

which, adopting our customary notation, we also denote simply by $X(\Phi_t^*(f))(p)$. Summarizing, we proved that, for every $t \in \mathbb{R}$ such that $M_t \neq \varnothing$,

$$\frac{d}{dt}\Phi_t^*(f) = X(\Phi_t^*(f)). \tag{6.12}$$

$$\diamond$$

Finally, we discuss the *Lie derivative* of a vector field along a(n other) vector field. So, let M be a manifold, and let X, Y be vector fields on M. The *Lie derivative* of Y along X is a new vector field, denoted $\mathcal{L}_X Y$ and defined, through its values, as follows:

$$(\mathcal{L}_X Y)_p := \frac{d}{dt}|_{t=0}\Phi_t^*(Y)_p. \tag{6.13}$$

This definition requires some explanations. First of all, what we actually mean by $\Phi_t^*(Y)$ is, more precisely, $\Phi_t^*(Y|_{M_{-t}})$: the pull-back of the vector field $Y|_{M_{-t}} \in \mathfrak{X}(M_{-t})$ along the diffeomorphism $\Phi_t : M_t \to M_{-t}$ (if X is complete, then $M_t = M$ and we don't need this level of precision). In particular, $\Phi_t^*(Y)$ is a vector field on M_t. Hence, the expression $\Phi_t^*(Y)_p$ under the t-derivative, does only make sense when $p \in M_t$, or, equivalently, $t \in I_p$, and, in this case, $\Phi_t^*(Y)_p$ is a t-dependent tangent vector in $T_p M$. We can only take the t-derivative if $\Phi_t^*(Y)_p$ depends smoothly on t (at least in an open neighborhood of 0). If this is the case, according to our interpretation of the velocity of a curve in a finite dimensional real vector space (see the end of Section 3.2), the derivative in (6.13) is again a vector in $T_p M$.

Proposition 6.22 (Lie Derivative of a Vector Field). *Let M be a manifold, and let $X, Y \in \mathfrak{X}(M)$ be vector fields. Then the Lie derivative of Y along X is a well-defined vector field simply given by*

$$\mathcal{L}_X Y = [X, Y]. \tag{6.14}$$

Proof. First we need to show that the tangent vector $\Phi_t^*(Y)_p$ depends smoothly on t in an open neighborhood of 0 (so that we can take the t-derivative (6.13)). To do this, we work in local coordinates as follows. Fix a point $p_0 \in M$, and let $(U, \varphi = (x^1, \ldots, x^n))$ be a chart around p_0. Then

$$X|_U = X^i \frac{\partial}{\partial x^i} \quad \text{and} \quad Y|_U = Y^i \frac{\partial}{\partial x^i}$$

for some smooth functions $X^i, Y^i \in C^\infty(U)$. The preimage $\Phi^{-1}(U)$ of U under the flow $\Phi : D \to M$ is an open subset of $\mathbb{R} \times M$ containing $(0, p_0)$. Hence, there is an open interval J containing 0, and a subchart (U_0, φ) of (U, φ) around p_0, such that $J \times U_0 \subseteq \Phi^{-1}(U)$, i.e., $\Phi(J \times U_0) \subseteq U$. In particular, we can consider the restriction

$$\Phi : J \times U_0 \to U \tag{6.15}$$

(yet in other words, for every $t \in J$, $U_0 \subseteq M_t$ and $\Phi_t(U_0) \subseteq U$). As usual, we denote by $\Phi^i = \Phi^*(x^i)$ the components of (6.15). We will need the following identities:

$$\frac{\partial \Phi^i}{\partial t}(0, p) = X^i(p) \tag{6.16}$$

and

$$\frac{\partial \Phi^i}{\partial x^k}(0, p) = \delta^i_k \tag{6.17}$$

for all $p \in U_0$. To prove them, note first that

$$\Phi^i(t, p) = \gamma^i_p(t)$$

for all $t \in J$ and all $p \in U_0$ (do you see it?). In particular,

$$\Phi^i(0, p) = \gamma^i_p(0) = x^i(\gamma_p(0)) = x^i(p), \quad \text{i.e.,} \quad \Phi^i(0, -) = x^i.$$

Hence

$$\frac{\partial \Phi^i}{\partial t}(0, p) = \frac{d}{dt}|_{t=0}\Phi^i(t, p) = \frac{d}{dt}|_{t=0}\gamma^i_p(t) = X^i(p)$$

and

$$\frac{\partial \Phi^i}{\partial x^k}(0, p) = \frac{\partial}{\partial x^k}|_p\Phi^i(0, -) = \frac{\partial}{\partial x^k}|_p x^i = \delta^i_k.$$

Now let $(t, p) \in J \times U_0$, and, using (5.5), compute

$$\Phi^*_t(Y)_p = d_{\Phi_t(p)}\Phi^{-1}_t(Y_{\Phi_t(p)})$$

$$= d_{\Phi_t(p)}\Phi_{-t}(Y_{\Phi_t(p)})$$

$$= \frac{\partial \Phi^i}{\partial x^j}(-t, \Phi(t, p))Y^j(\Phi(t, p))\frac{\partial}{\partial x^i}|_p.$$

The coefficients in the last expression depend smoothly on t. This shows that we can take the derivative in (6.13). It remains to compute

$$(\mathcal{L}_X Y)_p = \frac{d}{dt}\big|_{t=0} \Phi_t^*(Y)_p$$

$$= \left(\frac{d}{dt}\big|_{t=0} \left(\frac{\partial \Phi^i}{\partial t^k}(-t, \Phi(t,p)) Y^k(\Phi(t,p)) \right) \right) \frac{\partial}{\partial x^i}\big|_p.$$

The i-th coefficient is

$$\frac{d}{dt}\big|_{t=0} \left(\frac{\partial \Phi^i}{\partial x^j}(-t, \Phi(t,p)) Y^j(\Phi(t,p)) \right)$$

$$= \left(-\frac{\partial^2 \Phi^i}{\partial t \partial x^j}(0, \Phi(0,p)) + \frac{\partial^2 \Phi^i}{\partial x^k \partial x^j}(0, \Phi(0,p)) \frac{\partial \Phi^k}{\partial t}(0,p) \right) Y^j(\Phi(0,p))$$

$$+ \frac{\partial \Phi^i}{\partial x^j}(0, \Phi(0,p)) \frac{\partial Y^j}{\partial x^k}(\Phi(0,p)) \frac{\partial \Phi^k}{\partial t}(0,p)$$

$$= -\frac{\partial X^i}{\partial x^j}(p) Y^j(p) + \frac{\partial Y^i}{\partial x^j}(p) X^j(p)$$

$$= [X,Y]^i(p)$$

for all $p \in U_0$. This concludes the proof. $\qquad\square$

Sometimes Definition (6.13) (and Identity (6.14)) is also written as follows

$$[X,Y] = \mathcal{L}_X Y = \frac{d}{dt}\big|_{t=0} \Phi_t^*(Y).$$

It is also useful to compute the following *derivative at an arbitrary time* t:

$$\frac{d}{dt} \Phi_t^*(Y)_p = \frac{d}{ds}\big|_{s=t} \Phi_s^*(Y)_p. \qquad (6.18)$$

Similarly as for functions, (6.18) does only make sense when $t \in I_p$ or, equivalently, $p \in M_t$, and the correspondence

$$\frac{d}{dt} \Phi_t^*(Y) : M_t \to TM_t, \quad p \mapsto \frac{d}{dt} \Phi_t^*(Y)_p = \frac{d}{ds}\big|_{s=t} \Phi_s^*(Y)_p$$

is a well-defined (*a priori* non-necessarily smooth) section of TM_t. One can show that it is actually a smooth section, hence it corresponds to a vector field on M_t. To see this, let $p \in M_t$, and compute

$$\frac{d}{ds}\big|_{s=t}\Phi_s^*(Y)_p = \frac{d}{ds'}\big|_{s'=0}\Phi_{s'+t}^*(Y)_p \cdot \frac{d}{ds}\big|_{s=t}(s-t)$$

$$= \frac{d}{ds'}\big|_{s'=0}\Phi_t^*(\Phi_{s'}^*(Y))_p$$

$$= \frac{d}{ds'}\big|_{s'=0}d_{\Phi_t(p)}\Phi_t^{-1}(\Phi_{s'}^*(Y)_{\Phi_t(p)}). \qquad (6.19)$$

At this point, we need a

Lemma 6.23. *Let V, W be finite dimensional real vector spaces, let $\gamma : I \to V$ be a curve in V, and let $A : V \to W$ be a linear map. Then $A \circ \gamma : I \to W$ is a smooth curve and*

$$\frac{d}{ds}\big|_{s=s_0}(A \circ \gamma)(s) = A\left(\frac{d}{ds}\big|_{s=s_0}\gamma(s)\right)$$

for all $s_0 \in I$.

Proof. Let $\mathcal{R} = (e_1, \ldots, e_n)$ be a frame of V, $n = \dim V$. Then $\gamma(s) = \gamma^i(s)e_i$ for all $s \in I$, where the $\gamma^i : I \to \mathbb{R}$, $s \mapsto \gamma^i(s)$ are smooth functions. We have

$$\frac{d}{ds}\big|_{s=s_0}(A \circ \gamma)(s) = \frac{d}{ds}\big|_{s=s_0}A(\gamma^i(s)e_i)$$

$$= \frac{d}{ds}\big|_{s=s_0}\gamma^i(s)A(e_i)$$

$$= \left(\frac{d}{ds}\big|_{s=s_0}\gamma^i(s)\right)A(e_i)$$

$$= A\left(\frac{d}{ds}\big|_{s=s_0}\gamma^i(s)e_i\right)$$

$$= A\left(\frac{d}{ds}\big|_{s=s_0}\gamma(s)\right),$$

where, in the third step, we used Exercise 3.7. $\qquad \square$

Applying Lemma 6.23 to (6.19) (with $A = d_{\Phi_t(p)}\Phi_t^{-1}$), we find

$$\frac{d}{ds}\big|_{s=t}\Phi_s^*(Y)_p = \frac{d}{ds'}\big|_{s'=0}d_{\Phi_t(p)}\Phi_t^{-1}(\Phi_{s'}^*(Y)_{\Phi_t(p)})$$

$$= d_{\Phi_t(p)}\Phi_t^{-1}\left(\frac{d}{ds'}\big|_{s'=0}\Phi_{s'}^*(Y)_{\Phi_t(p)}\right)$$

$$= d_{\Phi_t(p)}\Phi_t^{-1}\left((\mathcal{L}_X Y)_{\Phi_t(p)}\right)$$

$$= d_{\Phi_t(p)}\Phi_t^{-1}\left([X,Y]_{\Phi_t(p)}\right)$$

$$= \Phi_t^*([X,Y])_p.$$

Summarizing, we proved that, for every $t \in \mathbb{R}$ (such that $M_t \neq \varnothing$),

$$\frac{d}{dt}\Phi_t^*(Y) = \Phi_t^*([X,Y]). \tag{6.20}$$

Remark 6.24. As for functions, we can also compute $\frac{d}{dt}\Phi_t^*(Y)$ following a different path. Namely,

$$\frac{d}{ds}\big|_{s=t}\Phi_s^*(Y)_p = \frac{d}{ds'}\big|_{s'=0}\Phi_{s'+t}^*(Y)_p$$

$$= \frac{d}{ds'}\big|_{s'=0}\Phi_{s'}^*(\Phi_t^*(Y))$$

$$= (\mathcal{L}_X\Phi_t^*(Y))_p$$

$$= [X, \Phi_t^*(Y)]_p,$$

for all $p \in M_t$. This shows that

$$\frac{d}{dt}\Phi_t^*(Y) = [X, \Phi_t^*(Y)].$$

We leave to the reader to check that every single step in the above computation makes sense. $\diamond$

6.3   Symmetries and Infinitesimal Symmetries

In Mathematics, a symmetry of an object $\mathcal{O}$ is always a *transformation preserving* $\mathcal{O}$. The precise meaning of the words *transformation* and *preserving* depends on the nature of the object $\mathcal{O}$. For instance, when $\mathcal{O}$ is a group, then a symmetry is an automorphism of $\mathcal{O}$. Similarly, when $\mathcal{O}$ is a topological space, then a symmetry is a homeomorphism $\Phi : \mathcal{O} \to \mathcal{O}$, and when $\mathcal{O}$ is a manifold, a symmetry is a diffeomorphism $\Phi : \mathcal{O} \to \mathcal{O}$. The *smoothness* of objects in Differential Geometry allows us to talk not only about symmetries but also about *infinitesimal symmetries*. As discussed in the previous section, it is natural to interpret vector fields on a manifold M as infinitesimal diffeomorphisms, hence they are also infinitesimal symmetries of M. We can also talk about symmetries and infinitesimal symmetries of a *structure* on a manifold M (see the previous section). In this section, we discuss in some details symmetries and infinitesimal symmetries of submanifolds, smooth functions, and vector fields on a manifold. In Chapter 8, we will quickly consider one more example: symmetries and infinitesimal symmetries of *differential forms*.

We begin with submanifolds. Let M be a manifold, and let $S \subseteq M$ be a submanifold. Recall that the image $\Phi(S)$ of S under a diffeomorphism $\Phi : M \to M$ is a submanifold as well (Exercise 2.13).

Definition 6.25 (Symmetry of a Submanifold). A *symmetry* of a submanifold $S \subseteq M$ in a manifold M is a diffeomorphism $\Phi : M \to M$ *preserving* S in the sense that $\Phi(S) = S$. A *local symmetry* of S is a diffeomorphism $\Phi : \mathcal{U} \to \Phi(\mathcal{U})$ between open submanifolds $\mathcal{U}, \Phi(\mathcal{U}) \subseteq M$, such that $\Phi(S \cap \mathcal{U}) = S \cap \Phi(\mathcal{U})$.

Example 6.26. In $\mathbb{R}^2$ with standard coordinates (x, y), consider a 1-dimensional subspace $\ell \subseteq \mathbb{R}^2$. Let $\lambda \in \mathbb{R}$ be a non-zero real number, and consider the dilation

$$\Phi : \mathbb{R}^2 \to \mathbb{R}^2, \quad (x, y) \mapsto \Phi(x, y) = (\lambda x, \lambda y).$$

Then Φ is clearly a symmetry of ℓ. ◆

Example 6.27. In $\mathbb{R}^2$ with standard coordinates (x, y), consider the circle $S^1 \subseteq \mathbb{R}^2$. Let $\theta \in \mathbb{R}$, and consider the rotation around the origin by an angle θ:

$$\Phi : \mathbb{R}^2 \to \mathbb{R}^2, \quad (x, y) \mapsto \Phi(x, y) = (x \cos \theta - y \sin \theta, x \sin \theta + y \cos \theta).$$

Then Φ is clearly a symmetry of S^1. ◆

Proposition 6.28. *Let M be a manifold, and let $S \subseteq M$ be a submanifold. The subset*

$$\mathrm{Diffeo}(M, S) := \{\text{symmetries of } S\} \subseteq \mathrm{Diffeo}(M)$$

of the group $\mathrm{Diffeo}(M)$ of diffeomorphisms of M is a subgroup.

Proof.   Left as Exercise 6.4.

$\square$

Exercise 6.4. Prove Proposition 6.28.

Definition 6.29 (Infinitesimal Symmetry of a Submanifold). An *infinitesimal symmetry* of a submanifold $S \subseteq M$ in a manifold M is a vector field X on M generating a flow $\{\Phi_t\}_t$ by local symmetries of S, i.e., $\Phi_t(S \cap M_t) = S \cap M_{-t}$ for all t (such that $M_t \neq \varnothing$).

Example 6.30. In $\mathbb{R}^2$ with standard coordinates (x, y), consider a 1-dimensional subspace $\ell \subseteq \mathbb{R}^2$ and the circle $S^1 \subseteq \mathbb{R}^2$. Consider also the vector fields

$$X_1 = x\frac{\partial}{\partial x} + y\frac{\partial}{\partial y} \quad \text{and} \quad X_2 = x\frac{\partial}{\partial y} - y\frac{\partial}{\partial x}.$$

It follows from Examples 6.17 and 6.26 that X_1 is an infinitesimal symmetry of ℓ. Similarly, it follows from Examples 6.18 and 6.27 that X_2 is an infinitesimal symmetry of S^1. ◆

Checking whether or not a vector field X is an infinitesimal symmetry of a submanifold S using the definition might be problematic, because it requires computing the flow of X, hence solving a system of ODEs for every possible initial data (which is far from being an easy task in general). However, infinitesimal symmetries of a submanifold can be (almost) characterized in a way which is (almost) independent of the flow.

Proposition 6.31. *Let M be a manifold, let $S \subseteq M$ be a submanifold, and let X be a vector field on M:*

- *If X is an infinitesimal symmetry of S, then X is tangent to S.*
- *Conversely, if X is tangent to S and, additionally, the restriction $X|_S$ is a complete vector field, then X is an infinitesimal symmetry of S.*

Proof.   Left as Exercise 6.5.

$\square$

Exercise 6.5. Prove Proposition 6.31. (**Hint:** *Use Proposition 6.10.*)

We now pass to symmetries of smooth functions, noting that the group $\text{Diffeo}(M)$ of diffeomorphisms of a manifold M *acts* on the algebra of smooth functions $C^\infty(M)$ (by algebra isomorphisms) via the pull-back, i.e., for any diffeomorphism $\Phi : M \to M$, we have an algebra isomorphism $\Phi^* : C^\infty(M) \to C^\infty(M)$. This suggests the following

Definition 6.32 (Symmetry of a Function). A *symmetry* of a smooth function $f \in C^\infty(M)$ on a manifold M is a diffeomorphism $\Phi : M \to M$

preserving f in the sense that $\Phi^*(f) = f$. A *local symmetry* of f is a diffeomorphism $\Phi : \mathcal{U} \to \Phi(\mathcal{U})$ between open submanifolds $\mathcal{U}, \Phi(\mathcal{U}) \subseteq M$ such that $\Phi^*(f|_{\Phi(\mathcal{U})}) = f|_{\mathcal{U}}$.

Example 6.33. On $\mathbb{R}^2$ with standard coordinates (x, y), consider a smooth function $f \in C^\infty(\mathbb{R}^2)$ depending on the sole y, i.e., there is a functions $g \in C^\infty(\mathbb{R})$ such that $f(x, y) = g(y)$ for all $(x, y) \in \mathbb{R}^2$. Let $x_0 \in \mathbb{R}$, and consider the translation along the x-axis

$$\Phi : \mathbb{R}^2 \to \mathbb{R}^2, \quad (x, y) \mapsto \Phi(x, y) = (x + x_0, y).$$

Then Φ is a symmetry of f, indeed, for every $(x, y) \in \mathbb{R}^2$:

$$\Phi^*(f)(x, y) = f(\Phi(x, y)) = f(x + x_0, y) = g(y) = f(x, y).$$

◆

Exercise 6.6. On $\mathbb{R}^2$ with standard coordinates (x, y), consider a smooth function $f \in C^\infty(\mathbb{R}^2)$ depending on the sole $x^2 + y^2$, i.e., there is a function $g \in C^\infty(\mathbb{R})$ such that $f(x, y) = g(x^2 + y^2)$ for all $(x, y) \in \mathbb{R}^2$. Let $\theta \in \mathbb{R}$, and consider the rotation around the origin by an angle θ:

$$\Phi : \mathbb{R}^2 \to \mathbb{R}^2, \quad (x, y) \mapsto \Phi(x, y) = (x \cos\theta - y \sin\theta, x \sin\theta + y \cos\theta).$$

Show that Φ is a symmetry of f.

Proposition 6.34. *Let M be a manifold, and let $f \in C^\infty(M)$ be a smooth function on M. The subset*

$$\mathrm{Diffeo}(M, f) := \{\text{symmetries of } f\} \subseteq \mathrm{Diffeo}(M)$$

of the group of diffeomorphisms of M is a subgroup.

Proof. Left as Exercise 6.7. ☐

Exercise 6.7. Prove Proposition 6.34.

Proposition 6.35. *Let M be a manifold, let $f, g \in C^\infty(M)$, and let $\Phi : M \to M$ be a diffeomorphism. If Φ is a symmetry of both f and g, then it is also a symmetry of $f + g$ and fg.*

Proof. Left as Exercise 6.8. ☐

> **Exercise 6.8.** Prove Proposition 6.35.

We now come to infinitesimal symmetries of functions.

Definition 6.36 (Infinitesimal Symmetry of a Function). An *infinitesimal symmetry* of a smooth function $f \in C^\infty(M)$ on a manifold M is a vector field X on M generating a flow $\{\Phi_t\}_t$ by local symmetries of f, i.e., $\Phi_t^*(f|_{M_{-t}}) = f|_{M_t}$ for all t (such that $M_t \neq \varnothing$).

Example 6.37. On $\mathbb{R}^2$ with standard coordinates (x, y), consider a smooth function f_1 depending on the sole y, and a smooth function f_2 depending on the sole $x^2 + y^2$. Consider also the vector fields

$$X_1 = \frac{\partial}{\partial x} \quad \text{and} \quad X_2 = x\frac{\partial}{\partial y} - y\frac{\partial}{\partial x}.$$

It follows from Examples 6.16 and 6.33 that X_1 is an infinitesimal symmetry of f_1. Similarly, it follows from Example 6.18 and Exercise 6.6 that X_2 is an infinitesimal symmetry of f_2. ♦

Similarly as for submanifolds, checking whether or not a vector field X is an infinitesimal symmetry of a smooth function f using the definition might be problematic. Luckily, infinitesimal symmetries of a smooth function can be characterized in a way which is totally independent of the flow.

Proposition 6.38. *Let M be a smooth manifold, let $f \in C^\infty(M)$ be a smooth function, and let X be a vector field on M. Then X is an infinitesimal symmetry of f if and only if $X(f) = 0$.*

We postpone a little bit the proof of Proposition 6.38. First, we propose an exercise and discuss a consequence. Both should convince the reader of the relevance of characterizing infinitesimal symmetries of smooth functions in a "flow-independent" way.

> **Exercise 6.9.** On $\mathbb{R}^2$ with standard coordinates (x, y) consider a smooth function $f \in C^\infty(\mathbb{R}^2)$ and the vector field
>
> $$X = \frac{\partial f}{\partial x}\frac{\partial}{\partial y} - \frac{\partial f}{\partial y}\frac{\partial}{\partial x}.$$
>
> Show that X is an infinitesimal symmetry of f.

Corollary 6.39. *Let M be a manifold, and let $f \in C^\infty(M)$ be a smooth function of M. The subset*

$$\mathfrak{X}(M, f) := \{\text{infinitesimal symmetries of } f\} \subseteq \mathfrak{X}(M)$$

of the Lie algebra of vector fields on M is a Lie subalgebra.

Proof. Left as Exercise 6.10. □

Exercise 6.10. Prove Corollary 6.39.

Proof of Proposition 6.38. Let M, f, X be as in the statement, and assume first that X is an infinitesimal symmetry of f. This means that

$$\Phi_t^*(f) = f|_{M_t}$$

for all $t \in \mathbb{R}$ (such that $M_t \neq \varnothing$). Then, for every $p \in M$,

$$X(f)(p) = \mathcal{L}_X f(p) = \frac{d}{dt}\Big|_{t=0}\Phi_t^*(f)(p) = \frac{d}{dt}\Big|_{t=0}f(p) = 0.$$

Conversely, let $X(f) = 0$, and let $t \in \mathbb{R}$ be such that $M_t \neq \varnothing$. Then, for every $p \in M_t$, we have

$$\frac{d}{dt}\Phi_t^*(f)(p) = \Phi_t^*(X(f))(p) = 0, \tag{6.21}$$

where we used (6.10). Identity (6.21) now shows that

$$\Phi_t^*(f)(p) = \text{const} = \Phi_0^*(f)(p) = f(p),$$

i.e., $\Phi_t^*(f) = f|_{M_t}$. This concludes the proof. □

Corollary 6.40. *Let M be a manifold, let $X \in \mathfrak{X}(M)$, and let $f, g \in C^\infty(M)$. If X is an infinitesimal symmetry of both f and g, then it is also an infinitesimal symmetry of $f + g$ and fg.*

Proof. Left as Exercise 6.11. □

Exercise 6.11. Prove Corollary 6.40.

We now pass to symmetries (and infinitesimal symmetries) of vector fields.

Definition 6.41 (Symmetry of a Vector Field). A *symmetry* of a vector field $Y \in \mathfrak{X}(M)$ on a manifold M is a diffeomorphism $\Phi : M \to M$ *preserving* Y in the sense that $\Phi^*(Y) = Y$. A *local symmetry* of Y is a diffeomorphism $\Phi : \mathcal{U} \to \Phi(\mathcal{U})$ between open submanifolds $\mathcal{U}, \Phi(\mathcal{U}) \subseteq M$ such that $\Phi^*(Y|_{\Phi(\mathcal{U})}) = Y|_{\mathcal{U}}$.

Example 6.42. On $\mathbb{R}^2$ with standard coordinates (x, y), consider the vector field $Y = \frac{\partial}{\partial y}$. Let $x_0 \in \mathbb{R}$, and consider the translation along the x-axis

$$\Phi : \mathbb{R}^2 \to \mathbb{R}^2, \quad (x, y) \mapsto \Phi(x, y) := (x + x_0, y).$$

Then Φ is a symmetry of Y. To see this, let's compute $\Phi^*(Y)$. We have

$$\Phi^*(Y) = A\frac{\partial}{\partial x} + B\frac{\partial}{\partial y},$$

for some functions $A, B \in C^\infty(\mathbb{R}^2)$ given by

$$A = \Phi^*(Y)(x) = \Phi^*(Y((\Phi^{-1})^*(x)))$$

and

$$B = \Phi^*(Y)(y) = \Phi^*(Y((\Phi^{-1})^*(y))).$$

Now $\Phi^{-1} : \mathbb{R}^2 \to \mathbb{R}^2$ is given by

$$\Phi^{-1} : \mathbb{R}^2 \to \mathbb{R}^2, \quad (x, y) \mapsto \Phi^{-1}(x, y) := (x - x_0, y),$$

and $(\Phi^{-1})^*(x)$ is the first component of Φ^{-1}, i.e., $x - x_0$. Hence

$$A = \Phi^*(Y((\Phi^{-1})^*(x))) = \Phi^*(Y(x - x_0)) = \Phi^*\left(\frac{\partial(x - x_0)}{\partial y}\right)$$

$$= \Phi^*(0) = 0.$$

Similarly, $(\Phi^{-1})^*(y)$ is the second component of $\Phi^{-1}(y)$, and

$$B = \Phi^*(Y)(y) = \Phi^*(Y((\Phi^{-1})^*(y))) = \Phi^*(Y(y)) = \Phi^*\left(\frac{\partial y}{\partial y}\right)$$

$$= \Phi^*(1) = 1.$$

We conclude that

$$\Phi^*(Y) = 0 \cdot \frac{\partial}{\partial x} + 1 \cdot \frac{\partial}{\partial y} = \frac{\partial}{\partial y} = Y,$$

i.e., Φ is a symmetry of Y as claimed. ◆

Exercise 6.12. On $\mathbb{R}^2$ with standard coordinates (x, y), consider the vector field

$$Y = x\frac{\partial}{\partial x} + y\frac{\partial}{\partial y}.$$

Additionally, let $\theta \in \mathbb{R}$, and consider the rotation around the origin of an angle θ:

$$\Phi : \mathbb{R}^2 \to \mathbb{R}^2, \quad (x, y) \mapsto \Phi(x, y) = (x\cos\theta - y\sin\theta, x\sin\theta + y\cos\theta).$$

Show that Φ is a symmetry of Y.

Proposition 6.43. *Let M be a manifold, and let $Y \in \mathfrak{X}(M)$ be a vector field on M. The subset*

$$\mathrm{Diffeo}(M, Y) := \{\text{symmetries of } Y\} \subseteq \mathrm{Diffeo}(M)$$

of the group of diffeomorphisms of M is a subgroup.

Proof.   Left as Exercise 6.13. $\square$

Exercise 6.13. Prove Proposition 6.43.

The following is a Corollary of Proposition 5.21.

Corollary 6.44. *Let M be a manifold, let $Y, Z \in \mathfrak{X}(M)$, let $f \in C^\infty(M)$, and let $\Phi : M \to M$ be a diffeomorphism. If Φ is a symmetry of Y, Z and f, then it is also a symmetry of $Y + Z$, $Y(f)$, fY, and $[Y, Z]$.*

Proof.   Left as Exercise 6.14. $\square$

Exercise 6.14. Prove Corollary 6.44.

We now come to infinitesimal symmetries of vector fields.

Definition 6.45 (Infinitesimal Symmetry of a Vector Field). An *infinitesimal symmetry* of a vector field $Y \in \mathfrak{X}(M)$ on a manifold M is a(nother) vector field X on M generating a flow $\{\Phi_t\}_t$ by local symmetries of Y, i.e., $\Phi_t^*(Y|_{M_{-t}}) = Y|_{M_t}$ for all t (such that $M_t \neq \varnothing$).

Beware that the two vector fields X, Y in Definition 6.45 play completely different roles.

Example 6.46. On $\mathbb{R}^2$ with standard coordinates (x, y), consider the vector fields

$$Y_1 = \frac{\partial}{\partial y} \quad \text{and} \quad Y_2 = x\frac{\partial}{\partial x} + y\frac{\partial}{\partial y}$$

and the vector fields

$$X_1 = \frac{\partial}{\partial x} \quad \text{and} \quad X_2 = x\frac{\partial}{\partial y} - y\frac{\partial}{\partial x}.$$

It follows from Examples 6.16 and 6.42 that X_1 is an infinitesimal symmetry of Y_1. Similarly, it follows from Example 6.18 and Exercise 6.12 that X_2 is an infinitesimal symmetry of Y_2. ◆

Similarly as for functions, checking whether or not a vector field X is an infinitesimal symmetry of an other vector field Y using the definition might be problematic but, as for functions, infinitesimal symmetries of a vector field can be characterized in a rather efficient way.

Proposition 6.47. *Let M be a smooth manifold, and let $Y, X \in \mathfrak{X}(M)$ be vector fields. Then X is an infinitesimal symmetry of Y if and only if $[Y, X] = 0$.*

Proof. The proof is very similar to that of Proposition 6.38 and we leave it as Exercise 6.15. ☐

Exercise 6.15. Prove Proposition 6.47.

Corollary 6.48. *Let M be a smooth manifold, let $Y, X \in \mathfrak{X}(M)$ be vector fields. Then X is an infinitesimal symmetry of Y if and only if Y is an infinitesimal symmetry of X.*

Proof. Obvious. ☐

Corollary 6.49. *Let M be a manifold, and let $Y \in \mathfrak{X}(M)$ be a vector field on M. The subset*

$$\mathfrak{X}(M, Y) := \{\text{infinitesimal symmetries of } Y\} \subseteq \mathfrak{X}(M)$$

of the Lie algebra of vector fields on M is a Lie subalgebra.

Proof. Left as Exercise 6.16. ☐

> **Exercise 6.16.** Prove Corollary 6.49.

The following infinitesimal version of Corollary 6.44 is a corollary of either Corollary 6.44 itself or (in the part about the commutator) Proposition 6.47 and the Jacobi identity for the commutator of vector fields.

Corollary 6.50. *Let M be a manifold, let $X, Y, Z \in \mathfrak{X}(M)$, and let $f \in C^\infty(M)$. If X is an infinitesimal symmetry of both Y, Z, and f, then it is also an infinitesimal symmetry of $Y + Z$, $Y(f)$, fY, and $[Y, Z]$.*

Proof. Left as Exercise 6.17. □

> **Exercise 6.17.** Prove Corollary 6.50.

We conclude this chapter discussing the flows of two *commuting* vector fields, i.e., two vector fields X, Y such that their commutator vanishes: $[X, Y] = 0$. Let M be a manifold, and let $X, Y \in \mathfrak{X}(M)$ be two vector fields on M. Denote by $\{\Phi_t\}_t$ and $\{\Psi_s\}_s$ the flows of X and Y, respectively. We want to show that X and Y commute if and only if their flows commute. First, we have to explain what does it mean that $\{\Phi_t\}_t$ and $\{\Psi_s\}_s$ commute. We will only discuss in details the case when both Φ and Ψ are defined on the whole $\mathbb{R} \times M$, however, we report a definition in the general case for completeness.

Definition 6.51 (Commuting Flows). The flows $\{\Phi_t\}_t$ and $\{\Psi_s\}_s$ commute if, for every $p \in M$, we have that, whenever $I, J \subseteq \mathbb{R}$ are open intervals containing 0 such that one of the expressions

$$\Phi_t(\Psi_s(p)) \quad \text{or} \quad \Psi_s(\Phi_t(p))$$

makes sense for all $(t, s) \in I \times J$, then the other one also makes sense and they are equal.

Proposition 6.52 (Commuting Vector Fields). *The vector fields X and Y commute if and only if so do their flows $\{\Phi_t\}_t$ and $\{\Psi_s\}_s$.*

Proof. We provide a proof only in the case when both X and Y are complete (for the general case, see, e.g., Lee, 2013). We begin remarking that, in this case, Definition 6.51 boils down to the condition

$$\Phi_t \circ \Psi_s = \Psi_s \circ \Phi_t \quad \text{for all } (s, t) \in \mathbb{R} \times \mathbb{R}.$$

Assume first that X and Y commute. Then, from Proposition 6.47, X is an infinitesimal symmetry of Y, i.e., Φ_t is a symmetry of Y for all $t \in \mathbb{R}$. It now follows from Proposition 6.10 that Φ_t maps integral curves of Y to integral curves. In particular, for all $p \in M$, the maximal integral curve of

Y starting from p, say $\gamma_p : \mathbb{R} \to M$, is mapped to an integral curve $\Phi_t \circ \gamma_p :$
$\mathbb{R} \to M$. Note that, being defined on the whole $\mathbb{R}$, not only $\Phi_t \circ \gamma_p$ is an
integral curve, but it is also maximal. As it starts from $\Phi_t(\gamma_p(0)) = \Phi_t(p)$,
it must coincide with the maximal integral curve $\gamma_{\Phi_t(p)}$, i.e., $\Phi_t \circ \gamma_p =$
$\gamma_{\Phi_t(p)} : \mathbb{R} \to M$. In other words,

$$\Phi_t(\gamma_p(s)) = \gamma_{\Phi_t(p)}(s),$$

for all $p \in M$ and all $(t, s) \in \mathbb{R} \times \mathbb{R}$. The left-hand side is

$$\Phi_t(\gamma_p(s)) = \Phi_t(\Psi_s(t)),$$

while the right-hand side is

$$\gamma_{\Phi_t(p)}(s) = \Psi_s(\Phi_t(p)).$$

This conclude the proof of the "only if" part of the statement.
    For the "if part", suppose that

$$\Phi_t \circ \Psi_s = \Psi_s \circ \Phi_t \quad \text{for all } (s, t) \in \mathbb{R} \times \mathbb{R}.$$

Then, similarly as in the proof of Proposition 6.10, for all $p \in M$ and all
$t \in \mathbb{R}$,

$$d_p\Phi_t(Y_p) = d_p\Phi_t\left(\frac{d}{ds}\Big|_{s=0}\Psi_s(p)\right) \qquad \text{($s \mapsto \Psi_s(p)$ is an integral}$$
$$\text{curve of Y)}$$

$$= \frac{d}{ds}\Big|_{s=0}\Phi_t(\Psi_s(p)) \qquad \text{(tangent map applied to the}$$
$$\text{velocity of a curve)}$$

$$= \frac{d}{ds}\Big|_{s=0}\Psi_s(\Phi_t(p)) \qquad \text{($\{\Phi_t\}_t$ and $\{\Psi_s\}_s$ commute)}$$

$$= Y_{\Phi_t(p)} \qquad \text{($s \mapsto \Psi_s(\Phi_t(p))$ is an integral}$$
$$\text{curve of Y).}$$

This shows that Φ_t is a symmetry of Y for all t, hence X is an infinitesimal
symmetry of Y and, from Proposition 6.47, X and Y commute. $\square$

Chapter 7

Covectors and Differential 1-Forms

In this chapter, we introduce (tangent) covectors and differential 1-forms. Covectors and tangent vectors are dual objects. Similarly, differential 1-forms and vector fields are dual objects. Covectors are the points of a new manifold, the *cotangent bundle*, and differential 1-forms are equivalent to sections of the cotangent bundle, in a very similar way as for tangent vectors and vector fields. However, there are important differences between vector fields and differential forms. Among the most relevant, unlike vector fields, differential forms on a manifold can be transformed into differential forms on another manifold along any smooth map, via the *pull-back construction*.

7.1 Covectors and the Cotangent Bundle

Let M be an n-dimensional manifold, let $p \in M$ be a point, and let $(U, \varphi = (x^1, \ldots, x^n))$ be a chart on M around p.

Definition 7.1 (Cotangent Space). The *cotangent space* to M at the point p is the dual space

$$T_p^* M := (T_p M)^*$$

of the tangent space. Vectors in $T_p^* M$ are called *tangent covectors*, or *cotangent vectors*, or, simply, *covectors*. In other words, covectors are $\mathbb{R}$-linear maps $\theta : T_p M \to \mathbb{R}$. The frame of $T_p^* M$ dual to the coordinate frame

$$\left(\frac{\partial}{\partial x^1} \Big|_p, \ldots, \frac{\partial}{\partial x^n} \Big|_p \right)$$

is denoted

$$\left(d_p x^1, \ldots, d_p x^n \right)$$

and it is called the *coordinate coframe*.

Remark 7.2. Let $\mathcal{U} \subseteq M$ be an open neighborhood of p. As $T_p\mathcal{U}$ can be identified with T_pM, the cotangent space $T_p^*\mathcal{U}$ can be identified with T_p^*M, and, in what follows, we will always take this point of view. $\diamond$

Similarly as for the coordinate frame, if we change coordinates around p, the coordinate coframe changes. So, let $(\tilde{U}, \tilde{\varphi} = (\tilde{x}^1, \ldots, \tilde{x}^n))$ be another chart on M around p. Denote by $t^1, \ldots, t^n$ the standard coordinates on $\varphi(U)$ and by $\tilde{t}^1, \ldots, \tilde{t}^n$ the standard coordinates on $\tilde{\varphi}(\tilde{U})$. Finally, let $P = \varphi(p)$. The transition matrix

$$M_{\mathcal{R}^*, \tilde{\mathcal{R}}^*} \in \mathrm{GL}(n, \mathbb{R})$$

between the coordinate coframes

$$\tilde{\mathcal{R}}^* = \left(d_p \tilde{x}^1, \ldots, d_p \tilde{x}^n \right) \quad \text{and} \quad \mathcal{R}^* = \left(d_p x^1, \ldots, d_p x^n \right)$$

is the transpose of the transition matrix $M_{\tilde{\mathcal{R}}, \mathcal{R}}$ between the coordinate frames

$$\mathcal{R} = \left(\frac{\partial}{\partial x^1}\Big|_p, \ldots, \frac{\partial}{\partial x^n}\Big|_p \right) \quad \text{and} \quad \tilde{\mathcal{R}} = \left(\frac{\partial}{\partial \tilde{x}^1}\Big|_p, \ldots, \frac{\partial}{\partial \tilde{x}^n}\Big|_p \right).$$

Hence, the former is

$$M_{\mathcal{R}^*, \tilde{\mathcal{R}}^*} = M_{\tilde{\mathcal{R}}, \mathcal{R}}^T = \left(\frac{\partial \tilde{x}^i}{\partial x^j}(p) \right)_{i=1,\ldots,n}^{j=1,\ldots,n} = \left(\frac{\partial (\tilde{\varphi} \circ \varphi^{-1})^i}{\partial t^j}(P) \right)_{i=1,\ldots,n}^{j=1,\ldots,n}.$$

In other words,

$$d_p \tilde{x}^i = \frac{\partial \tilde{x}^i}{\partial x^j}(p) d_p x^j = \frac{\partial (\tilde{\varphi} \circ \varphi^{-1})^i}{\partial t^j}(P) d_p x^j,$$

for all $i = 1, \ldots, n$.

The following notion of *differential of a smooth function f at a point p* encodes *partial derivatives of f at p* in a coordinate-free manner.

Definition 7.3 (Differential at a Point). The *differential* at the point p is the map:

$$d_p : C^\infty(M) \to T_p^* M, \quad f \mapsto d_p f$$

defined by

$$d_p f(v) = v(f), \quad \text{for all } v \in T_p M.$$

The differential d_p is well defined in the sense that $d_p f$ is a covector for every $f \in C^\infty(M)$. Indeed, let $v, w \in T_p M$, and let $a, b \in \mathbb{R}$. Then

$$d_p f(av + bw) = (av + bw)(f) = a \cdot v(f) + b \cdot w(f)$$
$$= a \cdot d_p f(v) + b \cdot d_p f(w),$$

showing that $d_p f$ is indeed $\mathbb{R}$-linear.

Proposition 7.4 (Properties of the Differential at a Point). *The differential* $d_p : C^\infty(M) \to T_p^* M$ *enjoys the following properties:*

- d_p *is $\mathbb{R}$-linear,*
- d_p *satisfies the following* Leibniz rule at the point p:

$$d_p(fg) = f(p) d_p g + g(p) d_p f, \quad \text{for all } f, g \in C^\infty(M),$$

- $d_p c = 0$ *for every constant function c,*
- d_p *is* local, *in the sense that $d_p f$ does only depend on the values of f around p, i.e.,*

$$d_p f = d_p g$$

whenever f and g agree on an open neighborhood of p.

Proof.   Left as Exercise 7.1   □

Exercise 7.1. Prove Proposition 7.4.

Let $f \in C^\infty(M)$, let $p \in M$, and let $(U, \varphi = (x^1, \ldots, x^n))$ be a chart on M around p. We now compute the components of $d_p f$ in the coordinate coframe. If $d_p f = \theta_i d_p x^i$, then

$$\theta_i = d_p f \left(\frac{\partial}{\partial x^i} |_p \right) = \frac{\partial}{\partial x_i} |_p f = \frac{\partial f}{\partial x^i}(p),$$

hence

$$d_p f = \frac{\partial f}{\partial x^i}(p) d_p x^i, \tag{7.1}$$

showing that $d_p f$ encodes, in a coordinate-free way, partial derivatives of f, as announced. It follows from Equation (7.1) that the elements of the coordinate coframe are exactly the differentials at the point p of the coordinate functions, motivating the notation $d_p x^i$.

Proposition 7.5. *Let M be a manifold and let $p \in M$ be a point. For any covector $\theta \in T_p^* M$, there exists a (non-unique) smooth function f such that $\theta = d_p f$.*

Proof.   Choose a chart $(U, \varphi = (x^1, \ldots, x^n))$ around p, and let $\theta \in T_p^* M$. There exist real numbers θ_i such that $\theta = \theta_i d_p x^i$. Consider the smooth function $g = \theta_i x^i \in C^\infty(U)$. In view of the Local Extension Lemma, there exists a smooth function $f \in C^\infty(M)$ agreeing with g in an open neighborhood of p, and we have

$$d_p f = d_p g = \frac{\partial g}{\partial x^i}(p) d_p x^i = \theta_i d_p x^i = \theta. \qquad \square$$

Similarly as tangent vectors, covectors on a manifold M can be organized into a manifold called the *cotangent bundle to M* and denoted $T^* M$. Let's construct $T^* M$. First of all, we put

$$T^* M := \coprod_{p \in M} T_p^* M = \left\{ (p, \theta) : p \in M \text{ and } \theta \in T_p^* M \right\}.$$

A point (p, θ) in $T^* M$ will be sometimes simply denoted by θ. The set $T^* M$ comes with a natural surjection $\pi : T^* M \to M$, $(p, \theta) \mapsto p$, and the pair $(T^* M, \pi)$ is the *cotangent bundle to M* (sometimes, $T^* M$ itself is referred to as the cotangent bundle). The smooth structure on M induces an atlas on $T^* M$. To see this, begin with a chart $(U, \varphi = (x^1, \ldots, x^n))$ on M, and define a chart $(T^* U, T^* \varphi)$ on $T^* M$ as follows. First of all, put

$$T^* U := \pi^{-1}(U) = \coprod_{p \in U} T_p^* M = \{ (p, \theta) \in T^* M : p \in U \}.$$

So, T^*U is exactly the cotangent bundle to U. Next, define a map

$$T^*\varphi : T^*U \to \widehat{U} \times \mathbb{R}^n$$

by

$$(p,\theta) \mapsto T^*\varphi(p,\theta) := \left(\varphi(p); \theta\left(\frac{\partial}{\partial x^1}|_p\right), \ldots, \theta\left(\frac{\partial}{\partial x^n}|_p\right)\right).$$

Note that the last n entries of $T^*\varphi(p,\theta)$ are the components of θ in the coordinate coframe

$$\left(d_p x^1, \ldots, d_p x^n\right).$$

Clearly, $(T^*U, T^*\varphi)$ is a $2n$-dimensional chart on T^*M. Every such chart is called a *standard chart*. The first n components of the coordinate map $T^*\varphi$ will be again denoted by $(x^1, \ldots, x^n)$, and the last n components will be denoted by $(p_1, \ldots, p_n)$.

We now show that any two standard charts on T^*M are compatible. So let $(U, \varphi = (x^1, \ldots, x^n))$ and $(\tilde{U}, \tilde{\varphi} = (\tilde{x}^1, \ldots, \tilde{x}^n))$ be charts on M, and let $(T^*U, T^*\varphi)$, $(T^*\tilde{U}, T^*\tilde{\varphi})$ be the associated standard charts on T^*M. If $U \cap \tilde{U} = \varnothing$, then $T^*U \cap T^*\tilde{U} = \varnothing$ and there is nothing else to prove. If $U \cap \tilde{U} \neq \varnothing$, then $T^*U \cap T^*\tilde{U} = T^*(U \cap \tilde{U}) \neq \varnothing$, and we can consider the transition map

$$T^*\varphi \circ (T^*\tilde{\varphi})^{-1} : T^*\tilde{\varphi}(T^*U \cap T^*\tilde{U}) \to T^*\varphi(T^*U \cap T^*\tilde{U}).$$

Now,

$$T^*\tilde{\varphi}(T^*U \cap T^*\tilde{U}) = (U \cap \tilde{U}) \times \mathbb{R}^n$$

is an open subset and similarly for $T^*\varphi(T^*U \cap T^*\tilde{U})$. Additionally, take a point $(\tilde{P}; a_1, \ldots, a_n) \in (U \cap \tilde{U}) \times \mathbb{R}^n$, and compute

$$(T^*\varphi \circ (T^*\tilde{\varphi})^{-1})(\tilde{P}; a_1, \ldots, a_n)$$

$$= T^*\varphi\left(\tilde{\varphi}^{-1}(\tilde{P}), a_i d_{\varphi^{-1}(\tilde{P})}\tilde{x}^i\right)$$

$$= \left((\varphi \circ \tilde{\varphi}^{-1})(\tilde{P}); a_i d_{\varphi^{-1}(\tilde{P})}\tilde{x}^i\left(\frac{\partial}{\partial x^1}|_{\tilde{\varphi}^{-1}(\tilde{P})}\right), \ldots, a_i d_{\varphi^{-1}(\tilde{P})}\tilde{x}^i\left(\frac{\partial}{\partial x^n}|_{\tilde{\varphi}^{-1}(\tilde{P})}\right)\right)$$

$$= \left((\varphi \circ \tilde{\varphi}^{-1})(\tilde{P}); a_i \frac{\partial \tilde{x}^i}{\partial x^1}(\varphi^{-1}(\tilde{P})), \ldots, a_i \frac{\partial \tilde{x}^i}{\partial x^n}(\varphi^{-1}(\tilde{P}))\right).$$

As $(U, \varphi), (\tilde{U}, \tilde{\varphi})$ are compatible, the first n entries depend smoothly on $\tilde{P}$. The last n entries depend smoothly on $\tilde{P}$ and linearly, hence smoothly, on $a_1, \ldots, a_n$. This shows that $T^*\varphi \circ (T^*\tilde{\varphi})^{-1}$ is smooth. By the same argument, its inverse is also smooth, so they are both diffeomorphisms and the charts $(T^*U, T^*\varphi), (T^*\tilde{U}, T^*\tilde{\varphi})$ are compatible. Finally, standard charts cover T^*M, so they form an atlas called the *standard atlas*. The standard atlas induces a Hausdorff and second countable topology on T^*M, and this can be proved exactly as in the case of the tangent bundle. So T^*M is a manifold, and one can also show that $\pi : T^*M \to M$ is a smooth map. We leave the details to the reader. Summarizing, we have proved the following

Proposition 7.6 (Smooth Structure on the Cotangent Bundle). *Standard charts on T^*M form an atlas. With the associated smooth structure, T^*M is a $2n$-dimensional manifold, and $\pi : T^*M \to M$ is a smooth map.*

Definition 7.7 (Section of the Cotangent Bundle). A *section* of T^*M is a map $s : M \to T^*M$ such that $\pi \circ s = \mathrm{id}_M$. In other words, a section s is the assignment of a covector $s(p)$ at p, for every point $p \in M$.

There is a natural $C^\infty(M)$-module structure on sections of T^*M defined exactly by the same formulas as for the tangent bundle (see Example 3.20). The *zero section* is the section mapping a point p to the zero covector $0 \in T_p^*M$, for all $p \in M$. Now, let $(U, \varphi = (x^1, \ldots, x^n))$ be a chart on M. Every section $s : M \to T^*M$ maps the coordinate domain U into the standard coordinate domain T^*U, and s is smooth if and only if the pullbacks $s_i := s^*(p_i)$ are smooth functions on U for every chart (U, φ) (in an atlas). Using these remarks, one can show that smooth sections form a submodule denoted $\Gamma(T^*M)$. All the claims in this paragraph can be proved exactly as for sections of the tangent bundle, and we leave the details to the reader.

7.2 Differential 1-Forms and Fields of Covectors

Let M be a manifold.

Definition 7.8 (Differential 1-Form). A *differential 1-form* on M, or, simply, a *1-form*, is a $C^\infty(M)$-linear map

$$\theta : \mathfrak{X}(M) \to C^\infty(M).$$

The space of 1-forms on M is denoted $\Omega^1(M)$.

The space $\Omega^1(M)$ of 1-forms on M is a $C^\infty(M)$-module. More precisely, it is the $C^\infty(M)$-module dual to the $C^\infty(M)$-module of vector fields.

Example 7.9 (Coordinate Coframe). Let $(U, \varphi = (x^1,\ldots,x^n))$ be a chart on M. Recall that the $C^\infty(U)$-module $\mathfrak{X}(U)$ of vector fields on U possesses a finite frame

$$\left(\frac{\partial}{\partial x^1}, \ldots, \frac{\partial}{\partial x^n} \right)$$

formed by coordinate vector fields. Accordingly, the dual module $\Omega^1(U)$ does also possess a finite frame, the dual frame, denoted by

$$\left(dx^1, \ldots, dx^n \right) \tag{7.2}$$

and uniquely determined by the conditions

$$dx^i \left(\frac{\partial}{\partial x^j} \right) = \delta^i_j.$$

The frame (7.2) of $\Omega^1(U)$ is (also) called the *coordinate coframe*. If

$$X = X^i \frac{\partial}{\partial x^i}$$

is a vector field on U, we have

$$dx^i(X) = X^i, \quad \text{for all } i = 1,\ldots,n.$$

Every 1-form θ on U can be uniquely written as

$$\theta = \theta_i dx^i,$$

for some smooth functions $\theta_i \in C^\infty(U)$ given by

$$\theta_i = \theta \left(\frac{\partial}{\partial x^i} \right).$$

If $(U, \tilde{\varphi} = (\tilde{x}^1,\ldots,\tilde{x}^n))$ is another chart with the same coordinate domain, then θ can be also written as

$$\theta = \tilde{\theta}_i d\tilde{x}^i$$

and

$$\theta_i = \theta \left(\frac{\partial}{\partial x^i} \right) = \theta \left(\frac{\partial \tilde{x}^j}{\partial x^i} \frac{\partial}{\partial \tilde{x}^j} \right) = \frac{\partial \tilde{x}^j}{\partial x^i} \theta \left(\frac{\partial}{\partial \tilde{x}^j} \right) = \frac{\partial \tilde{x}^j}{\partial x^i} \tilde{\theta}_j.$$

◆

Definition 7.10 (Differential). The *differential* is the map

$$d : C^\infty(M) \to \Omega^1(M), \quad f \mapsto df,$$

defined by

$$df(X) = X(f), \quad \text{for all } X \in \mathfrak{X}(M).$$

The differential d is well defined in the sense that df is a 1-form for every $f \in C^\infty(M)$. Check it as an exercise!

Proposition 7.11 (Properties of the Differential). *The differential* d : $C^\infty(M) \to \Omega^1(M)$ *enjoys the following properties:*

- *d is $\mathbb{R}$-linear,*
- *d satisfies the following* Leibniz rule:

$$d(fg) = f\,dg + g\,df, \quad \text{for all } f, g \in C^\infty(M),$$

- *$dc = 0$ for every constant function c.*

Proof.   Left as Exercise 7.2   □

Exercise 7.2. Prove Proposition 7.11.

Let $(U, \varphi = (x^1, \ldots, x^n))$ be a chart on M, and let $f \in C^\infty(U)$. We want to compute the components of $df \in \Omega^1(U)$ in the coordinate coframe. If $df = \theta_i dx^i$, then

$$\theta_i = df\left(\frac{\partial}{\partial x^i}\right) = \frac{\partial f}{\partial x^i},$$

hence

$$df = \frac{\partial f}{\partial x^i} dx^i.$$

So df encodes the gradient of f in a coordinate free manner. In particular, the differential of the i-th coordinate function is exactly the i-th coordinate 1-form, motivating the notation dx^i.

In the second part of this section, we show that differential 1-forms are equivalent to sections of the cotangent bundle. More precisely, there is a canonical $C^\infty(M)$-module isomorphism $\Omega^1(M) \cong \Gamma(T^*M)$. We begin proving that a differential 1-form $\theta \in \Omega^1(M)$ determines a covector $\theta_p \in T_p^*M$ for every point $p \in M$. In order to define θ_p, we need two lemmas.

Lemma 7.12. *Let $\theta \in \Omega^1(M)$, let $p_0 \in M$, and let $X \in \mathfrak{X}(M)$. Then the value*

$$\theta(X)(p_0)$$

of the function $\theta(X) \in C^\infty(M)$ at p_0, does only depend on (θ and) the value X_{p_0} of the vector field X at p_0.

Proof. The proof consists of various steps.

Step I: Differential 1-Forms are Local Operators. *For every open subset $\mathcal{U} \subseteq M$, the function $\theta(X)|_{\mathcal{U}}$ does only depend on (θ and) the restriction $X|_{\mathcal{U}}$.* To see this, let X' be another vector field on M such that $X'|_{\mathcal{U}} = X|_{\mathcal{U}}$, let $p \in \mathcal{U}$, and let $\beta \in C^\infty(M)$ be a bump function relative to the data $(\mathcal{U}, p)$. In other words, $\beta = 1$ in an open neighborhood of p, and $\mathrm{supp}\,\beta \subseteq \mathcal{U}$, in particular, $\beta = 0$ outside $\mathcal{U}$. Put $\eta = 1 - \beta$. Then

$$X - X' = \eta(X - X'),$$

hence

$$\theta(X - X')(p) = \theta(\eta(X - X'))(p) = \left(\eta\theta(X - X')\right)(p)$$
$$= \eta(p)\theta(X - X')(p) = 0$$

so that

$$\theta(X)(p) = \theta(X - X')(p) + \theta(X')(p) = \theta(X')(p).$$

It follows from the arbitrariness of $p \in \mathcal{U}$ that

$$\theta(X)|_{\mathcal{U}} = \theta(X')|_{\mathcal{U}}.$$

Step II: Local Extension Lemma for Vector Fields. *Let $S \subseteq M$ be a submanifold, and let X be a vector field on S. For every point p_0 in S, there exists an open neighborhood $U \subseteq M$ of p_0 and a vector field $\tilde{X}$ on M such that $\tilde{X}|_U$ is tangent to $S \cap U$, and $\tilde{X}|_U|_{S \cap U} = X|_{S \cap U}$.* We prove this claim in the case when $S = \mathcal{U}$ is an open submanifold and leave the general case as an exercise for the reader (the proof is not very different from that of the Local Extension Lemma for functions). So, let $\mathcal{U} \subseteq M$ be an open submanifold, let $p_0 \in \mathcal{U}$, and let $X \in \mathfrak{X}(\mathcal{U})$. Choose a bump function $\beta \in C^\infty(M)$ relative to the data $(\mathcal{U}, p_0)$, and consider the vector field $\tilde{X}$ defined (through its values) as follows:

$$\tilde{X}_p := \begin{cases} \beta(p)X_p & \text{if } p \in \mathcal{U} \\ 0 & \text{if } p \notin \mathcal{U} \end{cases}.$$

In view of the Gluing Lemma for Vector Fields, $\widetilde{X}$ is a well-defined vector field on M. Additionally, $\widetilde{X}$ agrees with X on an open neighborhood of p_0 (specifically, the open neighborhood where $\beta = 1$). This concludes the proof of the Local Extension Lemma (for $S = \mathcal{U}$ an open submanifold).

Step III: Vector Fields Vanishing at a Point. *Let X be a vector field vanishing at a point p, i.e., $X_p = 0$, then there exist vector fields $X_1, \dots, X_n$, and smooth functions $f_1, \dots, f_n$ such that (1) the f_i all vanish at p, i.e., $f_i(p) = 0$, and (2)*

$$X = f_1 X_1 + \cdots + f_n X_n$$

in an open neighborhood of p. To see this, consider $(U, \varphi = (x^1, \dots, x^n))$ a chart around p. Then

$$X|_U = g^i \frac{\partial}{\partial x^i}$$

for some smooth functions $g^i \in C^\infty(U)$. As

$$0 = X_p = g^i(p) \frac{\partial}{\partial x^i}\Big|_p,$$

we get $g^i(p) = 0$ for all $i = 1, \dots, n$. Now, let $f_1, \dots, f_n \in C^\infty(M)$ be smooth functions agreeing with $g^1, \dots, g^n$ in an open neighborhood of p. In particular, $f_i(p) = 0$, for all $i = 1, \dots, n$. Finally, use Step II to find vector fields $X_1, \dots, X_n \in \mathfrak{X}(M)$ agreeing with $\frac{\partial}{\partial x^1}, \dots, \frac{\partial}{\partial x^n}$ in an open neighborhood of p, and note that

$$X = f_1 X_1 + \cdots + f_n X_n$$

in an open neighborhood of p as required. This concludes Step III.

Step IV: Conclusion. Let $p_0 \in M$ be a point, and let $X, X' \in \mathfrak{X}(M)$ be vector fields such that $X_{p_0} = X'_{p_0}$. We want to show that $\theta(X)(p_0) = \theta(X')(p_0)$. To do this, note that

$$0 = X_{p_0} - X'_{p_0} = (X - X')_{p_0}.$$

Hence, it follows from Step III that there are functions $f_1, \dots, f_n \in C^\infty(M)$ and vector fields $X_1, \dots, X_n \in \mathfrak{X}(M)$ such that $f_i(p_0) = 0$, and

$$X - X' = f_1 X_1 + \cdots + f_n X_n$$

in an open neighborhood of p_0. It now follows from Step I that

$$\theta(X - X') = \theta(f_1 X_1 + \cdots + f_n X_n) = f_1 \theta(X_1) + \cdots + f_n \theta(X_n)$$

in an open neighborhood of p_0. In particular,

$$\begin{aligned}
\theta(X - X')(p_0) &= (f_1\theta(X_1) + \cdots + f_n\theta(X_n))(p_0) \\
&= f_1(p_0)\theta(X_1)(p_0) + \cdots + f_n(p_0)\theta(X_n)(p_0) \\
&= 0.
\end{aligned}$$

We conclude noting (again) that

$$\theta(X)(p_0) = \theta(X - X')(p_0) + \theta(X')(p_0) = \theta(X')(p_0). \qquad \square$$

Lemma 7.13. *Let $p \in M$, and let $v \in T_pM$. Then there exists a vector field $X \in \mathfrak{X}(M)$ such that $X_p = v$.*

Proof. Let $(U, \varphi = (x^1, \ldots, x^n))$ be a chart around p. Then

$$v = v^i \frac{\partial}{\partial x^i}\Big|_p$$

for some real numbers v^i. Interpreting the v^i as constant functions on U we can define a vector field on U:

$$Y = v^i \frac{\partial}{\partial x^i} \in \mathfrak{X}(U).$$

Clearly, $Y_p = v$. From the Local Extension Lemma for Vector Fields (Step II in the proof of Lemma 7.12), there exists a vector field $X \in \mathfrak{X}(M)$ on M such that X and Y agree in an open neighborhood of p. In particular,

$$X_p = Y_p = v.$$

This concludes the proof. $\qquad \square$

We are now ready to present the main result of this section. Begin with a differential 1-form θ on a manifold M. For any point $p \in M$, we want to define a covector $\theta_p \in T_p^*M$. So, let $v \in T_pM$, and let $X \in \mathfrak{X}(M)$ be any vector field such that $X_p = v$. The vector field X exists in view of Lemma 7.13. We put

$$\theta_p(v) := \theta(X)(p).$$

From Lemma 7.12, $\theta_p(v)$ does only depend on (θ and) v. Hence, θ_p is a well-defined map

$$\theta_p : T_pM \to \mathbb{R}.$$

It is easy to see that θ_p is a covector. Indeed, let $v, w \in T_pM$, let $a, b \in \mathbb{R}$, and let $X, Y \in \mathfrak{X}(M)$ be vector fields such that $X_p = v$ and $Y_p = w$. Then, from Exercise 5.5,

$$(aX + bY)_p = aX_p + bY_p = av + bw$$

and

$$\begin{aligned}
\theta_p(av + bw) &= \theta(aX + bY)(p) \\
&= (a\theta(X) + b\theta(Y))(p) \\
&= a\theta(X)(p) + b\theta(Y)(p) \\
&= a\theta_p(v) + b\theta_p(w).
\end{aligned}$$

The covector $\theta_p \in T_p^*M$ is called the *value at p* of the 1-form θ.

Exercise 7.3. Prove that, for every $\theta, \kappa \in \Omega^1(M)$, every $f \in C^\infty(M)$, and every $p \in M$,

$$\begin{aligned}
(\theta + \kappa)_p &= \theta_p + \kappa_p, \\
(f\theta)_p &= f(p)\theta_p, \\
(df)_p &= d_pf.
\end{aligned} \tag{7.3}$$

Additionally, let $(U, \varphi = (x^1, \ldots, x^n))$ be a chart around p. Prove that the value of the coordinate 1-form dx^i at p is exactly the coordinate covector d_px^i.

Let $\theta \in \Omega^1(M)$. The assignment $p \mapsto \theta_p$ is a(n *a priori* non-necessarily smooth) section s_θ of the cotangent bundle T^*M:

$$s_\theta : M \to T^*M, \quad p \mapsto s_\theta(p) := \theta_p.$$

Proposition 7.14. *The map*

$$\Omega^1(M) \to \Gamma(T^*M), \quad \theta \mapsto s_\theta \tag{7.4}$$

is a well-defined $C^\infty(M)$-module isomorphism.

Proof. First, we prove that for every $\theta \in \Omega^1(M)$, the section s_θ is actually smooth. As for vector fields, it is enough to check that, for every chart $(U, \varphi = (x^1, \ldots, x^n))$ on M, the *components* $(s_\theta)_i$ of s_θ in the chart (U, φ) (see the end of Section 7.1) are smooth functions on U. To do this, we fix $p_0 \in U$ and show that $(s_\theta)_i$ is smooth around p_0. So, let $p \in U$ and compute

$$(s_\theta)_i(p) = p_i(s_\theta(p)) = p_i(\theta_p)$$

$$= i\text{-th component of } \theta_p \text{ in the coordinate coframe}$$

$$= \theta_p \left(\frac{\partial}{\partial x^i} \Big|_p \right).$$

To continue the computation, we need to choose a vector field $X_i \in \mathfrak{X}(M)$ such that $(X_i)_p = \frac{\partial}{\partial x^i}\Big|_p$. It is convenient to choose X_i in a more precise way. Namely, we choose X_i such that it agrees with $\frac{\partial}{\partial x^i}$ in an open neighborhood $V \subseteq U$ of p_0. Such vector field exists in view of the Local Extension Lemma for Vector Fields, and we have

$$(s_\theta)_i(p) = \theta_p \left(\frac{\partial}{\partial x^i} \Big|_p \right) = \theta(X_i)(p)$$

for all $p \in V$. Clearly, $\theta(X_i)$ depends smoothly on p so that $(s_\theta)_i|_V$ is smooth.

We leave it as Exercise 7.4 (Point (1)) to check that the map (7.4) is $C^\infty(M)$-linear. It remains to check that (7.4) is invertible. To do this, we define its inverse. Given a section $s \in \Gamma(T^*M)$ of T^*M, we construct a 1-form $\theta_s \in \Omega^1(M)$ in such a way that

$$s_{\theta_s} = s \quad \text{and} \quad \theta_{s_\theta} = \theta, \tag{7.5}$$

for all $s \in \Gamma(T^*M)$, and all $\theta \in \Omega^1(M)$. For every $X \in \mathfrak{X}(M)$, we define

$$\theta_s(X) : M \to \mathbb{R}, \quad p \mapsto \theta_s(X)(p) := s(p)(X_p).$$

We have to show that

- $\theta_s(X) \in C^\infty(M)$ for all $s \in \Gamma(T^*M)$ and all $X \in \mathfrak{X}(M)$,
- $\theta_s : \mathfrak{X}(M) \to C^\infty(M)$ is a $C^\infty(M)$-linear map, hence a differential 1-form, and
- the identities (7.5) hold for all $s \in \Gamma(T^*M)$ and all $\theta \in \Omega^1(M)$.

We only discuss the first point and leave it to the reader to prove the remaining two points as Exercise 7.4 (Points (2) and (3)). Let $s \in \Gamma(T^*M)$ and $X \in \mathfrak{X}(M)$. In order to prove that $\theta_s(X)$ is smooth, we prove that the restriction $\theta_s(X)|_U$ is smooth for every chart $(U, \varphi = (x^1, \ldots, x^n))$. Now

$$X|_U = X^i \frac{\partial}{\partial x^i}$$

for some smooth functions X^i on U, and, for $p \in U$, we have

$$\theta_s(X)(p) = s(p)(X_p) = s(p)\left(X^i(p)\frac{\partial}{\partial x^i}|_p\right) = X^i(p)s_i(p) = (X^i s_i)(p).$$

As the s_i are smooth, the last expression depends smoothly on p. This concludes the proof. $\qquad\square$

Exercise 7.4. Complete the proof of Proposition 7.14 showing that

(1) The map $\theta \mapsto s_\theta$ is $C^\infty(M)$-linear,
(2) for every $s \in \Gamma(T^*M)$, the map $\theta_s : \mathfrak{X}(M) \to C^\infty(M)$ is $C^\infty(M)$-linear,
(3) the map $s \mapsto \theta_s$ inverts $\theta \mapsto s_\theta$.

As for vector fields, Proposition 7.14 has some important consequences. First of all, a 1-form θ on M can be "restricted" to an open submanifold $\mathcal{U} \subseteq M$ by restricting the associated section, exactly as for vector fields (we leave the details to the reader). We denote by $\theta|_\mathcal{U} \in \Omega^1(\mathcal{U})$ the restricted 1-form. Clearly, we have $\theta|_\mathcal{U}(X_\mathcal{U}) = \theta(X)|_\mathcal{U}$ for all $\theta \in \Omega^1(M)$ and all $X \in \mathfrak{X}(M)$. The second consequence of Proposition 7.14 is a Gluing Lemma for Differential 1-Forms. The statement and the proof are formally identical to that of the Gluing Lemma for Vector Fields. Finally, we can combine the Gluing Lemma for Differential 1-Forms with Example 7.9 to find that, for every 1-form θ on M,

(1) θ is completely determined by its restrictions $\theta|_U$ to some coordinate domains U covering M,
(2) for every chart $(U, \varphi = (x^1, \ldots, x^n))$, the restriction $\theta|_U$ can be uniquely written in the form

$$\theta|_U = \theta_i dx^i$$

for some smooth functions $\theta_i \in C^\infty(U)$, called the *components* of θ in the chart (U, φ),

(3) given a vector field $X \in \mathfrak{X}(M)$ locally given by

$$X|_U = X^i \frac{\partial}{\partial x^i},$$

we have

$$\theta(X)|_U = \theta|_U(X|_U) = \theta_i X^i.$$

The proof of Proposition 7.14 shows that the components θ_i of a 1-form in the chart $(U, \varphi = (x^1, \ldots, x^n))$ coincide with the components of the corresponding section. When the chart (U, φ) changes, the components of θ change. The *transformation rule* has been already discussed in Example 7.9.

Exercise 7.5. Consider $\mathbb{R}^2$ with standard coordinates x, y and the polar chart $(U, \psi = (r, \phi))$ on it. Recall that $U = \mathbb{R}^2 \smallsetminus \{y = 0, x \geq 0\}$, $\hat{U} = (0, +\infty) \times (0, 2\pi)$ and

$$\psi : U \to \hat{U}, \quad (x, y) \mapsto \psi(x, y) = (r, \phi) := \left(\sqrt{x^2 + y^2}, \phi(x, y) \right),$$

with

$$\phi(x, y) := \begin{cases} \arccos \dfrac{x}{\sqrt{x^2 + y^2}} & \text{if } y \geq 0 \\[3mm] -\arccos \dfrac{x}{\sqrt{x^2 + y^2}} & \text{if } y < 0 \end{cases}.$$

Recall also that the inverse map is

$$\psi^{-1} : \hat{U} \to U, \quad (r, \phi) \mapsto \psi^{-1}(r, \phi) = (r \cos \phi, r \sin \phi).$$

Consider the following differential 1-forms on $\mathbb{R}^2$:

$$\theta = x\,dy - y\,dx \quad \text{and} \quad \kappa = x\,dx + y\,dy.$$

Prove that

$$\theta|_U = r^2 d\phi \quad \text{and} \quad \kappa|_U = r\,dr.$$

Let $(U, \varphi = (x^1, \ldots, x^n))$ be a chart on M. Example 7.9 shows not only that the $C^\infty(U)$-module $\Omega^1(U)$ possesses a finite frame but also that it is generated by the differentials $dx^1, \ldots, dx^n$ of the coordinate functions

$x^1, \ldots, x^n$. In particular, it is generated by 1-forms of the type df for some $f \in C^\infty(U)$. The last fact is true even globally according to the following

Theorem 7.15. *Let M be a manifold. The $C^\infty(M)$-module $\Omega^1(M)$ of differential 1-forms on M is generated by differentials df of smooth functions $f \in C^\infty(M)$, i.e., every differential 1-form $\theta \in \Omega^1(M)$ can be written (in a possibly non-unique way) in the form*

$$\theta = f_1 dg_1 + \cdots + f_k dg_k$$

for some smooth functions $f_1, g_1, \ldots, f_k, g_k \in C^\infty(M)$.

Proof. The proof is non-trivial, in the sense that it is based on more involved results on smooth manifolds, and we omit it (but see Nestruev, 2020). □

7.3 Differential 1-Forms and Smooth Maps

Let M, N be manifolds and let $F : M \to N$ be a smooth map. While we can *move* tangent vectors from M to N via the tangent maps to F, there is no way to move vector fields from M to N, nor from N to M, unless F is a diffeomorphism. On the other hand, there is a natural way to move differential forms from N to M via a *pull-back construction* in many ways similar to the pull-back of functions.

Before defining the pull-back of differential forms along F, consider a point p in M and the tangent map

$$d_p F : T_p M \to T_{F(p)} N.$$

The *dual map* is denoted by

$$d_p^* F : T_{F(p)}^* N \to T_p^* M, \quad \theta \mapsto d_p^* F(\theta) := \theta \circ d_p F.$$

Example 7.16. Let $S \subseteq M$ be a submanifold, and let $p \in S$. The dual map

$$d_p^* i_S : T_p^* M \to T_p^* S$$

to the tangent map to the inclusion $i_S : S \hookrightarrow M$ is a surjection consisting in restricting a covector $\theta : T_p M \to \mathbb{R}$ on M at $p \in S$ to tangent vectors to S. ◆

Proposition 7.17 (Naturality of the Differential at a Point). *For every smooth function $f \in C^\infty(N)$, we have*

$$d_p (F^*(f)) = d_p^* F \left(d_{F(p)} f \right).$$

Proof.   Let $v \in T_pM$, then

$$d_p \left(F^*(f) \right)(v) = v \left(F^*(f) \right)$$
$$= d_pF(v)(f)$$
$$= d_{F(p)}f \left(d_pF(v) \right)$$
$$= d_p^*F \left(d_{F(p)}f \right)(v).$$ □

Now let $(U, \varphi = (x^1, \ldots, x^m))$ be a chart on M around p, and let $(V, \psi = (y^1, \ldots, y^n))$ be a chart on N around $F(p)$ such that $F(U) \subseteq V$. Put $P = \varphi(p)$. The representative matrix of the dual map $d_p^*F : T_{F(p)}^*N \to T_p^*M$ in the coordinate coframes

$$\left(d_{F(p)}y^1, \ldots, d_{F(p)}y^n \right) \quad \text{and} \quad \left(d_px^1, \ldots, d_px^m \right)$$

is the transpose of the representative matrix of the tangent map $d_pF :$ $T_pM \to T_{F(p)}N$ in the coordinate frames

$$\left(\frac{\partial}{\partial x^1}\Big|_p, \ldots, \frac{\partial}{\partial x^m}\Big|_p \right) \quad \text{and} \quad \left(\frac{\partial}{\partial y^1}\Big|_{F(p)}, \ldots, \frac{\partial}{\partial y^n}\Big|_{F(p)} \right).$$

Hence, it is

$$\left(\frac{\partial F^a}{\partial x^i}(p) \right)_{a=1,\ldots,n}^{i=1,\ldots,m} = \left(\frac{\partial \widehat{F}^a}{\partial t^i}(P) \right)_{a=1,\ldots,n}^{i=1,\ldots,m}.$$

Where, as usual, $F^a = F^*(y^a) = y^a \circ F$, $a = 1, \ldots, n$. In other words,

$$d_p^*F(d_{F(p)}y^a) = \frac{\partial F^a}{\partial x^i}(p)d_px^i = d_pF^a = d_pF^*(y^a),$$

for all $a \in \{1, \ldots, n\}$, which is duly consistent with Proposition 7.17. We now come to the main result of this section.

Proposition 7.18 (Pull-Back of 1-Forms). *Let M, N be smooth manifolds and let $F : M \to N$ be a smooth map. Then there exists a unique map*

$$F^* : \Omega^1(N) \to \Omega^1(M)$$

called the pull-back *of differential 1-forms along F such that, for all $\theta, \kappa \in \Omega^1(N)$, and all $f \in C^\infty(N)$,*

(1) $F^*(\theta + \kappa) = F^*(\theta) + F^*(\kappa)$,
(2) $F^*(f\theta) = F^*(f)F^*(\theta)$, and
(3) $F^*(df) = dF^*(f)$.

Proof. Let $\theta \in \Omega^1(N)$. We define $F^*(\theta) \in \Omega^1(M)$ through its values. So, for every $p \in M$, put

$$F^*(\theta)_p = d_p^* F(\theta_{F(p)}).$$

In other words, for every $v \in T_p M$,

$$F^*(\theta)_p(v) = \theta_{F(p)}(d_p F(v)).$$

This defines a section $s : p \mapsto F^*(\theta)_p$ of T^*M, and we have to show that it is a smooth section. To do this, we choose charts $(U, \varphi = (x^1, \ldots, x^m))$ and $(V, \psi = (y^1, \ldots, y^n))$ on M and N respectively, such that $F(U) \subseteq V$, and compute the components $s_i = s^*(p_i)$ of s in the chart (U, φ). We have

$$\theta|_V = \theta_a dy^a,$$

for some smooth functions $\theta_a \in C^\infty(V)$, $a = 1, \ldots, n$. Then, for every $p \in U$,

$$
\begin{aligned}
s_i(p) = s(p) \left(\frac{\partial}{\partial x^i}|_p \right) \\
= F^*(\theta)_p \left(\frac{\partial}{\partial x^i}|_p \right) \\
= \theta_{F(p)} \left(d_p F \left(\frac{\partial}{\partial x^i}|_p \right) \right) \\
= \theta_{F(p)} \left(\frac{\partial F^a}{\partial x^i}(p) \frac{\partial}{\partial y^a}|_{F(p)} \right) \\
= \frac{\partial F^a}{\partial x^i}(p) \theta_{F(p)} \left(\frac{\partial}{\partial y^a}|_{F(p)} \right) \\
= \frac{\partial F^a}{\partial x^i}(p) \theta_a(F(p)),
\end{aligned}
$$

for all $i \in \{1, \ldots, m\}$, where, as usual, $F^a = F^*(y^a) = y^a \circ F$, $a = 1, \ldots, n$. Both $\frac{\partial F^a}{\partial x^i}$ and $\theta_a \circ F = F^*(\theta_a)$ depend smoothly on $p \in U$, hence s_i is smooth. It follows from the arbitrariness of (U, φ) and (V, ψ) that s is smooth, and $F^*(\theta)$ is a well-defined 1-form on M.

We leave it as Exercise 7.6 to show that the map $F^* : \Omega^1(N) \to \Omega^1(M)$ enjoys Properties (1) and (2) in the statement, and we prove here Property (3). So, let $f \in C^\infty(N)$ and $p \in M$, and compute

$$F^*(df)_p = d_p^* F((df)_{F(p)})$$

$$= d_p^* F(d_{F(p)} f) \qquad \text{(Equation (7.3) in Exercise 7.3)}$$

$$= d_p(F^*(f)) \qquad \text{(Proposition 7.17)}$$

$$= (dF^*(f))_p. \qquad \text{(Equation (7.3) in Exercise 7.3)}$$

It follows from the arbitrariness of p that $F^*(df) = dF^*(f)$ as claimed.

It remains to show that Properties (1)–(3) uniquely define $F^* : \Omega^1(N) \to \Omega^1(M)$. To do this, we use Theorem 7.15. So, let $'F^* : \Omega^1(N) \to \Omega^1(M)$ be another map satisfying (1)–(3). Take an arbitrary $\theta \in \Omega^1(N)$. From Theorem 7.15, there exist $f_1, g_1, \ldots, f_k, g_k \in C^\infty(N)$ such that

$$\theta = f_1 dg_1 + \cdots + f_k dg_k.$$

Hence,

$$'F^*(\theta) = {}'F^*(f_1 dg_1 + \cdots + f_k dg_k)$$

$$= F^*(f_1)'F^*(dg_1) + \cdots + F^*(f_k)'F^*(dg_k)$$

$$= F^*(f_1)dF^*(g_1) + \cdots + F^*(f_k)dF^*(g_k)$$

$$= F^*(\theta).$$

This concludes the proof. $\qquad\qquad\qquad\qquad\qquad\qquad\qquad\qquad\qquad\square$

Exercise 7.6. Complete the proof of Proposition 7.18 showing that the map $F^* : \Omega^1(N) \to \Omega^1(M)$ enjoys Properties (1) and (2) in the statement.

Let $F : M \to N$ be a smooth map, and let (U, φ) and (V, ψ) be charts as in the proof of Proposition 7.18. It immediately follows from that proof that

$$F^*(\theta)|_U = F^*(\theta|_V) = F^*(\theta_a)\frac{\partial F^a}{\partial x^i}dx^i.$$

Example 7.19. Consider $\mathbb{R}^2$ with standard coordinates (u, v) and $\mathbb{R}^3$ with standard coordinates (x, y, z). Consider also the smooth map

$$F : \mathbb{R}^2 \to \mathbb{R}^3, \quad (u, v) \mapsto F(u, v) = (u^2, uv, v^2).$$

We want to compute the pull-back along F of the 1-form $\theta = y\,dx - x\,dy \in \Omega^1(\mathbb{R}^3)$:

$$F^*(\theta) = F^*(y\,dx - x\,dy) = F^*(y)dF^*(x) - F^*(x)dF^*(y)$$

$$= uv\,du^2 - u^2d(uv) = 2u^2v\,du - u^2v\,du - u^3\,dv = u^2v\,du - u^3\,dv.$$

$\blacklozenge$

Example 7.20 (Restriction of a 1-Form to a Submanifold). If $i_{\mathcal{U}} : \mathcal{U} \to M$ is the inclusion of an open submanifold, and $\theta \in \Omega^1(M)$ is a 1-form, then $i_{\mathcal{U}}^*(\theta) = \theta|_{\mathcal{U}}$ is the (usual) restriction of θ to $\mathcal{U}$.

More generally, we can "restrict" a 1-form $\theta \in \Omega^1(M)$ to *any* submanifold $S \subseteq M$ by pulling it back along the inclusion $i_S : S \hookrightarrow M$. The pull-back $i_S^*(\theta)$ is sometimes denoted by $\theta|_S$ and called the *restriction of θ to S*. In practice, it is obtained by restricting the values of θ at points of S to tangent vectors to S.

For instance, let M be $\mathbb{R}^3$ with standard coordinates (x, y, z), let $S = S^2 \subseteq \mathbb{R}^3$ be the 2-dimensional sphere, and let

$$\theta = x\,dx + y\,dy + z\,dz \in \Omega^1(\mathbb{R}^3).$$

We want to show that $\theta|_{S^2} = 0$. We do this in two different ways. First of all, let $P = (a, b, c) \in S^2$ and let

$$v = A\frac{\partial}{\partial x}\Big|_P + B\frac{\partial}{\partial y}\Big|_P + C\frac{\partial}{\partial z}\Big|_P \in T_PS^2.$$

This means that

$$aA + bB + cC = 0.$$

Hence,

$$\theta_P(v) = (ad_Px + bd_Py + cd_Pz)\left(A\frac{\partial}{\partial x}\Big|_P + B\frac{\partial}{\partial y}\Big|_P + C\frac{\partial}{\partial z}\Big|_P\right)$$

$$= aA + bB + cC = 0.$$

It follows from the arbitrariness of P and v that $\theta|_{S^2} = 0$.

Alternatively, note that

$$\theta = \frac{1}{2}d\left(x^2 + y^2 + z^2\right).$$

Hence,

$$\theta|_{S^2} = i_{S^2}^*(\theta) = i_{S^2}^* \left(\frac{1}{2} d \left(x^2 + y^2 + z^2 \right) \right) = \frac{1}{2} d i_{S^2}^* \left(x^2 + y^2 + z^2 \right)$$

$$= \frac{1}{2} d1 = 0.$$

♦

Exercise 7.7. Consider $\mathbb{R}^2$ with standard coordinates (x, y), and let (U_+, φ_+) be the stereographic chart (from the north) on the circle $S^1 \subseteq \mathbb{R}^2$. Consider the 1-form

$$\theta = y\, dx - x\, dy.$$

Write $\theta|_{U_+}$ in terms of the stereographic coordinate X_+. (**Hint:** *Note that*

$$\theta|_{U_+} = y|_{U_+} d\left(x|_{U_+}\right) - x|_{U_+} d\left(y|_{U_+}\right)$$

$$= \left(y|_{U_+} \frac{\partial x|_{U_+}}{\partial X_+} - x|_{U_+} \frac{\partial y|_{U_+}}{\partial X_+} \right) dX_+.$$

Finally, use the explicit expressions of $x|_{U_+}, y|_{U_+}$ *in terms of* X_+.)

Chapter 8

Differential Forms and Cartan Calculus

In this chapter, we define higher-degree differential forms. Differential forms are of great importance in modern Differential Geometry. Numerous interesting geometric structures (and physical systems) can be formalized via differential forms, and they play a key role in Algebraic Topology: the branch of Mathematics that uses algebraic constructions to study topological properties of spaces, in our case, smooth manifolds and related geometric objects (see, e.g., Bott and Tu, 1982). Additionally, differential forms serve as *integrands* in a coordinate-free *integration theory* (see Chapter 10). Finally, differential forms appear frequently in the mathematical formulation of physical laws (see, e.g., Nakahara, 2003). The basic rules of the calculus with vector fields and differential forms are collectively called *Cartan calculus*, after the French mathematician *Èlie Cartan* who first defined higher-degree differential forms as we know them in modern Differential Geometry (Cartan, 1899).

8.1 Algebraic Preliminaries: Alternating Forms

Let A be a real (associative, commutative, unital) algebra, let $\mathcal{M}$ be an A-module, and let k be a non-negative integer.

Definition 8.1 (Alternating Forms). An *alternating* (or *skew-symmetric*) *A-multilinear k-form* (or, simply, a *k-form*) on $\mathcal{M}$ is a map

$$\omega : \underbrace{\mathcal{M} \times \cdots \times \mathcal{M}}_{k \text{ times}} \to A$$

such that

- ω is A-multilinear,
- ω is alternating,

i.e., $\omega(\mu_1,\ldots,\mu_k)=0$ whenever two arguments among $\mu_1,\ldots,$ $\mu_k \in \mathcal{M}$ coincide. When $k = 0$, we interpret this definition requiring ω to be simply an element in A.

Remark 8.2. An alternating k-form $\omega : \mathcal{M} \times \cdots \times \mathcal{M} \to A$ is automatically skew-symmetric, i.e., for all $\mu_1,\ldots,\mu_k \in \mathcal{M}$ and for all permutations $\sigma \in S_k$,

$$\omega(\mu_{\sigma(1)},\ldots,\mu_{\sigma(k)}) = (-)^{\sigma}\omega(\mu_1,\ldots,\mu_k),$$

where we denoted by $(-)^{\sigma}$ the sign of the permutation σ. As we are working over a field of 0 characteristic, the converse is also true, i.e., *skew-symmetric forms are alternating.* $\diamond$

Example 8.3. Let $A = \mathbb{R}$ and $\mathcal{M} = \mathbb{R}^n$. The determinant

$$\det : \underbrace{\mathbb{R}^n \times \cdots \times \mathbb{R}^n}_{n \text{ times}} \to \mathbb{R}, \quad (A_1,\ldots,A_n) \mapsto \det(A_1 \cdots A_n)$$

is an n-form on $\mathbb{R}^n$.   ◆

We denote by $\mathrm{Alt}_k(\mathcal{M}, A)$ the space of alternating k-forms on $\mathcal{M}$. In particular, $\mathrm{Alt}_0(\mathcal{M}, A) = A$ and $\mathrm{Alt}_1(\mathcal{M}, A) = \mathrm{Hom}(\mathcal{M}, A)$ (the dual module of $\mathcal{M}$). If $\omega \in \mathrm{Alt}_k(\mathcal{M}, A)$, we also denote $|\omega| = k$ and call it the *degree* of ω. It is easy to see that $\mathrm{Alt}_k(\mathcal{M}, A)$ is an A-module for every k. The module operations are the following: for every $\omega, \omega_1, \omega_2 \in \mathrm{Alt}_k(\mathcal{M}, A)$ and all $\alpha \in A$, the sum $\omega_1 + \omega_2$ is given by

$$(\omega_1 + \omega_2)(\mu_1,\ldots,\mu_k) = \omega_1(\mu_1,\ldots,\mu_k) + \omega_2(\mu_1,\ldots,\mu_k),$$

and the product $\alpha\omega$ is given by

$$(\alpha\omega)(\mu_1,\ldots,\mu_k) = \alpha\omega(\mu_1,\ldots,\mu_k),$$

for all $\mu_1,\ldots,\mu_k \in \mathcal{M}$. When k is a negative integer, we put $\mathrm{Alt}_k(\mathcal{M}, A) = 0$: the zero module. In other words, by definition, *negative degree alternating forms vanish.*

There are more operations on alternating forms. In order to define them, we need one more

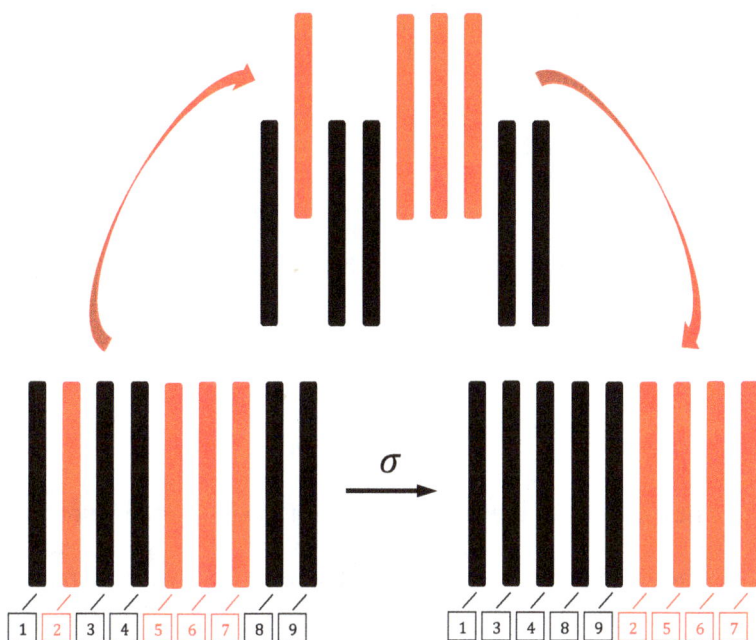

Figure 8.1. A $(5,4)$-unshuffle σ.

Definition 8.4 (Unshuffle). For any two non-negative integers k, l, a (k, l)-*unshuffle*, or simply an *unshuffle*, is a permutation $\sigma \in S_{k+l}$ such that

$$\sigma(1) < \cdots < \sigma(k) \quad \text{and} \quad \sigma(k+1) < \cdots < \sigma(k+l)$$

(see Figure 8.1 for an example). The subset in S_{k+l} consisting of (k, l)-unshuffles is denoted by $S_{k,l}$.

Now, for every k, l, there is a *wedge product* (also called *exterior product*):

$$\wedge : \mathrm{Alt}_k(\mathcal{M}, A) \times \mathrm{Alt}_l(\mathcal{M}, A) \to \mathrm{Alt}_{k+l}(\mathcal{M}, A), \quad (\omega, \rho) \mapsto \omega \wedge \rho. \tag{8.1}$$

The $(k+l)$-form $\omega \wedge \rho$ is defined by

$$(\omega \wedge \rho)(\mu_1, \ldots, \mu_{k+l})$$
$$= \sum_{\sigma \in S_{k,l}} (-)^{\sigma} \omega(\mu_{\sigma(1)}, \ldots, \mu_{\sigma(k)}) \rho(\mu_{\sigma(k+1)}, \ldots, \mu_{\sigma(k+l)}), \tag{8.2}$$

where, as indicated, the sum runs over the (k, l)-unshuffles. When $k = 0$, $\omega = \alpha$ is just an element in A, and we interpret Formula (8.2) by putting

$\omega \wedge \rho = \alpha \rho$. Similarly, when $l = 0$, $\rho = \beta \in A$, and we put $\omega \wedge \rho = \beta \omega$. Finally, when either k or l is negative, then either ω or ρ vanishes and we put $\omega \wedge \rho = 0$.

It is not too hard to see that $\omega \wedge \rho$ is indeed a well-defined $(k+l)$-form: the A-multilinearity of $\omega \wedge \rho$ immediately follows from the A-multilinearity of ω and ρ, and the skew-symmetry follows from both the skew-symmetry of ω and ρ and the sign $(-)^{\sigma}$ in the defining formula (8.2).

Remark 8.5. The wedge product $\omega \wedge \rho$ can be equivalently defined by the following formula:

$$(\omega \wedge \rho)(\mu_1, \ldots, \mu_{k+l})$$
$$= \frac{1}{k!l!} \sum_{\sigma \in S_{k+l}} (-)^{\sigma} \omega(\mu_{\sigma(1)}, \ldots, \mu_{\sigma(k)}) \rho(\mu_{\sigma(k+1)}, \ldots, \mu_{\sigma(k+l)}). \tag{8.3}$$

We stress that, in (8.3), the sum runs over *all* permutations, not just the unshuffles. However, as ω and ρ are alternating, hence skew-symmetric, several terms in (8.3) are counted several times, and we have to normalize by the factor $1/(k!l!)$. Using the unshuffles, instead, is a way to avoid such repetitions. Additionally, it makes it easier to prove several properties of the wedge product. ◇

Proposition 8.6. *The wedge product enjoys the following properties:*

- *(bilinearity) the wedge product is A-bilinear;*
- *(associativity) for any three alternating forms ω, ρ, τ,*

$$(\omega \wedge \rho) \wedge \tau = \omega \wedge (\rho \wedge \tau);$$

- *(graded commutativity) for any two alternating forms ω, ρ,*

$$\rho \wedge \omega = (-)^{|\omega||\rho|} \omega \wedge \rho.$$

Proof. The bilinearity should be clear. The associativity and the graded commutativity can be easily proved by playing with unshuffles. We omit the technical details. □

It immediately follows from the graded commutativity of the wedge product that, if ω is any alternating form of *odd* degree, then

$$\omega \wedge \omega = 0$$

(do you see it?).

The wedge products (8.1) can be combined into a single *interior opera-tion* in the *direct sum*

$$\text{Alt}_{\bullet}(\mathcal{M}, A) := \bigoplus_{k \in \mathbb{Z}} \text{Alt}_k(\mathcal{M}, A).$$

Before discussing this, we briefly recall what is the *direct sum* of a sequence of modules. So let $(\mathcal{N}_k)_{k \in \mathbb{Z}}$ be a sequence of A-modules. By def-inition, the *direct sum*

$$\mathcal{N}_{\bullet} = \bigoplus_{k \in \mathbb{Z}} \mathcal{N}_k \tag{8.4}$$

consists of sequences

$$(v_k)_{k \in \mathbb{Z}}, \quad v_k \in \mathcal{N}_k \quad \text{for all } k \in \mathbb{Z},$$

with the additional property that *only finitely many of the v_k are non-zero.* The direct sum $\mathcal{N}_{\bullet}$ is an A-module with the following two *term-wise* operations:

$$(v_k)_{k \in \mathbb{Z}} + (v'_k)_{k \in \mathbb{Z}} := (v_k + v'_k)_{k \in \mathbb{Z}}$$

and

$$\alpha (v_k)_{k \in \mathbb{Z}} := (\alpha v_k)_{k \in \mathbb{Z}},$$

for all $(v_k)_{k \in \mathbb{Z}}, (v'_k)_{k \in \mathbb{Z}} \in \mathcal{N}_{\bullet}$, and all $\alpha \in A$. Note that the *zero vector* in $\mathcal{N}_{\bullet}$ is the *zero sequence* $(0)_{k \in \mathbb{Z}}$, and the *opposite* of an element $(v_k)_{k \in \mathbb{Z}}$ is the *opposite sequence* $(-v_k)_{k \in \mathbb{Z}}$. A module of the form (8.4) is also called a *graded module.* If $A = \mathbb{R}$, we rather speak about a *graded vector space.*

Example 8.7. The real vector space $\mathbb{R}[t^1, \ldots, t^n]$ of real polynomials in the real indeterminates $(t^1, \ldots, t^n)$ is a graded vector space. Indeed, denote by $\mathbb{R}[t^1, \ldots, t^n]_k$ the space of homogeneous polynomials of degree k. We also put $\mathbb{R}[t^1, \ldots, t^n]_k = 0$ for $k < 0$. Then, $\mathbb{R}[t^1, \ldots, t^n]$ is (canonically isomorphic to) the direct sum of the $\mathbb{R}[t^1, \ldots, t^n]_k$:

$$\mathbb{R}[t^1, \ldots, t^n] = \bigoplus_{k \in \mathbb{Z}} \mathbb{R}[t^1, \ldots, t^n]_k$$

(do you see it?). ◆

The graded module $\mathcal{N}_\bullet = \bigoplus_{k \in \mathbb{Z}} \mathcal{N}_k$ comes with canonical monomorphisms of A-modules:

$$\iota_{k_0} : \mathcal{N}_{k_0} \to \mathcal{N}_\bullet, \quad v \mapsto \iota_{k_0}(v) := (v_k)_{k \in \mathbb{Z}},$$

where

$$v_k := \begin{cases} v & \text{if } k = k_0 \\ 0 & \text{if } k \neq k_0 \end{cases}.$$

The image of ι_{k_0} consists of sequences whose terms all vanish except, possibly, for the k_0-th one. If $(v_k)_{k \in \mathbb{Z}} \in \mathcal{N}_\bullet$ is a sequence whose only non-zero entries are $v_{k_1}, \ldots, v_{k_\ell}$, then, clearly,

$$(v_k)_{k \in \mathbb{Z}} = \iota_{k_1}(v_{k_1}) + \cdots + \iota_{k_\ell}(v_{k_\ell}). \tag{8.5}$$

In the following, we will always interpret the $\mathcal{N}_k$ as subspaces of $\mathcal{N}$, identifying them with their images under ι_k (if we do so, then $\mathcal{N}_\bullet$ is exactly the direct sum of its subspaces $\mathcal{N}_k$, in the usual sense of the direct sum of subspaces). For instance, rather than (8.5), we will simply write

$$(v_k)_{k \in \mathbb{Z}} = v_{k_1} + \cdots + v_{k_\ell}.$$

In this way, $\mathcal{N}_\bullet$ will equivalently consist of finite sums of elements in the $\mathcal{N}_k$.

An element v of $\mathcal{N}_\bullet$ in (the image of) $\mathcal{N}_k$ (under ι_k) is called *homogeneous*, the integer k is called the *degree* of v, and we write $|v| = k$. So, $\mathcal{N}_\bullet$ is generated by homogeneous elements.

Now, we go back to alternating forms. So, let $\mathcal{M}$ be an A-module. The $\mathrm{Alt}_k(\mathcal{M}, A)$ will play the role of $\mathcal{N}_k$. Elements in the graded module

$$\mathrm{Alt}_\bullet(\mathcal{M}, A) = \bigoplus_{k \in \mathbb{Z}} \mathrm{Alt}_k(\mathcal{M}, A)$$

will be also referred to as *alternating forms*. *Homogeneous forms* are then those with a "definite number of entries".

Proposition 8.8 (Wedge Product). *There exists a unique $\mathbb{R}$-bilinear map*

$$\wedge : \mathrm{Alt}_\bullet(\mathcal{M}, A) \times \mathrm{Alt}_\bullet(\mathcal{M}, A) \to \mathrm{Alt}_\bullet(\mathcal{M}, A), \quad (\omega, \rho) \mapsto \omega \wedge \rho \tag{8.6}$$

extending the wedge products (8.1), also called the wedge product. *The wedge product (8.6) enjoys the following properties:*

- (bilinearity) *the wedge product is A-bilinear;*
- (associativity) *for any three alternating forms* ω, ρ, τ,

$$(\omega \wedge \rho) \wedge \tau = \omega \wedge (\rho \wedge \tau);$$

- (graded commutativity) *for any two* homogeneous *alternating forms* ω, ρ,

$$\rho \wedge \omega = (-)^{|\omega||\rho|} \omega \wedge \rho;$$

- (unitality) *for any alternating form* ω

$$1 \wedge \omega = \omega \wedge 1 = \omega.$$

Proof. Let $\omega = (\omega_k)_{k \in \mathbb{Z}}$ and $\rho = (\rho_l)_{l \in \mathbb{Z}}$ be two (non-necessarily homogeneous) alternating forms. By definition, their wedge product is

$$\omega \wedge \rho = (\tau_m)_{m \in \mathbb{Z}},$$

with

$$\tau_m := \sum_{k+l=m} \omega_k \wedge \rho_l,$$

where, in the last formula, we used the wedge products (8.1). It is easy to see that the operation in $\text{Alt}_\bullet(\mathcal{M}, A)$ so defined enjoys all the required properties. We leave the simple details to the reader. $\qquad \square$

The vector space $\text{Alt}_\bullet(\mathcal{M}, A)$ of alternating forms, together with the wedge product (8.6), is an important instance of a *graded algebra*.

Definition 8.9 (Graded Algebra). A *real, associative, graded commutative algebra with unit*, or simply a *graded* $\mathbb{R}$-*algebra*, or just a *graded algebra*, is a graded vector space

$$(\mathcal{A}_\bullet = \bigoplus_{k \in \mathbb{Z}} \mathcal{A}_k, +, \cdot)$$

equipped with an additional interior operation $\wedge : \mathcal{A}_\bullet \times \mathcal{A}_\bullet \to \mathcal{A}_\bullet$, such that

(1) $\wedge$ is $\mathbb{R}$-bilinear,
(2) $\wedge$ is associative,

in particular, $(\mathcal{A}_\bullet, +, \wedge)$ is a ring and, additionally,

(3) $\wedge$ *preserves the degree*, i.e., for every two homogeneous elements $\alpha, \beta \in \mathcal{A}_\bullet$, the product $\alpha \wedge \beta$ is also homogeneous and

$$|\alpha \wedge \beta| = |\alpha| + |\beta|,$$

(4) $\wedge$ is *graded commutative*, i.e., for every two homogeneous elements $\alpha, \beta \in \mathcal{A}_\bullet$,

$$\beta \wedge \alpha = (-)^{|\alpha||\beta|}\alpha \wedge \beta,$$

(5) $\wedge$ possesses a *unit* $1 \in \mathcal{A}_0$.

If $\mathcal{A}_\bullet$ and $\mathcal{B}_\bullet$ are graded algebras, a *graded algebra homomorphism* between $\mathcal{A}_\bullet$ and $\mathcal{B}_\bullet$ is an $\mathbb{R}$-linear map $H : \mathcal{A}_\bullet \to \mathcal{B}_\bullet$ such that

(I) *H preserves the degree*, i.e., for every homogeneous element $\alpha \in \mathcal{A}_\bullet$, the image $H(\alpha) \in \mathcal{B}_\bullet$ is also homogeneous of the same degree:

$$|H(\alpha)| = |\alpha|,$$

(II) *H preserves the product*, i.e., for every two elements $\alpha, \beta \in \mathcal{A}_\bullet$,

$$H(\alpha \wedge \beta) = H(\alpha) \wedge H(\beta),$$

(III) *H preserves the unit*, i.e.,

$$H(1) = 1.$$

As usual, the identity is a graded algebra homomorphism, and the composition of graded algebra homomorphisms is a graded algebra homomorphism.

Theorem 8.10. *Let A be an algebra, let $\mathcal{M}$ be an A-module possessing a finite frame*

$$(\epsilon_1, \ldots, \epsilon_n),$$

and let

$$(\epsilon^1, \ldots, \epsilon^n)$$

be the dual frame of $\mathrm{Hom}(\mathcal{M}, A) = \mathrm{Alt}_1(\mathcal{M}, A)$. *Then, for every* $0 \leq k \leq n$, *the A-module* $\mathrm{Alt}_k(\mathcal{M}, A)$ *of alternating k-forms on $\mathcal{M}$ is generated by*

$$\epsilon^{i_1} \wedge \cdots \wedge \epsilon^{i_k}, \quad i_1, \ldots, i_k \in \{1, \ldots, n\}.$$

More precisely, the k-forms

$$\left(\epsilon^{i_1} \wedge \cdots \wedge \epsilon^{i_k}\right)_{i_1 < \cdots < i_k}$$

ordered (for instance) lexicographically form a frame of $\mathrm{Alt}_k(\mathcal{M}, A)$ *of cardinality*

$$\binom{n}{k}.$$

On the other hand, for $k > n$, $\mathrm{Alt}_k(\mathcal{M}, A) = 0$.

In the hypothesis of Theorem 8.10, let $i_1, \ldots, i_k \in \{1, \ldots, n\}$ be any k-tuple of indexes, and consider the multiple wedge product

$$\epsilon^{i_1} \wedge \cdots \wedge \epsilon^{i_k}.$$

We note preliminarily that, from the graded commutativity of the wedge product, when at least two indexes among $i_1, \ldots, i_k$ coincide, then $\epsilon^{i_1} \wedge \cdots \wedge \epsilon^{i_k} = 0$. So, $\epsilon^{i_1} \wedge \cdots \wedge \epsilon^{i_k}$ can only be non-trivial when the i_j are all different, $j = 1, \ldots, k$, for instance, when $i_1 < \cdots < i_k$.

Proof of Theorem 8.10. The proof generalizes the proof of the existence of the dual frame and we only sketch it. It is enough to prove the "more precise version" of the statement. Let $k \geq 0$. First, we prove that the system

$$\left(\epsilon^{i_1} \wedge \cdots \wedge \epsilon^{i_k} \right)_{i_1 < \cdots < i_k} \tag{8.7}$$

generates $\text{Alt}_k(\mathcal{M}, A)$. So, let ω be a k-form. For every $1 \leq i_1 < \cdots < i_k \leq n$, put

$$b_{i_1 \cdots i_k} := \omega(\epsilon_{i_1}, \ldots, \epsilon_{i_k}) \in A.$$

We want to show that

$$\omega = \sum_{i_1 < \cdots < i_k} b_{i_1 \cdots i_k} \epsilon^{i_1} \wedge \cdots \wedge \epsilon^{i_k}. \tag{8.8}$$

Denote by ω' the right-hand side of (8.8). By multilinearity, it is enough to show that ω' acts as ω on elements in the frame $(\epsilon_1, \ldots, \epsilon_n)$. By skew-symmetry, it is enough to check that

$$\omega'(\epsilon_{j_1}, \ldots, \epsilon_{j_k}) = \omega(\epsilon_{j_1}, \ldots, \epsilon_{j_k}) = b_{j_1 \cdots j_k}$$

for all $1 \leq j_1 < \cdots < j_k \leq n$. So, let $1 \leq j_1 < \cdots < j_k \leq n$ and compute

$$\omega'(\epsilon_{j_1}, \ldots, \epsilon_{j_k})$$

$$= \left(\sum_{i_1 < \cdots < i_k} b_{i_1 \cdots i_k} \epsilon^{i_1} \wedge \cdots \wedge \epsilon^{i_k} \right) (\epsilon_{j_1}, \ldots, \epsilon_{j_k})$$

$$= \sum_{i_1 < \cdots < i_k} b_{i_1 \cdots i_k} \left(\epsilon^{i_1} \wedge \cdots \wedge \epsilon^{i_k} (\epsilon_{j_1}, \ldots, \epsilon_{j_k}) \right).$$

As both $i_1, \ldots, i_k$ and $j_1, \ldots, j_k$ are increasingly ordered, the only non-trivial contributions to

$$\epsilon^{i_1} \wedge \cdots \wedge \epsilon^{i_k} (\epsilon_{j_1}, \ldots, \epsilon_{j_k})$$

come from

$$\epsilon^{i_1}(\epsilon_{j_1}) \cdots \epsilon^{i_k}(\epsilon_{j_k}) = \delta^{i_1}_{j_1} \cdots \delta^{i_k}_{j_k}.$$

Hence,

$$\omega'(\epsilon_{j_1}, \ldots, \epsilon_{j_k})$$

$$= \sum_{i_1 < \cdots < i_k} b_{i_1 \cdots i_k} \left(\epsilon^{i_1} \wedge \cdots \wedge \epsilon^{i_k}(\epsilon_{j_1}, \ldots, \epsilon_{j_k}) \right)$$

$$= \sum_{i_1 < \cdots < i_k} b_{i_1 \cdots i_k} \delta^{i_1}_{j_1} \cdots \delta^{i_k}_{j_k} = b_{j_1 \cdots j_k} = \omega(\epsilon_{j_1}, \ldots, \epsilon_{j_k}).$$

It remains to show that the system (8.7) consists of linearly independent k-forms. So, let $b_{i_1 \cdots i_k} \in A$ be scalars such that

$$\omega' = \sum_{i_1 < \cdots < i_k} b_{i_1 \cdots i_k} \epsilon^{i_1} \wedge \cdots \wedge \epsilon^{i_k} = 0.$$

In particular, ω' vanishes on elements in the frame $(\epsilon_1, \ldots, \epsilon_n)$ and, for every $1 \le j_1 < \cdots < j_k \le n$, the same computation as above, reveals that

$$0 = \omega'(\epsilon_{j_1}, \ldots, \epsilon_{j_k}) = b_{j_1 \cdots j_k}.$$

This concludes the proof.   □

Theorem 8.10 says that, when $\mathcal{M}$ possesses a finite frame $(\epsilon^1, \ldots, \epsilon^n)$, then every k-form $\omega \in \mathrm{Alt}_k(\mathcal{M}, A)$ can be written in the form

$$\omega = a_{i_1 \cdots i_k} \epsilon^{i_1} \wedge \cdots \wedge \epsilon^{i_k} \tag{8.9}$$

for some $a_{i_1 \cdots i_k} \in A$, where the sum (omitted according to the Einstein convention) runs over *all* indexes $i_1, \ldots, i_k$. On the other hand, ω can be *uniquely* written in the form

$$\omega = \sum_{i_1 < \cdots < i_k} b_{i_1, \ldots, i_k} \epsilon^{i_1} \wedge \cdots \wedge \epsilon^{i_k}. \tag{8.10}$$

Note that the $a_{i_1 \cdots i_k}$ can be uniquely chosen in such a way that they enjoy the following skew-symmetry property:

$$a_{i_{\sigma(1)} \cdots i_{\sigma(k)}} = (-)^\sigma a_{i_1 \cdots i_k} \tag{8.11}$$

for every permutation $\sigma \in S_k$ (do you see it?). In this case, the relationship between the $a_{i_1 \cdots i_k}$ and the $b_{i_1 \cdots i_k}$ is easily obtained by appropriately reordering the factors in (8.9), and it is

$$b_{i_1 \cdots i_k} = k! a_{i_1 \cdots i_k} = \omega(\epsilon_{i_1}, \ldots, \epsilon_{i_k}), \quad 1 \le i_1 < \cdots < i_k \le n.$$

It is sometimes convenient to expand ω as in (8.9) (with skew-symmetric coefficients $a_{i_1 \cdots i_k}$) rather than as in (8.10).

Corollary 8.11 (Alternating Forms on a Module with a Finite Frame).
Let A be an algebra, let $\mathcal{M}$ be an A-module possessing a finite frame

$$(\epsilon_1, \dots, \epsilon_n),$$

and let

$$(\epsilon^1, \dots, \epsilon^n)$$

be the dual frame of $\operatorname{Hom}(\mathcal{M}, A) = \operatorname{Alt}_1(\mathcal{M}, A)$. *Then, the graded $\mathbb{R}$-algebra* $\operatorname{Alt}_{\bullet}(\mathcal{M}, A)$ *of alternating forms is* generated *by*

(1) *0-forms, i.e., elements in* $\operatorname{Alt}_0(\mathcal{M}, A) = A$, *and*
(2) *the 1-forms* $\epsilon^1, \dots, \epsilon^n$.

This means that every form $\omega \in \operatorname{Alt}_{\bullet}(\mathcal{M}, A)$ *can be written in the form*

$$\omega = \sum_{k \in \mathbb{Z}} a_{i_1 \cdots i_k} \epsilon^{i_1} \wedge \cdots \wedge \epsilon^{i_k}, \tag{8.12}$$

for some $a_{i_1 \cdots i_k} \in A$, *where only finitely many of the* $a_{i_1 \cdots i_k}$ *are non-zero.*

Proof. Obvious. □

Note that, from the above discussion, the $a_{i_1 \cdots i_k}$ can be uniquely chosen so that they enjoy the skew-symmetry property (8.11). Additionally, any form $\omega \in \operatorname{Alt}_{\bullet}(\mathcal{M}, A)$ can be uniquely written in the form

$$\omega = \sum_{k \in \mathbb{Z}} \sum_{i_1 < \cdots < i_k} b_{i_1 \cdots i_k} \epsilon^{i_1} \wedge \cdots \wedge \epsilon^{i_k},$$

where only finitely many of the $b_{i_1 \cdots i_k}$ are non-zero.

Example 8.12 (Vector Product). Let $A = \mathbb{R}$ and $\mathcal{M} = \mathbb{R}^3$. Let $(\epsilon^1, \epsilon^2, \epsilon^3)$ be the canonical frame in $(\mathbb{R}^3)^* \cong \mathbb{R}^3$, and consider two 1-forms

$$A = A_1 \epsilon^1 + A_2 \epsilon^2 + A_3 \epsilon^3 \quad \text{and} \quad B = B_1 \epsilon^1 + B_2 \epsilon^2 + B_3 \epsilon^3.$$

The space $\operatorname{Alt}_2(\mathbb{R}^3, \mathbb{R})$ of 2-forms on $\mathbb{R}^3$ is 3-dimensional, and it is spanned by

$$\epsilon^2 \wedge \epsilon^3, \quad \epsilon^3 \wedge \epsilon^1, \quad \epsilon^1 \wedge \epsilon^2.$$

A direct computation shows that the wedge product $A \wedge B$ is given by

$$A \wedge B = \begin{vmatrix} A_2 & A_3 \\ B_2 & B_3 \end{vmatrix} \epsilon^2 \wedge \epsilon^3 - \begin{vmatrix} A_1 & A_3 \\ B_1 & B_3 \end{vmatrix} \epsilon^3 \wedge \epsilon^1 + \begin{vmatrix} A_1 & A_2 \\ B_1 & B_2 \end{vmatrix} \epsilon^1 \wedge \epsilon^2,$$

whose coefficients are the components of the vector product of the coordinate vectors (A_1, A_2, A_3) and (B_1, B_2, B_3). This shows that the wedge product generalizes the standard vector product in $\mathbb{R}^3$. ◆

8.2 Higher-Degree Differential Forms

Let M be a manifold, let $p \in M$ be a point, and let $(U, \varphi = (x^1, \ldots, x^n))$ be a chart on M around p. We will apply the language developed in the previous section to two cases:

(1) $A = \mathbb{R}$ and $\mathcal{M} = T_pM$,
(2) $A = C^\infty(M)$ and $\mathcal{M} = \mathfrak{X}(M)$.

We begin with the first one. Let k be a non-negative integer.

Definition 8.13 (k-Covector). The *space of k-covectors* of M at the point p is the space

$$\wedge^k T_p^* M := \mathrm{Alt}_k(T_pM, \mathbb{R})$$

of alternating k-forms on the tangent space T_pM. Forms in $\wedge^k T_p^* M$ are *k-covectors*. In other words, a k-covector at the point p is an $\mathbb{R}$-multilinear, alternating map:

$$\omega : \underbrace{T_pM \times \cdots \times T_pM}_{k \text{ times}} \to \mathbb{R}.$$

The frame

$$\left(d_p x^{i_1} \wedge \cdots \wedge d_p x^{i_k} \right)_{i_1 < \cdots < i_k}$$

of $\wedge^k T_p^* M$ is called the *coordinate frame*.

Note that 0-covectors are just real numbers, while 1-covectors are just covectors.

Every k-covector $\omega \in \wedge^k T_p^* M$ can be uniquely written as

$$\omega = a_{i_1 \cdots i_k} d_p x^{i_1} \wedge \cdots \wedge d_p x^{i_k},$$

for some $a_{i_1 \cdots i_k} \in \mathbb{R}$ with the skew-symmetry property (8.11), where the (omitted) sum runs over *all* multi-indexes. The k-covector ω can also be uniquely written as

$$\omega = \sum_{i_1 < \cdots < i_k} b_{i_1 \cdots i_k} d_p x^{i_1} \wedge \cdots \wedge d_p x^{i_k},$$

for some $b_{i_1 \cdots i_k} \in \mathbb{R}$ related to the $a_{i_1 \cdots i_k}$ by the formula

$$b_{i_1 \cdots i_k} = k! a_{i_1 \cdots i_k} = \omega \left(\frac{\partial}{\partial x^{i_1}} \Big|_p, \ldots, \frac{\partial}{\partial x^{i_k}} \Big|_p \right), \quad i_1 < \cdots < i_k.$$

Remark 8.14. Let $\mathcal{U} \subseteq M$ be an open neighborhood of p. Then, $\wedge^k T_p^* \mathcal{U}$ can be identified with $\wedge^k T_p^* M$ in the obvious way, and we will always do this. $\diamondsuit$

As already for tangent vectors and covectors, k-covectors can be organized into a manifold called the *bundle of k-covectors of M* and denoted $\wedge^k T^* M$, defined as

$$\wedge^k T^* M := \coprod_{p \in M} \wedge^k T_p^* M = \left\{ (p, \omega) : p \in M \text{ and } \omega \in \wedge^k T_p^* M \right\}.$$

As usual, a point $(p, \omega) \in \wedge^k T^* M$ will be often simply denoted by ω. The set $\wedge^k T^* M$ comes with an obvious surjection $\pi : \wedge^k T^* M \to M$, $(p, \omega) \mapsto p$, and the pair $(\wedge^k T^* M, \pi)$ is the *bundle of k-covectors of M*. The smooth structure on M induces an atlas on $\wedge^k T^* M$. To see this, begin with a chart $(U, \varphi = (x^1, \ldots, x^n))$ on M and define a chart $(\wedge^k T^* U, \wedge^k T^* \varphi)$ on $\wedge^k T^* M$ as follows. First of all, put

$$\wedge^k T^* U := \pi^{-1}(U) = \coprod_{p \in U} \wedge^k T_p^* M = \left\{ (p, \omega) \in \wedge^k T^* M : p \in U \right\}.$$

Next, put

$$N(k, n) = \binom{n}{k},$$

and define a map:

$$\wedge^k T^* \varphi : \wedge^k T^* U \to \widehat{U} \times \mathbb{R}^{N(k,n)}$$

by

$$(p, \omega) \mapsto \wedge^k T^* \varphi(p, \omega) := \left(\varphi(p); \left(b_{i_1 \cdots i_k} \right)_{i_1 < \cdots < i_k} \right),$$

where

$$b_{i_1 \cdots i_k} := \omega \left(\frac{\partial}{\partial x^{i_1}} \Big|_p, \ldots, \frac{\partial}{\partial x^{i_k}} \Big|_p \right), \quad i_1 < \cdots < i_k.$$

Recall that the $b_{i_1 \cdots i_k}$ are the components of ω in the coordinate frame

$$\left(d_p x^{i_1} \wedge \cdots \wedge d_p x^{i_k} \right)_{i_1 < \cdots < i_k}.$$

Clearly, $(\wedge^k T^* U, \wedge^k T^* \varphi)$ is an $(n + N(k, n))$-dimensional chart on $\wedge^k T^* M$, called a *standard chart*. One can show, in a very similar way as for the cotangent bundle, that any two standard charts on $\wedge^k T^* M$ are compatible. So, standard charts form an atlas called the *standard atlas*. Exactly as in the case of the tangent and the cotangent bundles, the standard atlas induces a Hausdorff and II-countable topology on $\wedge^k T^* M$. So, $\wedge^k T^* M$ is a manifold, and one can show that $\pi : \wedge^k T^* M \to M$ is a smooth map. We leave the details to the reader. Summarizing, we have (more or less) proved the following

Proposition 8.15 (Smooth Structure on the Bundle of k-Covectors).
Standard charts on $\wedge^k T^* M$ form an atlas. With the associated smooth structure, $\wedge^k T^* M$ is a manifold of dimension

$$n + \binom{n}{k},$$

and $\pi : \wedge^k T^* M \to M$ is a smooth map.

Remark 8.16. Note that a point in the bundle $\wedge^0 T^* M$ of 0-covectors is simply a pair (p, a), where $p \in M$ and $a \in \mathbb{R}$. Accordingly, as a set, $\wedge^0 T^* M$ is simply the cross product $M \times \mathbb{R}$, and it is easy to see that the manifold structure on $\wedge^0 T^* M$ agrees with the product manifold structure on $M \times \mathbb{R}$. Additionally, the discussion preceding Proposition 8.15 shows that the bundle of 1-covectors, with its manifold structure, is exactly the cotangent bundle. $\diamond$

Definition 8.17 (Field of k-Covectors). A *section* of $\wedge^k T^* M$ is a map $s : M \to \wedge^k T^* M$ such that $\pi \circ s = \mathrm{id}_M$. In other words, a section s is the assignment of a k-covector $s(p)$ at p, for every point $p \in M$.

There is a natural $C^\infty(M)$-module structure on sections of $\wedge^k T^* M$ defined exactly by the same formulas as for the tangent and the cotangent bundle. The *zero section* is the section mapping a point p to the zero k-covector $0 \in \wedge^k T^*_p M$, for all $p \in M$. Finally, smooth sections form a submodule denoted $\Gamma(\wedge^k T^* M)$.

We now come to alternating forms on the $C^\infty(M)$-module $\mathfrak{X}(M)$ of vector fields. Let k be a non-negative integer.

Definition 8.18 (Differential k-Form). A *differential k-form* on M, or, simply, a *k-form*, is an alternating k-form

$$\omega : \underbrace{\mathfrak{X}(M) \times \cdots \times \mathfrak{X}(M)}_{k \text{ times}} \to C^\infty(M).$$

The $C^\infty(M)$-module of k-forms is denoted $\Omega^k(M)$. For $k < 0$, we also put $\Omega^k(M) = 0$. The graded algebra of alternating forms on $\mathfrak{X}(M)$ is denoted

$$\Omega^\bullet(M) = \bigoplus_{k \in \mathbb{Z}} \Omega^k(M).$$

Elements in $\Omega^\bullet(M)$ are called *differential forms*.

Differential 0-forms on M are just functions on M. By definition, negative degree differential forms vanish.

Example 8.19 (Basis k-Forms on a Coordinate Domain). Let $(U, \varphi = (x^1, \ldots, x^n))$ be a chart on M. As the $C^\infty(U)$-module $\mathfrak{X}(U)$ possesses a finite frame, the space $\Omega^k(U)$ of k-forms on U does also possess a finite frame, the *coordinate frame*:

$$\left(dx^{i_1} \wedge \cdots \wedge dx^{i_k} \right)_{i_1 < \cdots < i_k}.$$

Every k-form ω on U can be uniquely written as

$$\omega = a_{i_1 \cdots i_k} dx^{i_1} \wedge \cdots \wedge dx^{i_k},$$

for some $a_{i_1 \cdots i_k} \in C^\infty(U)$ with the skew-symmetry property (8.11), where the (omitted) sum runs over *all* multi-indexes. The k-form ω can also be uniquely written as

$$\omega = \sum_{i_1 < \cdots < i_k} b_{i_1 \cdots i_k} dx^{i_1} \wedge \cdots \wedge dx^{i_k},$$

for some $b_{i_1 \cdots i_k} \in C^\infty(U)$ related to the $a_{i_1 \cdots i_k}$ by the formula

$$b_{i_1 \cdots i_k} = k! a_{i_1 \cdots i_k} = \omega \left(\frac{\partial}{\partial x^{i_1}}, \ldots, \frac{\partial}{\partial x^{i_k}} \right), \quad i_1 < \cdots < i_k.$$

◆

Differential k-forms are equivalent to sections of the bundle of k-covectors. More precisely, there is a canonical $C^\infty(M)$-module isomorphism $\Omega^k(M) \cong \Gamma(\wedge^k T^*M)$. To see this, first note that a k-form $\omega \in \Omega^k(M)$ determines a k-covector $\omega_p \in \wedge^k T_p^*M$, for every point $p \in M$. In order to define ω_p, we need the degree k analog of Lemma 7.12.

Lemma 8.20. *Let $\omega \in \Omega^k(M)$, let $p_0 \in M$, and let $X_1, \ldots, X_k \in \mathfrak{X}(M)$. Then, the value*

$$\omega(X_1, \ldots, X_k)(p_0)$$

of the function $\omega(X_1, \ldots, X_k) \in C^\infty(M)$ at p_0 does only depend on (ω and) the values

$$(X_1)_{p_0}, \ldots, (X_k)_{p_0}$$

of the vector fields $X_1, \ldots, X_k$ at p_0.

Proof. The proof is essentially the same as that of Lemma 7.12 and so we omit it. □

Now, consider a differential k-form ω on M. For any point $p \in M$, we define a k-covector $\omega_p \in \wedge^k T_p^* M$ as follows. Let $v_1, \ldots, v_k \in T_p M$, and let $X_1, \ldots, X_k \in \mathfrak{X}(M)$ be vector fields such that $(X_i)_p = v_i$. The X_i exist in view of Lemma 7.13. Put

$$\omega_p(v_1, \ldots, v_k) := \omega(X_1, \ldots, X_k)(p).$$

It is easy to see, exactly as in the case of 1-covectors, that ω_p is a well-defined k-covector

$$\omega_p : \underbrace{T_p M \times \cdots \times T_p M}_{k \text{ times}} \to \mathbb{R}$$

called the *value of ω at p.*

Exercise 8.1. Let k, l be non-negative integers. Prove that, for every ω, $\omega' \in \Omega^k(M)$, every $\rho \in \Omega^l(M)$, and every $p \in M$,

$$(\omega + \omega')_p = \omega_p + \omega'_p,$$

$$(\omega \wedge \rho)_p = \omega_p \wedge \rho_p. \tag{8.13}$$

Remark 8.21. Consider Exercise 8.1. Note that when $l = 0$, then $\rho = f$ is a function. Then, Equation (8.13) says that

$$(f\omega)_p = f(p)\omega_p.$$

Additionally, if $(U, \varphi = (x^1, \ldots, x^n))$ is a chart around p, then the values at p of the k-forms in the coordinate frame are

$$(dx^{i_1} \wedge \cdots \wedge dx^{i_k})_p = d_p x^{i_1} \wedge \cdots \wedge d_p x^{i_k}, \quad i_1 < \cdots < i_k.$$

$\diamondsuit$

The assignment $p \mapsto \omega_p$ is a section s_ω of the bundle of k-covectors $\wedge^k T^* M$:

$$s_\omega : M \to \wedge^k T^* M, \quad p \mapsto s_\omega(p) := \omega_p.$$

Proposition 8.22. *The map*

$$\Omega^k(M) \to \Gamma(\wedge^k T^* M), \quad \omega \mapsto s_\omega$$

is a well-defined $C^\infty(M)$-module isomorphism.

Proof. The proof is essentially the same as that of Proposition 7.14 and we omit it. □

As for vector fields and 1-forms, Proposition 8.22 has important consequences. Differential forms can be restricted to open submanifolds, by restricting the associated sections, exactly as vector fields and 1-forms, and we use the same notation $\omega|_{\mathcal{U}}$ for the restriction of a differential form ω to an open submanifold $\mathcal{U}$. Second, there is a(n obvious) Gluing Lemma for Differential Forms, and we leave it to the reader to state and prove it. Finally, given a differential k-form $\omega \in \Omega^k(M)$, we have the following:

(1) ω is completely determined by its restrictions $\omega|_U$ to some coordinate domains U covering M.
(2) For every chart $(U, \varphi = (x^1, \ldots, x^n))$, the restriction $\omega|_U$ (as already remarked) can be uniquely written as

$$\omega|_U = a_{i_1 \cdots i_k} dx^{i_1} \wedge \cdots \wedge dx^{i_k},$$

for some $a_{i_1 \cdots i_k} \in C^\infty(U)$ with the skew-symmetry property (8.11), where the (omitted) sum runs over *all* multi-indexes. The k-form $\omega|_U$ can also be uniquely written as

$$\omega|_U = \sum_{i_1 < \cdots < i_k} b_{i_1 \cdots i_k} dx^{i_1} \wedge \cdots \wedge dx^{i_k},$$

for some $b_{i_1 \cdots i_k} \in C^\infty(U)$.
(3) Given vector fields $X_1, \ldots, X_k \in \mathfrak{X}(M)$ locally given by

$$X_\alpha|_U = X_\alpha^i \frac{\partial}{\partial x^i}, \quad \alpha = 1, \ldots, k,$$

we have

$$\omega(X_1, \ldots, X_k)|_U = \omega|_U(X_1|_U, \ldots, X_k|_U)$$

$$= k! a_{i_1 \cdots i_k} X_1^{i_1} \cdots X_k^{i_k}$$

$$= \sum_{i_1 < \cdots < i_k} b_{i_1 \cdots i_k} X_1^{i_1} \cdots X_k^{i_k}.$$

(4) If $k > n = \dim M$, then $\omega|_U = 0$ for all charts (U, φ), hence $\omega = 0$, i.e., there are no non-trivial differential forms on M of degree greater than $\dim M$.

Exercise 8.2. Investigate how the local coordinate expression of a differential form changes under a change of coordinates.

Now, let $(U, \varphi = (x^1, \ldots, x^n))$ be a chart on M. From Example 8.19, the graded $\mathbb{R}$-algebra $\Omega^\bullet(U)$ is generated by functions $C^\infty(U)$ and the 1-forms $dx^1, \ldots, dx^n$. In particular, it is generated by functions and 1-forms of the type df for some $f \in C^\infty(U)$. This is true even globally according to the following analog of Theorem 7.15.

Theorem 8.23. *Let M be an n-dimensional manifold. The graded $\mathbb{R}$-algebra $\Omega^\bullet(M)$ is generated by functions $C^\infty(M)$ and differentials df of functions $f \in C^\infty(M)$, i.e., every differential form $\omega \in \Omega^k(M)$ can be written (in a possibly non-unique way) as a finite sum of forms of the type*

$$f\, dg_1 \wedge \cdots \wedge dg_k, \quad 0 \le k \le n,$$

where $f, g_1, \ldots, g_k \in C^\infty(M)$.

Proof.   Omitted (for a detailed proof based on the algebraic approach to calculus on manifolds, see Nestruev, 2020).   □

We conclude this section showing that the pull-back of functions and differential 1-forms along a smooth map $F : M \to N$ extends uniquely to a homomorphism of graded algebras

$$F^* : \Omega^\bullet(N) \to \Omega^\bullet(M),$$

also called the *pull-back*, according to the following.

Proposition 8.24 (Pull-Back of Differential Forms). *Let M, N be smooth manifolds and let $F : M \to N$ be a smooth map. Then, there exists a unique homomorphism of graded algebras*

$$F^* : \Omega^\bullet(N) \to \Omega^\bullet(M)$$

such that, on functions and 1-forms, F^* agrees with the pull-backs already defined.

Proof.   First of all, let $\omega \in \Omega^k(N)$ be a homogeneous differential form (of degree k). When $k = 0$, then $\omega = f$ is a functions and we put $F^*(f) = f \circ F$. When $k > 0$, we define $F^*(\omega) \in \Omega^k(M)$ through its values as follows. For every $p \in M$, and every $v_1, \ldots, v_k \in T_pM$, we put

$$F^*(\omega)_p(v_1, \ldots, v_k) = \omega_{F(p)}(d_pF(v_1), \ldots, d_pF(v_k)). \tag{8.14}$$

In particular, when $k = 1$, we recover the pull-back of 1-forms already defined. For generic k, Formula (8.14) defines a section $s : p \mapsto F^*(\omega)_p$ of $\wedge^k T^*M$. One can prove that s is a smooth section in a very similar way as for 1-forms and we leave the details to the reader. In this way, we defined the pull-back of homogeneous forms. We extend the definition to possibly non-homogeneous forms by *additivity* in the obvious way, i.e., for $(\omega_k)_{k\in\mathbb{Z}} \in \Omega^\bullet(M)$ a generic form, we put

$$F^* \left((\omega_k)_{k\in\mathbb{Z}}\right) = (F^*(\omega_k))_{k\in\mathbb{Z}}. \tag{8.15}$$

Next, we have to prove that

$$F^* : \Omega^\bullet(N) \to \Omega^\bullet(M)$$

is $\mathbb{R}$-linear and preserves the wedge product of differential forms. We leave this easy check as Exercise 8.3. Finally, we have to prove uniqueness. This immediately follows from Theorem 8.23 (do you see it?). □

Exercise 8.3. Complete the proof of Proposition 8.24 showing that the pull-back

$$F^* : \Omega^\bullet(N) \to \Omega^\bullet(M)$$

defined via (8.14) and (8.15) is $\mathbb{R}$-linear and preserves the wedge product. (**Hint:** *To show that F^* is $\mathbb{R}$-linear, discuss first linear combinations of homogeneous differential forms of the same degree. To show that F^* preserves the wedge product, discuss first the wedge product of homogeneous forms (of possibly different degrees). To do this, use Exercise 8.1. Then, discuss the general case.*)

Exercise 8.4. Let $F : M \to N$ and $G : N \to Q$ be smooth maps between manifolds. Show that, for every $\omega \in \Omega^\bullet(Q)$,

$$\mathrm{id}_Q^*(\omega) = \omega$$

and

$$(G \circ F)^*(\omega) = F^*(G^*(\omega)).$$

Example 8.25. Consider $\mathbb{R}^2$ with standard coordinates (u, v) and $\mathbb{R}^3$ with standard coordinates (x, y, z). Consider also the smooth map

$$F : \mathbb{R}^2 \to \mathbb{R}^3, \quad (u, v) \mapsto F(u, v) = (u^2, uv, v^2)$$

and the differential 2-form

$$\omega = dx \wedge dy + dy \wedge dz + dz \wedge dx \in \Omega^2(\mathbb{R}^3).$$

We want to compute the pull-back of ω along F. We have

$$
\begin{aligned}
F^*(\omega) &= F^*(dx \wedge dy + dy \wedge dz + dz \wedge dx) \\
&= F^*(dx) \wedge F^*(dy) + F^*(dy) \wedge F^*(dz) + F^*(dz) \wedge F^*(dx) \\
&= dF^*(x) \wedge dF^*(y) + dF^*(y) \wedge dF^*(z) + dF^*(z) \wedge dF^*(y) \\
&= du^2 \wedge d(uv) + d(uv) \wedge dv^2 + dv^2 \wedge du^2 \\
&= (2udu) \wedge (udv + vdu) + (udv + vdu) \wedge (2vdv) + (2vdv) \wedge (2udu) \\
&= (2u^2 + 2v^2 - 4uv)du \wedge dv \\
&= 2(u - v)^2 du \wedge dv,
\end{aligned}
$$

where we used, among other things, that $du \wedge du = dv \wedge dv = 0$ and $dv \wedge du = -du \wedge dv$ (can you motivate any single step in the above computation?).   ◆

Example 8.26 (Restriction of a k-Form to a Submanifold). If $i_{\mathcal{U}} : \mathcal{U} \to M$ is the inclusion of an open submanifold, and $\omega \in \Omega^k(M)$ is a k-form, then $i_{\mathcal{U}}^*(\omega) = \omega|_{\mathcal{U}}$ is the restriction of ω to $\mathcal{U}$.

More generally, exactly as for 1-forms (and functions), we can "restrict" a generic k-form $\omega \in \Omega^k(M)$ to *any* submanifold $S \subseteq M$ by pulling it back along the inclusion $i_S : S \hookrightarrow M$. The pull-back $i_S^*(\omega)$ is sometimes denoted $\omega|_S$ and called the *restriction of ω to S*. In practice, it is obtained by restricting the values of ω at points of S to tangent vectors to S.   ◆

Exercise 8.5. Consider $\mathbb{R}^3$ with standard coordinates (x, y, z) and the differential 2-form

$$
\omega = zdx \wedge dy + xdy \wedge dz + ydz \wedge dx \in \Omega^2(\mathbb{R}^3).
$$

Let $(U_+, \varphi_+ = (X_+, Y_+))$ be the stereographic chart on the 2-dimensional sphere $S^2 \subseteq \mathbb{R}^3$. Compute

$$
\omega|_{U_+} = \omega|_{S^2}|_{U_+}
$$

in terms of the stereographic coordinates (X_+, Y_+).

8.3   Cartan Calculus

We conclude this chapter presenting some natural operations on differential forms, besides the wedge product. These operations and the associated computational rules are sometimes collectively called *Cartan calculus* in honor of Èlie Cartan, who first discovered (some of) them.

We begin with some algebraic preliminaries. Let $\mathcal{A}_\bullet = \bigoplus_{k\in\mathbb{Z}} \mathcal{A}_k$ be a graded algebra, and let $l \in \mathbb{Z}$ be an integer.

Definition 8.27 (Graded Derivation). A degree l, *graded derivation* of $\mathcal{A}_\bullet$, or, simply, a *graded derivation*, is an $\mathbb{R}$-linear map $\Delta : \mathcal{A}_\bullet \to \mathcal{A}_\bullet$ such that

(1) $\Delta(\mathcal{A}_k) \subseteq \mathcal{A}_{k+l}$, for all $k \in \mathbb{Z}$,
(2) Δ satisfies the following *graded Leibniz rule*: for all homogeneous elements $\alpha, \beta \in \mathcal{A}_\bullet$,

$$\Delta(\alpha \wedge \beta) = \Delta(\alpha) \wedge \beta + (-)^{l|\alpha|}\alpha \wedge \Delta(\beta).$$

The (possibly negative) integer l is called the *degree* of Δ and is also denoted by $|\Delta|$.

Proposition 8.28. *Degree l graded derivations of $\mathcal{A}_\bullet$ form a real vector space under the sum and the product by a scalar defined in the following (obvious) way:*

$$(\Delta + \Delta')(\alpha) := \Delta(\alpha) + \Delta'(\alpha),$$

$$(a\Delta)(\alpha) := a\left(\Delta(\alpha)\right),$$

for all degree l graded derivations Δ, Δ', all $a \in \mathbb{R}$, and all $\alpha \in \mathcal{A}_\bullet$.

Proof. A straightforward computation that we omit. $\qquad\square$

Now, let Δ, ∇ be graded derivations (of possibly different degrees), and let $\beta \in \mathcal{A}_\bullet$ be a homogeneous element. We define new operators:

$$\beta \wedge \Delta : \mathcal{A}_\bullet \to \mathcal{A}_\bullet \quad \text{and} \quad [\Delta, \nabla] : \mathcal{A}_\bullet \to \mathcal{A}_\bullet$$

by putting

$$(\beta \wedge \Delta)(\alpha) := \beta \wedge \Delta(\alpha),$$

$$[\Delta, \nabla](\alpha) := \Delta(\nabla(\alpha)) - (-)^{|\Delta||\nabla|}\nabla(\Delta(\alpha)).$$

Proposition 8.29. *Both $\beta \wedge \Delta$ and $[\Delta, \nabla]$ are graded derivations. Their degrees are*

$$|\beta \wedge \Delta| = |\beta| + |\Delta| \quad \text{and} \quad |[\Delta, \nabla]| = |\Delta| + |\nabla|.$$

Additionally, the bracket $[-,-]$

(1) *is* $\mathbb{R}$*-bilinear,*
(2) *is* graded skew-symmetric, *i.e.,*

$$[\nabla, \Delta] = -(-)^{|\Delta||\nabla|}[\Delta, \nabla]$$

for all graded derivations Δ, ∇,

(3) *satisfies the following* graded Jacobi identity:

$$[\Delta, [\nabla, \square]] + (-)^{|\Delta|(|\nabla|+|\square|)}[\nabla, [\square, \Delta]] + (-)^{|\square|(|\Delta|+|\nabla|)}[\square, [\Delta, \nabla]] = 0,$$

for all graded derivations $\Delta, \nabla, \square$.

Finally,

(4) *the wedge product of a homogeneous element of* $\mathcal{A}_\bullet$ *and a graded derivation is an* $\mathbb{R}$*-bilinear operation,*
(5)

$$(\beta \wedge \gamma) \wedge \Delta = \beta \wedge (\gamma \wedge \Delta),$$

(6)

$$[\Delta, \beta \wedge \nabla] = \Delta(\beta) \wedge \nabla + (-)^{|\Delta||\beta|}\beta \wedge [\Delta, \nabla],$$

for all graded derivations Δ, ∇ *and all homogeneous elements* $\beta, \gamma \in \mathcal{A}_\bullet$.

Proof.   A long but straightforward computation left as Exercise 8.6. $\square$

Exercise 8.6. Prove Proposition 8.29.

The derivation $[\Delta, \nabla]$ is called the *graded commutator* of Δ and ∇. Note that, while the graded commutator of graded derivations is a graded derivation, the usual commutator $\Delta \circ \nabla - \nabla \circ \Delta$ is *not* a graded derivation in general.

Remark 8.30. The reader has probably already understood that the wedge product in a graded algebra and the graded commutator of graded derivations are graded analogs of the product in an algebra and the usual commutator of usual derivations. These operations enjoy similar properties as their standard cousins. However, the "graded formulas" differ from the usual ones by signs depending on the degrees of the objects involved. There is a (mnemonic) *meta-mathematical rule* called the *Koszul sign rule* that might help the reader remembering these signs:

Koszul Sign Rule: *A graded formula differs from its standard (non-graded) analog by signs appearing every time two (homogeneous) graded objects swap. Namely, every time that two graded objects O, O' swap, there appears a sign $(-)^{|O||O'|}$.*

The reader is invited to check that the Koszul sign rule applies to all cases discussed so far: commutativity rule of the product, Leibniz rule, definition of the commutator, skew-symmetry of the commutator, Jacobi identity, etc. ◇

Proposition 8.31. *A graded derivation Δ of the graded algebra $\mathcal{A}_\bullet$ is completely determined by its action on homogeneous generators of $\mathcal{A}_\bullet$, i.e., if $\{\alpha_i\}_{i \in I}$ is a family of homogeneous generators of $\mathcal{A}_\bullet$ (indexed by some set I) and Δ' is another graded derivation of $\mathcal{A}_\bullet$ such that $|\Delta'| = |\Delta|$ and*

$$\Delta'(\alpha_i) = \Delta(\alpha_i), \quad \text{for all } i \in I,$$

then $\Delta = \Delta'$.

Proof. First of all, we clarify/recall what we mean by a *family of homogeneous generators* $\{\alpha_i\}_{i \in I}$ of $\mathcal{A}_\bullet$. This means that α_i are homogeneous elements of $\mathcal{A}_\bullet$, and every other element $\alpha \in \mathcal{A}_\bullet$ can be written as a finite sum of products of the form

$$\alpha_{i_1} \wedge \cdots \wedge \alpha_{i_k}, \quad i_1, \ldots, i_k \in I, \quad k \in \mathbb{Z}.$$

Now, assume Δ, Δ' are as in the statement, and let $\alpha \in \mathcal{A}_\bullet$. Then,

$$\alpha = \sum \alpha_{i_1} \wedge \cdots \wedge \alpha_{i_k} \quad i_1, \ldots, i_k \in I, \quad k \in \mathbb{Z},$$

and

$$\Delta'(\alpha) = \Delta'\left(\sum \alpha_{i_1} \wedge \cdots \wedge \alpha_{i_k}\right)$$

$$= \sum \sum_{j=1}^{k} (-)^{|\Delta'|(|\alpha_{i_1}| + \cdots |\alpha_{i_{j-1}}|)} \alpha_{i_1} \wedge \cdots \wedge \alpha_{i_{j-1}} \wedge \Delta'(\alpha_{i_j}) \wedge \alpha_{i_{j+1}} \wedge \cdots \wedge \alpha_{i_k}$$

$$= \sum \sum_{j=1}^{k} (-)^{|\Delta|(|\alpha_{i_1}| + \cdots |\alpha_{i_{j-1}}|)} \alpha_{i_1} \wedge \cdots \wedge \alpha_{i_{j-1}} \wedge \Delta(\alpha_{i_j}) \wedge \alpha_{i_{j+1}} \wedge \cdots \wedge \alpha_{i_k}$$

$$= \Delta(\alpha),$$

where we used the graded Leibniz rule. This concludes the proof. □

Theorem 8.32 (Interior Product). *Let M be a manifold, and let $X \in \mathfrak{X}(M)$ be a vector field on M. There exists a unique degree -1 graded derivation ι_X of the graded algebra $\Omega^\bullet(M)$ of differential forms on M such that*

- $\iota_X f = 0,$
- $\iota_X df = X(f),$

for all smooth functions $f \in C^\infty(M) = \Omega^0(M)$.

Proof.   First, we define ι_X on homogeneous forms ω. If $|\omega| = 0$, then $\omega = f \in C^\infty(M)$, and we put

$$\iota_X f := 0.$$

If $|\omega| = k > 0$, then

$$\iota_X \omega := \omega(X, -, \dots, -),$$

i.e., $\iota_X \omega$ is the $(k-1)$-form defined by

$$\iota_X \omega(X_1, \dots, X_{k-1}) := \omega(X, X_1, \dots, X_{k-1}), \qquad (8.16)$$

for all $X_1, \dots, X_{k-1} \in \mathfrak{X}(M)$. It is clear that $\iota_X \omega$ is a $(k-1)$-form, indeed the expression in the right-hand side of (8.16) is obviously $C^\infty(M)$-multilinear and alternating in the arguments $X_1, \dots, X_{k-1}$. Moreover, $\iota_X \omega$ is $\mathbb{R}$-linear (actually even $C^\infty(M)$-linear) in the argument ω. Now, we extend ι_X to all forms by additivity in the following obvious way: for all $(\omega_k)_{k \in \mathbb{Z}} \in \Omega^\bullet(M)$, we put

$$\iota_X(\omega_k)_{k \in \mathbb{Z}} = (\iota_X \omega_k)_{k \in \mathbb{Z}}.$$

The map

$$\iota_X : \Omega^\bullet(M) \to \Omega^\bullet(M)$$

defined in this way is $\mathbb{R}$-linear and maps k-forms to $(k-1)$-form. We now prove that it is a graded derivation (of degree -1). So, let $\omega \in \Omega^k(M)$ and $\rho \in \Omega^l(M)$. We assume $k, l > 0$ and we leave the simpler cases $k = 0$ or $l = 0$ to the reader. Let $X_1, \dots, X_{k+l-1} \in \mathfrak{X}(M)$ and compute

$$\iota_X(\omega \wedge \rho)(X_1, \dots, X_{k+l-1})$$
$$= \omega \wedge \rho(X, X_1, \dots, X_{k+l-1})$$
$$= \sum_{\sigma \in S_{k-1,l}} (-)^\sigma \omega(X, X_{\sigma(1)}, \dots, X_{\sigma(k-1)}) \rho(X_{\sigma(k)}, \dots, X_{\sigma(k+l-1)})$$
$$+ \sum_{\sigma \in S_{k,l-1}} (-)^\sigma (-)^k \omega(X_{\sigma(1)}, \dots, X_{\sigma(k)}) \rho(X, X_{\sigma(k)}, \dots, X_{\sigma(k+l-1)})$$

$$= \sum_{\sigma \in S_{k-1,l}} (-)^{\sigma} \iota_X \omega (X_{\sigma(1)}, \ldots, X_{\sigma(k-1)}) \rho (X_{\sigma(k)}, \ldots, X_{\sigma(k+l-1)})$$

$$+ (-)^k \sum_{\sigma \in S_{k,l-1}} (-)^{\sigma} \omega (X_{\sigma(1)}, \ldots, X_{\sigma(k)}) \iota_X \rho (X_{\sigma(k)}, \ldots, X_{\sigma(k+l-1)})$$

$$= \left((\iota_X \omega) \wedge \rho + (-)^{|\omega|} \omega \wedge \iota_X \rho \right) (X_1, \ldots, X_{k+l-1}).$$

Finally, let $f \in C^{\infty}(M)$ and compute

$$\iota_X df = df(X) = X(f).$$

This shows that ι_X satisfies all the required properties. Uniqueness immediately follows from Proposition 8.31 and Theorem 8.23. $\qquad\square$

The graded derivation ι_X is called the *insertion of X*, or *contraction with X*, or also *interior product with X*. We want to describe it in local coordinates. So, let $(U, \varphi = (x^1, \ldots, x^n))$ be a chart on M, let $\omega \in \Omega^k(M)$ be locally given by

$$\omega|_U = a_{i_1 \cdots i_k} dx^{i_1} \wedge \cdots \wedge dx^{i_k},$$

for some $a_{i_1 \cdots i_k} \in C^{\infty}(U)$ enjoying the usual skew-symmetry property (8.11), and let $X \in \mathfrak{X}(M)$ be locally given by

$$X|_U = X^i \frac{\partial}{\partial x^i}.$$

We want to show that

$$(\iota_X \omega)|_U = k a_{i_1 i_2 \cdots i_k} X^{i_1} dx^{i_2} \wedge \cdots \wedge dx^{i_k}.$$

First of all, it is easy to see using the definition of ι_X that $(\iota_X \omega)|_U = \iota_{X|_U} \omega|_U$ (do you see it?). Now, compute

$$\iota_{X|_U} \omega|_U$$

$$= \iota_{X|_U} \left(a_{i_1 \cdots i_k} dx^{i_1} \wedge \cdots \wedge dx^{i_k} \right)$$

$$= \sum_{j=1}^{k} (-)^{j-1} a_{i_1 \cdots i_k} dx^{i_1} \wedge \cdots \wedge \iota_X dx^{i_j} \wedge \cdots \wedge dx^{i_k}$$

$$= \sum_{j=1}^{k} (-)^{j-1} a_{i_1 \cdots i_k} dx^{i_1} \wedge \cdots \wedge X(x^{i_j}) \wedge \cdots \wedge dx^{i_k}$$

$$= \sum_{j=1}^{k} (-)^{j-1} a_{i_1 \cdots i_k} X^{i_j} dx^{i_1} \wedge \cdots \wedge \widehat{dx^{i_j}} \wedge \cdots \wedge dx^{i_k}$$

$$= \sum_{j=1}^{k} a_{i_j i_1 \cdots \widehat{i_j} \cdots i_k} X^{i_j} dx^{i_1} \wedge \cdots \wedge \widehat{dx^{i_j}} \wedge \cdots \wedge dx^{i_k} \quad \text{(reordering the indexes)}$$

$$= \sum_{j=1}^{k} a_{i_1 i_2 \cdots i_k} X^{i_1} dx^{i_2} \wedge \cdots \wedge dx^{i_k} \quad \text{(renaming the indexes)}$$

$$= k a_{i_1 i_2 \cdots i_k} X^{i_1} dx^{i_2} \wedge \cdots \wedge dx^{i_k}.$$

Example 8.33. On $\mathbb{R}^3$ with standard coordinates (x, y, z), consider the 2-form

$$\omega = x \, dy \wedge dz + y \, dz \wedge dx + z \, dx \wedge dy$$

and the vector field

$$X = x \frac{\partial}{\partial x} + y \frac{\partial}{\partial y} + z \frac{\partial}{\partial z}.$$

Compute the interior product of ω with X:

$$\begin{aligned}
\iota_X \omega &= \iota_X (x \, dy \wedge dz + y \, dz \wedge dx + z \, dx \wedge dy) \\
&= x \iota_X (dy \wedge dz) + y \iota_X (dz \wedge dx) + z \iota_X (dx \wedge dy) \\
&= x (\iota_X dy) \wedge dz - x \, dy \wedge \iota_X dz + y (\iota_X dz) \wedge dx \\
&\quad - y \, dz \wedge \iota_X dx + z (\iota_X dx) \wedge dy - z \, dx \wedge \iota_X dy \\
&= x X(y) dz - x X(z) dy + y X(z) dx \\
&\quad - y X(x) dz + z X(x) dy - z X(y) dx \\
&= xy \, dz - xz \, dy + yz \, dx - yx \, dz + zx \, dy - zy \, dx \\
&= 0.
\end{aligned}$$

◆

Proposition 8.34. *Let* $X, Y \in \mathfrak{X}(M)$ *and let* $f \in C^\infty(M)$. *Then,*

$$\begin{aligned}
\iota_{X+Y} &= \iota_X + \iota_Y, \\
\iota_{fX} &= f \iota_X.
\end{aligned}$$
(8.17)

Proof. Both the left-hand sides and the right-hand sides of (8.17) are degree -1 graded derivations of $\Omega^\bullet(M)$. In order to check that they agree, it is enough to check that they agree on generators, and this is an easy consequence of Theorem 8.32. We leave the details to the reader. $\qquad\square$

Theorem 8.35 (Lie Derivative). *Let M be a manifold and let $X \in \mathfrak{X}(M)$ be a vector field on M. There exists a unique degree 0 graded derivation $\mathcal{L}_X$ of the graded algebra $\Omega^\bullet(M)$ of differential forms on M such that*

- $\mathcal{L}_X f = X(f)$,
- $\mathcal{L}_X df = dX(f)$,

for all smooth functions $f \in C^\infty(M) = \Omega^0(M)$.

Proof. First, we define $\mathcal{L}_X$ on homogeneous forms ω. So, let $|\omega| = k$. Denote by $\{\Phi_t\}_t$ the flow of X. We define $\mathcal{L}_X \omega$ through its values as follows:

$$(\mathcal{L}_X \omega)_p := \frac{d}{dt}\Big|_{t=0} \Phi_t^*(\omega)_p, \tag{8.18}$$

for all $p \in M$. As already for the Lie derivative of a vector field, we have to show that Formula (8.18) defines correctly a smooth section of the bundle $\wedge^k T^* M$ of k-covectors. We work in local coordinates. So, fix a point $p_0 \in M$, and let $(U, \varphi = (x^1, \ldots, x^n))$ be a chart around p_0. Then,

$$X|_U = X^i \frac{\partial}{\partial x^i} \quad \text{and} \quad \omega|_U = a_{i_1 \cdots i_k} dx^{i_1} \wedge \cdots \wedge dx^{i_k},$$

for some smooth functions X^i, and $a_{i_1 \cdots i_k} \in C^\infty(U)$ (enjoying the usual skew-symmetry property). Additionally, let $J \subseteq \mathbb{R}$ be an open interval containing 0 and let (U_0, φ) be a subchart of (U, φ) around p_0 exactly as in the proof of Proposition 6.22, i.e., such that $\Phi(J \times U_0) \subseteq U$. Now, let $(t, p) \in J \times U_0$ and, using the same notation and formulas as in the proof of Proposition 6.22, compute

$$\Phi_t^*(\omega)_p = \Phi_t^*\left(a_{i_1 \cdots i_k} dx^{i_1} \wedge \cdots \wedge dx^{i_k}\right)_p$$

$$= \Phi_t^*(a_{i_1 \cdots i_k})(p) d_p \Phi_t^{i_1} \wedge \cdots \wedge d_p \Phi_t^{i_k}$$

$$= a_{i_1 \cdots i_k}(\Phi(t, p)) \frac{\partial \Phi^{i_1}}{\partial x^{j_1}}(t, p) \cdots \frac{\partial \Phi^{i_k}}{\partial x^{j_k}}(t, p) d_p x^{j_1} \wedge \cdots \wedge d_p x^{j_k}.$$

$$\tag{8.19}$$

The coefficients in the last expression depend smoothly on t, and we can take the derivative in (8.18) (see Exercise 3.7):

$$(\mathcal{L}_X\omega)_p = \frac{d}{dt}\big|_{t=0}\Phi_t^*(\omega)_p$$

$$= \frac{d}{dt}\big|_{t=0}\left(a_{i_1\cdots i_k}(\Phi(t,p))\frac{\partial\Phi^{i_1}}{\partial x^{j_1}}(t,p)\cdots\frac{\partial\Phi^{i_k}}{\partial x^{j_k}}(t,p)\right)d_px^{j_1}\wedge\cdots\wedge d_px^{j_k}.$$

The coefficients are all smooth in the argument p, and this shows that the section

$$p \mapsto (\mathcal{L}_X\omega)_p$$

of $\wedge^k T_p^* M$ is well defined and smooth. For future use, we write explicitly the coefficient labeled by $(j_1,\ldots,j_k)$. A direct computation exploiting (6.16) and (6.17) shows that

$$\frac{d}{dt}\big|_{t=0}\left(a_{i_1\cdots i_k}(\Phi(t,p))\frac{\partial\Phi^{i_1}}{\partial x^{j_1}}(t,p)\cdots\frac{\partial\Phi^{i_k}}{\partial x^{j_k}}(t,p)\right)$$

$$= \left(X^i\frac{\partial a_{j_1\cdots j_k}}{\partial x^i} + \frac{\partial X^i}{\partial x^{j_1}}a_{ij_2\cdots j_k} + \cdots + \frac{\partial X^i}{\partial x^{j_k}}a_{j_1\cdots j_{k-1}i}\right)(p). \tag{8.20}$$

We leave the computational details to the reader.

Next, we extend $\mathcal{L}_X$ to all forms by additivity as follows: for all $(\omega_k)_{k\in\mathbb{Z}} \in \Omega^\bullet(M)$, we put

$$\mathcal{L}_X(\omega_k)_{k\in\mathbb{Z}} = (\mathcal{L}_X\omega_k)_{k\in\mathbb{Z}}.$$

The map

$$\mathcal{L}_X : \Omega^\bullet(M) \to \Omega^\bullet(M)$$

defined in this way is $\mathbb{R}$-linear and maps k-forms to k-forms. We now prove that it is a graded derivation (of degree 0). So, let ω,ρ be homogeneous forms, let $p \in M$, and compute

$$(\mathcal{L}_X(\omega\wedge\rho))_p = \frac{d}{dt}\big|_{t=0}\Phi_t^*(\omega\wedge\rho)_p$$

$$= \frac{d}{dt}\big|_{t=0}(\Phi_t^*(\omega)\wedge\Phi_t^*(\rho))_p$$

$$= \frac{d}{dt}\big|_{t=0}\Phi_t^*(\omega)_p\wedge\Phi_t^*(\rho)_p.$$

Now, we apply Lemma 8.36 below to the case $B = \wedge$ to find

$$
\begin{aligned}
(\mathcal{L}_X(\omega \wedge \rho))_p &= \frac{d}{dt}\big|_{t=0}\Phi_t^*(\omega)_p \wedge \Phi_t^*(\rho)_p \\
&= \left(\frac{d}{dt}\big|_{t=0}\Phi_t^*(\omega)_p\right) \wedge \rho_p + \omega_p \wedge \left(\frac{d}{dt}\big|_{t=0}\Phi_t^*(\rho)_p\right) \\
&= (\mathcal{L}_X\omega)_p \wedge \rho_p + \omega_p \wedge (\mathcal{L}_X\rho)_p \\
&= ((\mathcal{L}_X\omega) \wedge \rho + \omega \wedge (\mathcal{L}_X\rho))_p \, .
\end{aligned}
$$

The graded Leibniz rule now follows from the arbitrariness of p.

Next, let $f \in C^\infty(M)$. Property $\mathcal{L}_X f = X(f)$ is provided by Proposition 6.19. In order to compute $\mathcal{L}_X df$, fix $p_0 \in M$, and let $(U, \varphi = (x^1, \ldots, x^n))$ and U_0 be as in the first part of the proof. Then, for all $p \in U_0$,

$$
\begin{aligned}
(\mathcal{L}_X df)_p &= \frac{d}{dt}\big|_{t=0}\Phi_t^*(df)_p \\
&= \frac{d}{dt}\big|_{t=0}d_p\Phi_t^*(f) \\
&= \frac{d}{dt}\big|_{t=0}\frac{\partial \Phi_t^*(f)}{\partial x^i}(p)d_px^i \\
&= \left(\frac{\partial^2}{\partial t\partial x^i}\big|_{t=0,q=p}f(\Phi(t,q))\right)d_px^i \\
&= \frac{\partial}{\partial x^i}\big|_p\frac{d}{dt}\big|_{t=0}\Phi_t^*(f)d_px^i \\
&= (dX(f))_p
\end{aligned}
$$

(can you reproduce all the above steps?) and, from the arbitrariness of p, we have $\mathcal{L}_X df = dX(f)$. Finally, uniqueness immediately follows from Proposition 8.31 and Theorem 8.23. □

Lemma 8.36. *Let V_1, V_2, V be finite-dimensional real vector spaces, and let*

$$
B : V_1 \times V_2 \to V
$$

be a bilinear map. Let $I \subseteq \mathbb{R}$ be an open interval. Then, for any two smooth curves $\gamma_1 : I \to V_1, \gamma_2 : I \to V_2,$

(1) *the map* $I \to V, t \mapsto B(\gamma_1(t), \gamma_2(t))$ *is a smooth curve and*
(2) *for all $t_0 \in I$, the following Leibniz rule holds:*

$$\frac{d}{dt}\Big|_{t=t_0} B(\gamma_1(t), \gamma_2(t))$$

$$= B\left(\frac{d}{dt}\Big|_{t=t_0}\gamma_1(t), \gamma_2(t_0)\right) + B\left(\gamma_1(t_0), \frac{d}{dt}\Big|_{t=t_0}\gamma_2(t)\right).$$

Proof. Left as Exercise 8.7. □

Exercise 8.7. Prove Lemma 8.36.

The graded derivation $\mathcal{L}_X$ is called the *Lie derivative along X*. We want to describe it in local coordinates. So, let $(U, \varphi = (x^1, \ldots, x^n))$ be a chart on M, let $\omega \in \Omega^k(M)$ be locally given by

$$\omega|_U = a_{i_1 \cdots i_k} dx^{i_1} \wedge \cdots \wedge dx^{i_k},$$

with the $a_{i_1 \cdots i_k}$ skew-symmetric, and let $X \in \mathfrak{X}(M)$ be locally given by

$$X|_U = X^i \frac{\partial}{\partial x^i}.$$

It immediately follows from (8.20) that

$$(\mathcal{L}_X\omega)|_U = \mathcal{L}_{X|_U}\omega|_U$$

$$= \left(X^i \frac{\partial a_{j_1 \cdots j_k}}{\partial x^i} + \frac{\partial X^i}{\partial x^{j_1}} a_{ij_2 \cdots j_k} + \cdots + \frac{\partial X^i}{\partial x^{j_k}} a_{j_1 \cdots j_{k-1}i}\right) dx^{j_1} \wedge \cdots \wedge dx^{j_k}.$$

$$(8.21)$$

Example 8.37. On the 3-dimensional standard Euclidean space with standard coordinates (x, y, z), consider the 2-form

$$\omega = (dx + 2zdy) \wedge dz$$

and the vector field

$$X = y\frac{\partial}{\partial x} - x\frac{\partial}{\partial y}.$$

Compute the Lie derivative of ω along X:

$$\mathcal{L}_X\omega = \mathcal{L}_X((dx + 2dy) \wedge dz)$$
$$= (\mathcal{L}_X(dx + 2dy)) \wedge dz + (dx + 2zdy) \wedge \mathcal{L}_X(dz)$$
$$= (dX(x) + 2X(z)dy + 2zdX(y)) \wedge dz + (dx + 2zdy) \wedge dX(z)$$
$$= (dy - 2zdx) \wedge dz.$$

♦

Proposition 8.38. *Let* $X, Y \in \mathfrak{X}(M)$ *and let* $f \in C^{\infty}(M)$. *Then,*

$$\mathcal{L}_{X+Y} = \mathcal{L}_X + \mathcal{L}_Y,$$
$$\mathcal{L}_{fX} = f\mathcal{L}_X + df \wedge \iota_X. \tag{8.22}$$

Proof. Both the left-hand sides and the right-hand sides of (8.22) are degree 0 graded derivations of $\Omega^{\bullet}(M)$. In order to check that they agree, it is enough to check that they agree on generators, and this is an easy consequence of Theorem 8.35. $\qquad\square$

Theorem 8.39 (Exterior Differential). *Let* M *be a manifold. There exists a unique degree 1 graded derivation d of the graded algebra* $\Omega^{\bullet}(M)$ *of differential forms on M such that*

- df *is exactly the differential of* f,
- $ddf = 0$,

for all smooth functions $f \in C^{\infty}(M) = \Omega^0(M)$.

Proof. First, we define d on homogeneous forms ω. If $|\omega| = 0$, then $\omega = f \in C^{\infty}(M)$, and, by definition, df is simply the usual differential of a function. If $|\omega| = k > 0$, then $d\omega$ is the $(k+1)$-form defined by the following *Chevalley–Eilenberg formula*:

$$d\omega(X_1, \ldots, X_{k+1})$$
$$= \sum_i (-)^{i+1} X_i \left(\omega(X_1, \ldots, \widehat{X_i}, \ldots, X_{k+1}) \right)$$
$$+ \sum_{i<j} (-)^{i+j} \omega([X_i, X_j], X_1, \ldots, \widehat{X_i}, \ldots, \widehat{X_j}, \ldots, X_{k+1}), \tag{8.23}$$

for all $X_1, \ldots, X_{k+1} \in \mathfrak{X}(M)$. It is clear that $d\omega$ is $\mathbb{R}$-multilinear. The fact that $d\omega$ is alternating easily follows from (the fact that ω is alternating and) the signs $(-)^i$ and $(-)^{i+j}$ in (8.23) and we leave the details to the reader. Next, we have to show that $d\omega$ is $C^{\infty}(M)$-multilinear. It is enough to show that $d\omega$ is $C^{\infty}(M)$-linear in the first argument. The $C^{\infty}(M)$-linearity in the other arguments then follows from skew-symmetry. So, we have to show that

$$d\omega(fX_1, X_2, \ldots, X_{k+1}) = fd\omega(X_1, X_2, \ldots, X_{k+1}),$$

for all $X_1, X_2, \ldots, X_{k+1} \in \mathfrak{X}(M)$ and all $f \in C^\infty(M)$. To see this, we compute

$$d\omega(fX_1, X_2, \ldots, X_{k+1})$$

$$= fX_1\left(\omega(X_2, \ldots, X_{k+1})\right)$$

$$+ \sum_{1 < i}(-)^{i+1} X_i\left(\omega(fX_1, X_2, \ldots, \widehat{X_i}, \ldots, X_{k+1})\right)$$

$$+ \sum_{1 < j}(-)^{j+1}\omega([fX_1, X_j], X_1, \ldots, \widehat{X_j}, \ldots, X_{k+1})$$

$$+ \sum_{1 < i < j}(-)^{i+j}\omega([X_i, X_j], fX_1, X_2, \ldots, \widehat{X_i}, \ldots, \widehat{X_j}, \ldots, X_{k+1})$$

$$= fX_1\left(\omega(X_2, \ldots, X_{k+1})\right) + \sum_{1 < i}(-)^{i+1} X_i\left(f\omega(X_1, X_2, \ldots, \widehat{X_i}, \ldots, X_{k+1})\right)$$

$$+ \sum_{1 < j}(-)^{j+1}\omega(f[X_1, X_j] - X_j(f)X_1, X_2, \ldots, \widehat{X_j}, \ldots, X_{k+1})$$

$$+ \sum_{1 < i < j}(-)^{i+j}f\omega([X_i, X_j], X_1, X_2, \ldots, \widehat{X_i}, \ldots, \widehat{X_j}, \ldots, X_{k+1})$$

$$= fX_1\left(\omega(X_2, \ldots, X_{k+1})\right)$$

$$+ \sum_{1 < i}(-)^{i+1} X_i(f)\omega(X_1, X_2, \ldots, \widehat{X_i}, \ldots, X_{k+1})$$

$$+ \sum_{1 < i}(-)^{i+1} fX_i\left(\omega(X_1, X_2, \ldots, \widehat{X_i}, \ldots, X_{k+1})\right)$$

$$+ \sum_{1 < j}(-)^{j+1} f\omega([X_1, X_j], X_1, \ldots, \widehat{X_j}, \ldots, X_{k+1})$$

$$- \sum_{1 < j}(-)^{j+1} X_j(f)\omega(X_1, X_2, \ldots, \widehat{X_j}, \ldots, X_{k+1})$$

$$+ \sum_{1 < i < j}(-)^{i+j} f\omega([X_i, X_j], X_1, X_2, \ldots, \widehat{X_i}, \ldots, \widehat{X_j}, \ldots, X_{k+1})$$

$$= fd\omega(X_1, X_2, \ldots, X_{k+1}).$$

We extend d to all forms by additivity in the usual way. The map

$$d : \Omega^\bullet(M) \to \Omega^\bullet(M)$$

defined in this way is clearly $\mathbb{R}$-linear and maps k-forms to $(k+1)$-forms. We now prove that it is a graded derivation (of degree 1). So, let $\omega \in \Omega^k(M)$ and $\rho \in \Omega^l(M)$. If $k = 0$, then $\omega = f \in C^\infty(M)$, and $\omega \wedge \rho = f\rho$. So, for every $X_1, \ldots, X_{l+1} \in \mathfrak{X}(M)$,

$$
\begin{aligned}
d(\omega \wedge \rho)&(X_1, \ldots, X_{l+1}) \\
&= d(f\rho)(X_1, \ldots, X_{l+1}) \\
&= \sum_i (-)^{i+1} X_i \left(f\rho(X_1, \ldots, \widehat{X_i}, \ldots, X_{l+1}) \right) \\
&\quad + \sum_{i<j} (-)^{i+j} f\rho([X_i, X_j], X_1, \ldots, \widehat{X_i}, \ldots, \widehat{X_j}, \ldots, X_{l+1}) \\
&= \sum_i (-)^{i+1} X_i(f)\rho(X_1, \ldots, \widehat{X_i}, \ldots, X_{l+1}) \\
&\quad + \sum_i (-)^{i+1} f X_i \left(\rho(X_1, \ldots, \widehat{X_i}, \ldots, X_{l+1}) \right) \\
&\quad + \sum_{i<j} (-)^{i+j} f\rho([X_i, X_j], X_1, \ldots, \widehat{X_i}, \ldots, \widehat{X_j}, \ldots, X_{l+1}) \\
&= \sum_i (-)^{i+1} df(X_i)\rho(X_1, \ldots, \widehat{X_i}, \ldots, X_{l+1}) + f d\rho(X_1, \ldots, X_{l+1}) \\
&= (df \wedge \rho + f d\rho)(X_1, \ldots, X_{l+1}).
\end{aligned}
$$

This shows that the graded Leibniz rule for $d(\omega \wedge \rho)$ works when $|\omega| = 0$. We leave it to the reader to show that it also works when $|\omega| = 1$ as Exercise 8.8 (it's a long computation but, if well organized, it is not conceptually complicated). This allows us to prove the graded Leibniz rule for $d(\omega \wedge \rho)$ by induction on $|\omega|$ (note that the graded Leibniz rule could also be proved by a direct, but cumbersome, computation). So, assume that the graded Leibniz rule for $d(\omega \wedge \rho)$ works when $|\omega| < r - 1$, and let $|\omega| = r > 1$. By Theorem 7.15, ω can be written as a sum of terms of the form

$$
\theta_1 \wedge \cdots \wedge \theta_r,
$$

where $\theta_1, \ldots, \theta_r \in \Omega^1(M)$. Then,

$$
\begin{aligned}
d(\omega \wedge \rho) \\
&= \sum d(\theta_1 \wedge \theta_2 \wedge \cdots \wedge \theta_r \wedge \rho) \\
&= \sum (d\theta_1) \wedge \theta_2 \wedge \cdots \wedge \theta_r \wedge \rho - \sum \theta_1 \wedge d(\theta_2 \wedge \cdots \wedge \theta_r \wedge \rho) \quad \text{(base of induction)}
\end{aligned}
$$

$$= \sum (d\theta_1) \wedge \theta_2 \wedge \cdots \wedge \theta_r \wedge \rho - \sum \theta_1 \wedge d(\theta_2 \wedge \cdots \wedge \theta_r) \wedge \rho$$

$$- \sum (-)^{r-1} \theta_1 \wedge \theta_2 \wedge \cdots \wedge \theta_r \wedge d\rho \qquad \text{(induction hypothesis)}$$

$$= (d\omega) \wedge \rho + (-)^r \omega \wedge d\rho \qquad\qquad \text{(base of induction)}.$$

Finally, specializing the Chevalley–Eilenberg formula (8.23) to $k = 1$, we get

$$ddf(X, Y) = X(df(Y)) - Y(df(X)) - df([X, Y])$$
$$= X(Y(f)) - Y(X(f)) - [X, Y](f)$$
$$= 0,$$

for all $X, Y \in \mathfrak{X}(M)$. This concludes the proof of the existence. The uniqueness again follows from Theorem 7.15. $\qquad\qquad\qquad\qquad\qquad\qquad\square$

Exercise 8.8. Complete the proof of Theorem 8.39 showing that

$$d(\omega \wedge \rho) = (d\omega) \wedge \rho - \omega \wedge d\rho,$$

for all $\omega \in \Omega^1(M)$ and all homogeneous forms ρ.

The graded derivation d is called the *exterior differential*, or *de Rham differential*, or just the *differential*. We want to describe it in local coordinates. So, let $(U, \varphi = (x^1, \ldots, x^n))$ be a chart on M, and let $\omega \in \Omega^k(M)$ be locally given by

$$\omega|_U = a_{i_1 \cdots i_k} dx^{i_1} \wedge \cdots \wedge dx^{i_k},$$

for some skew-symmetric $a_{i_1 \cdots i_k}$. We want to show that

$$(d\omega)|_U = \frac{\partial}{\partial x^{i_1}} a_{i_2 \cdots i_{k+1}} dx^{i_1} \wedge dx^{i_2} \wedge \cdots \wedge dx^{i_{k+1}}. \qquad (8.24)$$

Before proving this formula, we stress that, while it is simple and useful, the coefficients in it are not skew-symmetric in the indexes $i_1 \cdots i_{k+1}$. As we know, they can be uniquely replaced by skew-symmetric coefficients $d_{i_1 \cdots i_{k+1}}$ given by

$$d_{i_1 \cdots i_{k+1}} = \frac{1}{k+1} \sum_{j=1}^{k+1} (-)^{j+1} \frac{\partial}{\partial x^{i_j}} \left(a_{i_1 \cdots \hat{i_j} \cdots i_{k+1}} \right).$$

We leave the details to the reader. We now prove (8.24). First of all, it is easy to see, using the definition of the exterior differential, that

$(d\omega)|_U = d(\omega|_U)$. Now,

$$d(\omega|_U) = d\left(a_{i_1\cdots i_k}dx^{i_1} \wedge \cdots \wedge dx^{i_k}\right)$$

$$= da_{i_1\cdots i_k} \wedge dx^{i_1} \wedge \cdots \wedge dx^{i_k}$$

$$= \frac{\partial}{\partial x^i}a_{i_1\cdots i_k}dx^i \wedge dx^{i_1} \wedge \cdots \wedge dx^{i_k},$$

where we used that

$$da_{i_1\cdots i_k} = \frac{\partial}{\partial x^i}a_{i_1\cdots i_k}dx^i$$

and that $ddx^i = 0$. After renaming the indexes, this proves (8.24).

Example 8.40. On $\mathbb{R}^3$ with standard coordinates (x,y,z), consider the 2-form

$$\omega = xdy \wedge dz + ydz \wedge dx + zdx \wedge dy.$$

Compute the exterior differential of ω:

$$d\omega = dx \wedge dy \wedge dz + dy \wedge dz \wedge dx + dz \wedge dx \wedge dy = 3dx \wedge dy \wedge dz.$$

♦

Example 8.41 (Gradient, Rotor, Divergence). On $\mathbb{R}^3$ with standard coordinates (x^1, x^2, x^3), consider a generic function

$$f = f(x^1, x^2, x^3),$$

a generic 1-form

$$A = A_1dx^1 + A_2dx^2 + A_3dx^3,$$

and a generic 2-form

$$B = B_1dx^2 \wedge dx^3 + B_2dx^3 \wedge dx^1 + B_3dx^1 \wedge dx^2.$$

Then,

$$df = \frac{\partial f}{\partial x^1}dx^1 + \frac{\partial f}{\partial x^2}dx^2 + \frac{\partial f}{\partial x^3}dx^3$$

encodes, as we already know, the *gradient* grad f of f,

$$dA = \left(\frac{\partial A_3}{\partial x^2} - \frac{\partial A_2}{\partial x^3}\right) dx^2 \wedge dx^3 + \left(\frac{\partial A_1}{\partial x^3} - \frac{\partial A_3}{\partial x^1}\right) dx^3 \wedge dx^1$$
$$- \left(\frac{\partial A_2}{\partial x^1} - \frac{\partial A_1}{\partial x^2}\right) dx^1 \wedge dx^2$$

encodes the *rotor* rot $\mathbf{A}$ of $\mathbf{A} = (A_1, A_2, A_3)$, and

$$dB = \left(\frac{\partial B_1}{\partial x^1} + \frac{\partial B_2}{\partial x^2} + \frac{\partial B_3}{\partial x^3}\right) dx^1 \wedge dx^2 \wedge dx^3$$

encodes the *divergence* div $\mathbf{B}$ of $\mathbf{B} = (B_1, B_2, B_3)$, showing that the exterior differential generalizes the standard differential operators grad, rot, div. ♦

Now, we want to compute the graded commutators of interior products, Lie derivatives, and the exterior differential. These formulas are the main formulas in Cartan calculus. We collect them in Table 8.1 of graded commutators.

Theorem 8.42 (Cartan Calculus). *Let M be a manifold and let $X, Y \in \mathfrak{X}(M)$ be vector fields on M. The interior products ι_X, ι_Y, the Lie derivatives $\mathcal{L}_X, \mathcal{L}_Y$, and the exterior differential d fit in Table 8.1 of graded commutators.*

Proof. All the graded commutators in Table 8.1 can be computed in the same way. We discuss one example and leave all the rest to the reader as Exercise 8.9. The formula

$$[d, \iota_X] = \mathcal{L}_X \tag{8.25}$$

is known as *Cartan Magic Formula* and can be proved as follows. Both sides of (8.25) are graded derivations of degree 0. To show that they coincide, it is enough to show that they agree on generators. So, let $f \in C^\infty(M)$, and compute

$$[d, \iota_X]f = d\iota_X f + \iota_X df = \iota_X df = X(f) = \mathcal{L}_X f,$$

Table 8.1.   Distinguished graded commutators.

$[-,-]$	ι_Y	$\mathcal{L}_Y$	d
ι_X	0	$\iota_{[X,Y]}$	$\mathcal{L}_X$
$\mathcal{L}_X$	$\iota_{[X,Y]}$	$\mathcal{L}_{[X,Y]}$	0
d	$\mathcal{L}_Y$	0	0

and

$$[d, \iota_X]df = d\iota_X df + \iota_X ddf = dX(f) = \mathcal{L}_X df.$$

This concludes the proof of the Cartan Magic Formula. □

Exercise 8.9. Complete the proof of Theorem 8.42 computing the remaining graded commutators in Table 8.1.

Note that the formula $[d, d] = 0$, in the bottom-right corner of Table 8.1, is not at all trivial. Indeed, as d is a degree 1 derivation, we have

$$[d, d] = d \circ d + d \circ d = 2d \circ d.$$

Hence, $[d, d] = 0$ is the same as

$$d \circ d = 0,$$

which is in turn equivalent to

$$\mathrm{im}\, d \subseteq \ker d.$$

A form in the kernel of d is called a *closed form* or a *cocycle*. A form in the image of d is called an *exact form* or a *coboundary*. So, $[d, d] = 0$ means that exact forms are also closed. The converse might not be true. The quotient vector space

$$H_{\mathrm{dR}}(M) := \frac{\ker d}{\mathrm{im}\, d}$$

is a graded vector space called the *de Rham cohomology* of M and contains important topological information about M (see, e.g., Lee, 2013; Bott and Tu, 1982).

Remark 8.43. It should be clear how to define symmetries and infinitesimal symmetries of a differential form. A *symmetry* of a differential form ω on a manifold M is a diffeomorphism $\Phi : M \to M$ *preserving* ω in the sense that $\Phi^*(\omega) = \omega$. A *local symmetry* of ω is a diffeomorphism $\Phi : \mathcal{U} \to \Phi(\mathcal{U})$ between open submanifolds $\mathcal{U}, \Phi(\mathcal{U}) \subseteq M$ such that $\Phi^*(\omega|_{\Phi(\mathcal{U})}) = \omega|_{\mathcal{U}}$. An *infinitesimal symmetry* of ω is a vector field X on M generating a flow $\{\Phi_t\}_t$ by local symmetries of ω, i.e., $\Phi_t^*(\omega|_{M_{-t}}) = \omega|_{M_t}$ for all t. The subset

$$\mathrm{Diffeo}(M, \omega) := \{\text{symmetries of } \omega\} \subseteq \mathrm{Diffeo}(M)$$

of the group of diffeomorphisms of M is a subgroup. The subset

$$\mathfrak{X}(M, \omega) := \{\text{infinitesimal symmetries of } \omega\} \subseteq \mathfrak{X}(M)$$

of the Lie algebra of vector fields is a Lie subalgebra. To prove this last claim, it is convenient to characterize infinitesimal symmetries of ω in a suitable way. Namely, *a vector field X on M is an infinitesimal symmetry of ω if and only if $\mathcal{L}_X \omega = 0$.* This statement can be proved, exactly as Proposition 6.38, after proving the following formulas:

$$\frac{d}{dt}\Phi_t^*(\omega) = \Phi_t^*(\mathcal{L}_X \omega) = \mathcal{L}_X(\Phi_t^*(\omega)), \tag{8.26}$$

where $\{\Phi_t\}_t$ is the flow of X. By definition, given t such that $M_t \neq \varnothing$, the left-hand side of (8.26) is the differential form on M_t defined through its values as follows: for all $p \in M_t$, the value of $\frac{d}{dt}\Phi_t^*(\omega)$ at p is

$$\frac{d}{dt}\Phi_t^*(\omega)_p = \frac{d}{ds}\big|_{s=t}\Phi_t^*(\omega)_p.$$

One can show that this a well-defined form and that Formulas (8.26) indeed hold. We leave the details to the reader. As already mentioned, from those formulas, one can easily prove that X is an infinitesimal symmetry of ω if and only if $\mathcal{L}_X \omega = 0$. Finally, to see that $\mathfrak{X}(M, \omega) \subseteq \mathfrak{X}(M)$ is a Lie subalgebra, take $X, Y \in \mathfrak{X}(M, \omega)$ and $a \in \mathbb{R}$. Then,

$$\mathcal{L}_{X+Y}\omega = \mathcal{L}_X \omega + \mathcal{L}_Y \omega = 0.$$

Additionally,

$$\mathcal{L}_{aX}\omega = a\mathcal{L}_X \omega = 0.$$

The last two formulas show that $\mathfrak{X}(M, \omega)$ is a vector subspace of $\mathfrak{X}(M)$. We conclude computing

$$\mathcal{L}_{[X,Y]}\omega = [\mathcal{L}_X, \mathcal{L}_Y]\omega = \mathcal{L}_X \mathcal{L}_Y \omega - \mathcal{L}_Y \mathcal{L}_X \omega = 0,$$

where we used the commutator in the center of Table 8.1. Hence, $\mathfrak{X}(M, \omega)$ is also a Lie subalgebra of $\mathfrak{X}(M)$, as claimed. $\diamond$

Besides the definition provided in the proof of Theorem 8.35, and the one provided by the Cartan Magic Formula, there is yet another equivalent definition of the Lie derivative.

Proposition 8.44. *Let M be a manifold, and let $X \in \mathfrak{X}(M)$. The Lie derivative along X satisfies the following Leibniz rule with respect to insertions of vector* fields: *for any degree k differential form ω, and any $X_1, \ldots, X_k \in \mathfrak{X}(M)$,*

$$(\mathcal{L}_X \omega)(X_1, \ldots, X_k) = X\left(\omega(X_1, \ldots, X_k)\right) - \sum_{i=1}^{k} \omega(X_1, \ldots, \underbrace{[X, X_i]}_{i\text{-th place}}, \ldots, X_k).$$

(8.27)

Proof. The proof is by induction on k. The case $k = 1$ is easily obtained applying the formula

$$[\iota_{X_1}, \mathcal{L}_X] = \iota_{[X_1, X]}$$

to a 1-form (do you see it?). Now, rewrite the left-hand side of (8.27) as follows:

$$\begin{aligned}(\mathcal{L}_X \omega)(X_1, \ldots, X_k) &= (\iota_{X_1} \mathcal{L}_X \omega)(X_2, \ldots, X_k) \\ &= ([\iota_{X_1}, \mathcal{L}_X]\omega)(X_2, \ldots, X_k) + (\mathcal{L}_X \iota_{X_1} \omega)(X_2, \ldots, X_k) \\ &= (\iota_{[X_1, X]}\omega)(X_2, \ldots, X_k) + (\mathcal{L}_X \iota_{X_1} \omega)(X_2, \ldots, X_k) \\ &= \omega([X_1, X], X_2, \ldots, X_k) + (\mathcal{L}_X \iota_{X_1} \omega)(X_2, \ldots, X_k)\end{aligned}$$

and use induction on the last summand. $\qquad\square$

We conclude this section and this chapter discussing how do the interior product, the Lie derivative, and the exterior differential interact with smooth maps. So, let M, N be smooth manifolds and let $F : M \to N$ be a smooth map. We begin with the most important case: that of the exterior differential.

Proposition 8.45 (Naturality of the Exterior Differential). *The exterior differential "commutes" with pull-backs, i.e.,*

$$d \circ F^* = F^* \circ d. \tag{8.28}$$

Before proposing a proof, we stress that the reader should not be confused by the terminology "the exterior differential commutes with pull-backs". Namely, the exterior differential in the left-hand side of (8.28) is that in $\Omega^\bullet(M)$ while the one in the right-hand side is that in $\Omega^\bullet(N)$.

Proof of Proposition 8.45 (a sketch). The statement could be proved by a direct computation. We propose an alternative proof exploiting Theorem 8.23. Consider the $\mathbb{R}$-linear map

$$\mathfrak{D} : \Omega^\bullet(N) \to \Omega^\bullet(M), \quad \omega \mapsto \mathfrak{D}(\omega) := dF^*(\omega) - F^*(d\omega).$$

We have to prove that $\mathfrak{D} = 0$. To do this, we first list some properties of $\mathfrak{D}$. Besides the $\mathbb{R}$-linearity, $\mathfrak{D}$ enjoys the following properties:

(1) $\mathfrak{D}$ is a degree 1 map, i.e., $\mathfrak{D}(\Omega^k(N)) \subseteq \Omega^{k+1}(M)$, for all $k \in \mathbb{Z}$,
(2) $\mathfrak{D}$ satisfies the following (*F-relative*) version of the *graded Leibniz rule*: for all homogeneous forms $\omega, \rho \in \Omega^\bullet(N)$,

$$\mathfrak{D}(\omega \wedge \rho) = \mathfrak{D}(\omega) \wedge F^*(\rho) + (-)^{|\omega|} F^*(\omega) \wedge \mathfrak{D}(\rho).$$

Property (1) is obvious. Property (2) can be easily checked using that d is a graded derivation and that F^* is a graded algebra homomorphism, and we leave the details to the reader. In a very similar way as in the proof of Proposition 8.31, it follows from Properties (1) and (2) that, if $\mathfrak{D}$ vanishes on generators, then $\mathfrak{D} = 0$. So, it remains to check that

$$\mathfrak{D}(f) = 0,$$
$$\mathfrak{D}(df) = 0,$$

i.e.,

$$dF^*(f) = F^*(df), \tag{8.29}$$
$$dF^*(df) = F^*(ddf), \tag{8.30}$$

for all $f \in C^\infty(M)$. Now, Identity (8.29) has already been proved as Proposition 7.18 Point (3), while Identity (8.30) follows from Proposition 7.18 Point (3), and $d \circ d = 0$. This concludes the proof. $\qquad\square$

We now pass to interior products and Lie derivatives.

Proposition 8.46. *Let $X \in \mathfrak{X}(M)$ and $Y \in \mathfrak{X}(N)$ be F-related vector fields. Then,*

$$\iota_X \circ F^* = F^* \circ \iota_Y,$$
$$\mathcal{L}_X \circ F^* = F^* \circ \mathcal{L}_Y.$$

In particular, if $\Phi : M \to N$ is a diffeomorphism, and $Y \in \mathfrak{X}(N)$, then

$$\iota_{\Phi^*(Y)} \circ \Phi^* = \Phi^* \circ \iota_Y,$$
$$\mathcal{L}_{\Phi^*(Y)} \circ \Phi^* = \Phi^* \circ \mathcal{L}_Y.$$

Proof.   Left as Exercise 8.10. $\qquad\square$

> **Exercise 8.10.** Prove Proposition 8.46. (**Hint:** *There is a proof very similar to the proof of Proposition 8.45.*)

Corollary 8.47. *Let M be a manifold, let $Y \in \mathfrak{X}(M)$, let $\omega, \omega_1, \omega_2 \in \Omega^\bullet(M)$, and let $\Phi : M \to M$ be a diffeomorphism. If Φ is a symmetry of $\omega, \omega_1, \omega_2$, then it is also a symmetry of $d\omega, \omega_1 + \omega_2, \omega_1 \wedge \omega_2$. If, additionally, Φ is a symmetry of Y, then it is also a symmetry of $\iota_Y \omega$ and $\mathcal{L}_Y \omega$.*

Proof. Left as Exercise 8.11. $\qquad\qquad\qquad\qquad\qquad\qquad\qquad$ $\square$

> **Exercise 8.11.** Prove Corollary 8.47.

The following infinitesimal version of Corollary 8.47 is a corollary of either Corollary 8.47 itself or Theorem 8.42.

Corollary 8.48. *Let M be a manifold, let $X, Y \in \mathfrak{X}(M)$, and let $\omega, \omega_1, \omega_2 \in \Omega^\bullet(M)$. If X is an infinitesimal symmetry of $\omega, \omega_1, \omega_2$, then it is also an infinitesimal symmetry of $d\omega, \omega_1 + \omega_2, \omega_1 \wedge \omega_2$. If, additionally, X is an infinitesimal symmetry of Y, then it is also an infinitesimal symmetry of $\iota_Y \omega$ and $\mathcal{L}_Y \omega$.*

Proof. Left as Exercise 8.12. $\qquad\qquad\qquad\qquad\qquad\qquad\qquad$ $\square$

> **Exercise 8.12.** Prove Corollary 8.48.

Chapter 9

Vector Bundles

In this chapter, we discuss a class of geometric objects playing an important role in many branches of Differential Geometry: *vector bundles*. Roughly, a smooth vector bundle is a family $\{E_p\}_{p \in M}$ of vector spaces smoothly parameterized by a point p in a smooth manifold M. A little bit more precisely, this means that the union $E = \coprod_{p \in M} E_p$ is a manifold and the fiber-wise (vector space) operations are smooth. The tangent bundle, the cotangent bundle, and, more generally, the bundle of k-covectors are noteworthy instances of vector bundles.

9.1 Vector Bundles and Vector Bundle Maps

Let M be an n-dimensional manifold, let m be a non-negative integer, and let V be an m-dimensional real vector space. Consider a family $\{E_p\}_{p \in M}$ of m-dimensional real vector spaces E_p parameterized by a point p in M. We denote by E their disjoint union

$$E := \coprod_{p \in M} E_p = \{(p, e) : p \in M \text{ and } e \in E_p\}$$

and by

$$\pi : E \to M, \quad (p, e) \mapsto p,$$

the natural projection. Clearly, the preimage $\pi^{-1}(p)$ of a point $p \in M$ under π identifies canonically with the vector space E_p. For an open subset

$\mathcal{U} \subseteq M$, we denote by $E_{\mathcal{U}}$ the preimage of $\mathcal{U}$ under π:

$$E_{\mathcal{U}} := \pi^{-1}(\mathcal{U}) = \coprod_{p \in \mathcal{U}} E_p.$$

Definition 9.1 (Vector Bundle). A *rank m vector bundle* over M is a family $\{E_p\}_{p \in M}$ of m-dimensional real vector spaces E_p parameterized by a point p in M equipped with a smooth $(n + m)$-dimensional manifold structure on their disjoint union E in such a way that the following two conditions are satisfied:

(1) the natural surjection $\pi : E \to M$ is a smooth map,
(2) for every point $p_0 \in M$ there exist an open neighborhood $\mathcal{U} \in M$ of p_0 and a diffeomorphism $\Phi_{\mathcal{U}} : E_{\mathcal{U}} \to \mathcal{U} \times V$ such that

  • the diagram

$$
\begin{array}{ccc}
E_{\mathcal{U}} & \xrightarrow{\ \Phi_{\mathcal{U}}\ } & \mathcal{U} \times V \\
& {\scriptstyle \pi} \searrow \quad \swarrow {\scriptstyle \mathrm{pr}_{\mathcal{U}}} & \\
& \mathcal{U} &
\end{array}
\tag{9.1}
$$

  commutes (here $\mathrm{pr}_{\mathcal{U}} : \mathcal{U} \times V \to \mathcal{U}$ is the projection onto the first factor); in other words, for any $p \in \mathcal{U}$, $\Phi_{\mathcal{U}}$ maps E_p to $\{p\} \times V$, and
  • for every $p \in \mathcal{U}$, the restriction

$$\Phi_{\mathcal{U}} : E_p \to \{p\} \times V \cong V \tag{9.2}$$

  is a vector space isomorphism (where we are identifying $\{p\} \times V$ with V in the obvious way).

In this situation, we also say that (E, π, M) (or simply E) is a *vector bundle* (over M). Then E is called the *total space*, M is the *base manifold*, V is the *abstract fiber*, and $\pi : E \to M$ is the *vector bundle projection*. The vector space $E_p = \pi^{-1}(p)$ is the *fiber* of E over $p \in M$, and the pair $(\mathcal{U}, \Phi_{\mathcal{U}})$ is called a *local trivialization* of E. If $\mathcal{U} = M$, then we call $(\mathcal{U}, \Phi_{\mathcal{U}})$ a *global trivialization*. Every vector bundle possessing a global trivialization is called *trivializable*.

A local trivialization $\Phi_{\mathcal{U}}$ of a vector bundle E over M allows us to identify the fiber E_p of E over $p \in \mathcal{U}$ with the abstract fiber V (via (9.2)), hence the fibers over different points in $\mathcal{U}$ with each other, while, in general, there is no canonical way to identify two fibers of a vector bundle (if a trivialization is not provided).

Let (E, π, M) be a vector bundle and let $V \subseteq M$ be an open submanifold. Then (E_V, π, V) (equivalently, the family of vector spaces $\{E_p\}_{p \in V}$)
is a vector bundle (over V) with the same abstract fiber, indeed, for every
point $p \in V$, it is always possible to find a local trivialization $(\mathcal{U}, \Phi_{\mathcal{U}})$ of E
such that $\mathcal{U} \subseteq V$ (do you see it?). It follows that $(\mathcal{U}, \Phi_{\mathcal{U}})$ is also a local trivialization for E_V. In particular, when $V = \mathcal{U}$ for some local trivialization
$(\mathcal{U}, \Phi_{\mathcal{U}})$ of E, then $(E_V, \pi, V) = (E_{\mathcal{U}}, \pi, \mathcal{U})$ is a trivializable vector bundle
and $\Phi_{\mathcal{U}}$ is a global trivialization of it. In this sense, every vector bundle is
locally trivializable.

Example 9.2 (Trivial Vector Bundle). The product $M \times V$ together with the
projection $\mathrm{pr}_M : M \times V \to M$ onto the first factor is a rank m trivializable
vector bundle that we denote V_M (it corresponds to the constant family
$\{E_p = V\}_{p \in M}$). This is obvious: there is a global trivialization $V_M \to M \times$
V given by the identity map. Every vector bundle of the form V_M is called
a *trivial vector bundle*. ◆

Example 9.3 ((Co)tangent Bundle as a Vector Bundle). The tangent bun-
dle (TM, τ) is a rank n vector bundle with abstract fiber $\mathbb{R}^n$. For any point
$p_0 \in M$, we can use a little modification of a standard chart as a local trivialization. Namely, let $(U, \varphi = (x^1, \ldots, x^n))$ be a chart on M around p_0.
Define a map $\Phi_U : TU \to U \times \mathbb{R}^n$ by putting

$$\Phi_U(p, v) := \left(p; \dot{x}^1(v), \ldots, \dot{x}^n(v)\right),$$

where, as usual, $(\dot{x}^1(v), \ldots, \dot{x}^n(v))$ are the components of v in the coordinate frame

$$\left(\frac{\partial}{\partial x^1}\Big|_p, \ldots, \frac{\partial}{\partial x^n}\Big|_p\right).$$

Clearly, the Φ_U are local trivializations of a vector bundle structure (do you
see it?). Similarly, the bundle of k-covectors $(\wedge^k T^* M, \pi)$ is a rank $N(k, n) =$
$\binom{k}{n}$ vector bundle with local trivializations $\Phi_U : \wedge^k T^* U \to U \times \mathbb{R}^{N(k,n)}$
given by

$$\Phi_U(p, \omega) := \left(p; (b_{i_1 \cdots i_k})_{i_1 < \cdots < i_k}\right),$$

where

$$b_{i_1 \cdots i_k} = \omega\left(\frac{\partial}{\partial x^{i_1}}\Big|_p, \ldots, \frac{\partial}{\partial x^{i_k}}\Big|_p\right)$$

are the components of $\omega \in \wedge^k T^* M$ in the coordinate frame $(d_p x^{i_1} \wedge \cdots \wedge d_p x^{i_k})_{i_1 < \cdots < i_k}$. In particular, the cotangent bundle $(T^* M, \pi)$ is a rank n vector bundle. ♦

Note that the abstract fiber of a rank m vector bundle (E, π, M) can always be chosen to be $\mathbb{R}^m$. Indeed, let V be the abstract fiber of E. Choose a frame $\mathcal{R} = (e_1, \ldots, e_m)$ in V and let $\varphi_{\mathcal{R}} : V \to \mathbb{R}^m$ be the coordinate isomorphism. Finally, let $\Phi_{\mathcal{U}} : E_{\mathcal{U}} \to \mathcal{U} \times V$ be a local trivialization of E. Then it is clear that the composition

$$E_{\mathcal{U}} \xrightarrow{\ \Phi_{\mathcal{U}}\ } \mathcal{U} \times V \xrightarrow{\ \mathrm{id} \times \varphi_{\mathcal{R}}\ } \mathcal{U} \times \mathbb{R}^m$$

is a local trivialization as well (i.e., it is a diffeomorphism, and $\mathrm{pr}_{\mathcal{U}} \circ (\mathrm{id} \times \varphi_{\mathcal{R}}) \circ \Phi_{\mathcal{U}} = \pi$). Here $\mathrm{id} \times \varphi_{\mathcal{R}} : \mathcal{U} \times V \to \mathcal{U} \times \mathbb{R}^m$ is the diffeomorphism given by $\mathrm{id} \times \varphi_{\mathcal{R}}(p, v) = (p, \varphi_{\mathcal{R}}(v))$.

Remark 9.4 (Subtrivializations). Let (E, π, M) be a vector bundle with abstract fiber V, let $(\mathcal{U}, \Phi_{\mathcal{U}})$ be a local trivialization of E, and let $\mathcal{V} \subseteq \mathcal{U}$ be an open subset. It is clear that the pair $(\mathcal{V}, \Phi_{\mathcal{U}} : E_{\mathcal{V}} \to \mathcal{V} \times V)$ is again a local trivialization, sometimes called a *subtrivialization*. In this way, the base $\mathcal{U}$ of a local trivialization $(\mathcal{U}, \Phi_{\mathcal{U}})$ can be made arbitrarily small around a given point, if necessary, passing to a subtrivialization. We will often exploit this trick in what follows, without further comments. ◇

The following theorem is often useful to construct vector bundles when we are given just a family of vector spaces parameterized by a point in a manifold (see Example 9.6, see also Section 9.3).

Theorem 9.5 (Vector Bundle Chart Theorem). *Let M be an n-dimensional manifold, let V be an m-dimensional real vector space, and let $\{E_p\}_{p \in M}$ be a family of m-dimensional real vector spaces parameterized by a point in M. Suppose that we are given an open cover $\mathcal{C} = \{\mathcal{U}\}$ of M and, for every $\mathcal{U} \in \mathcal{C}$, a bijection*

$$\Phi_{\mathcal{U}} : E_{\mathcal{U}} \to \mathcal{U} \times V,$$

such that

(1) *for any $\mathcal{U} \in \mathcal{C}$, the diagram (9.1) commutes (i.e., for any $p \in \mathcal{U}$, $\Phi_{\mathcal{U}}$ maps E_p to $\{p\} \times V$),*
(2) *for any $\mathcal{U} \in \mathcal{C}$ and any $p \in \mathcal{U}$, the restriction $\Phi_{\mathcal{U}} : E_p \to \{p\} \times V \cong V$ is a vector space isomorphism,*

(3) *for any* $\mathcal{U}, \mathcal{V} \in \mathcal{C}$, *the map*

$$\Phi_{\mathcal{U}\mathcal{V}} := \Phi_{\mathcal{V}} \circ \Phi_{\mathcal{U}}^{-1} : (\mathcal{U} \cap \mathcal{V}) \times V \to (\mathcal{U} \cap \mathcal{V}) \times V \qquad (9.3)$$

is of the form

$$\Phi_{\mathcal{U}\mathcal{V}}(p, v) = (p, A(p)(v)) \qquad (9.4)$$

for some smooth function $A : \mathcal{U} \cap \mathcal{V} \to \mathrm{GL}(V)$.

Then there exists a unique smooth structure $\mathcal{A}_E$ *on* E *such that* E *is an* $(n+m)$-*dimensional manifold and a vector bundle over* M *with the* $(\mathcal{U}, \Phi_{\mathcal{U}})$ *as local trivializations.*

Proof. For completeness, we remark that, given an open cover $\mathcal{C} = \{\mathcal{U}\}$ of M and bijections $\Phi_{\mathcal{U}} : E_{\mathcal{U}} \to \mathcal{U} \times V$ satisfying Conditions (1) and (2) in the statement, then the maps (9.3) are necessarily of the form (9.4) for some maps $A : \mathcal{U} \cap \mathcal{V} \to \mathrm{GL}(V)$ (do you see it?) but, in general, the A need not be smooth. Now, assume that A is a smooth map for all $\mathcal{U}, \mathcal{V} \in \mathcal{C}$ as in the statement. We begin constructing a smooth $(n+m)$-dimensional atlas on $E = \coprod_{p \in M} E_p$. So, let $p_0 \in M$, and let $\mathcal{U} \in \mathcal{C}$ contain p_0. Choose a chart (U, φ) on M around p_0 such that $U \subseteq \mathcal{U}$. Choose also a frame $\mathcal{R}$ in V and denote by φ^E the composition

$$E_U \xrightarrow{\Phi_{\mathcal{U}}} U \times V \xrightarrow{\varphi \times \varphi_{\mathcal{R}}} \widehat{U} \times \mathbb{R}^m ,$$

where $\varphi \times \varphi_{\mathcal{R}} : U \times V \to \widehat{U} \times \mathbb{R}^m$ is the map given by $\varphi \times \varphi_{\mathcal{R}}(p, v) = (\varphi(p), \varphi_{\mathcal{R}}(v))$. Clearly, (E_U, φ^E) is an $(n+m)$-dimensional chart on E that we call a *vector bundle chart* (we also say that (E_U, φ^E) is a vector bundle chart *over* (U, φ)). Any two such charts are compatible. Indeed, let $(\tilde{U}, \tilde{\varphi})$ be another chart on M, let $\tilde{\mathcal{U}} \in \mathcal{C}$ contain $\tilde{U}$, let $\tilde{\mathcal{R}}$ be another frame in V, and let $(E_{\tilde{U}}, \tilde{\varphi}^E)$ be the corresponding vector bundle chart. Either $U \cap \tilde{U} = \varnothing$, hence $E_U \cap E_{\tilde{U}} = \varnothing$, or $U \cap \tilde{U} \neq \varnothing$ hence

$$E_U \cap E_{\tilde{U}} = E_{U \cap \tilde{U}} \neq \varnothing.$$

In the latter case, the transition map

$$\tilde{\varphi}^E \circ (\varphi^E)^{-1} : \varphi^E(E_U \cap E_{\tilde{U}}) \to \tilde{\varphi}^E(E_U \cap E_{\tilde{U}})$$

is a diffeomorphism between open subsets of $\mathbb{R}^{n+m}$. Indeed, first of all,

$$\varphi^E(E_U \cap E_{\tilde{U}}) = \varphi^E(E_{U \cap \tilde{U}}) = \varphi(U \cap \tilde{U}) \times \mathbb{R}^m$$

is indeed an open subset, and similarly $\tilde{\varphi}^E(E_{U \cap \tilde{U}}) = \tilde{\varphi}(U \cap \tilde{U}) \times \mathbb{R}^m$. Now, check the smoothness of $\tilde{\varphi}^E \circ (\varphi^E)^{-1}$: take a point $(P; x) \in \varphi(U \cap \tilde{U}) \times \mathbb{R}^m$ and compute

$$\tilde{\varphi}^E \circ (\varphi^E)^{-1}(P, x) = \left((\tilde{\varphi} \times \varphi_{\tilde{\mathcal{R}}}) \circ \Phi_{\tilde{U}} \circ \Phi_U^{-1} \circ (\varphi \times \varphi_{\mathcal{R}})^{-1} \right)(P, x)$$

$$= \left((\tilde{\varphi} \times \varphi_{\tilde{\mathcal{R}}}) \circ \Phi_{U\tilde{U}} \circ (\varphi \times \varphi_{\mathcal{R}})^{-1} \right)(P, x)$$

$$= \left((\tilde{\varphi} \times \varphi_{\tilde{\mathcal{R}}}) \circ \Phi_{U\tilde{U}} \right)(\varphi^{-1}(P), \varphi_{\mathcal{R}}^{-1}(x))$$

$$= (\tilde{\varphi} \times \varphi_{\tilde{\mathcal{R}}}) \left(\varphi^{-1}(P), A(\varphi^{-1}(P))(\varphi_{\mathcal{R}}^{-1}(x)) \right)$$

$$= \left((\tilde{\varphi} \circ \varphi^{-1})(P), (\varphi_{\tilde{\mathcal{R}}} \circ A(\varphi^{-1}(P)) \circ \varphi_{\mathcal{R}}^{-1})(x) \right).$$

As $(U, \varphi), (\tilde{U}, \tilde{\varphi})$ are compatible charts, the first n entries depend smoothly on P. As A is a smooth map by Condition (3) in the statement, then the last m entries depend smoothly on P. They also depend linearly, hence smoothly, on x. So $\tilde{\varphi}^E \circ (\varphi^E)^{-1}$ is smooth. Changing the roles of the two charts reveals that the inverse $\varphi^E \circ (\tilde{\varphi}^E)^{-1}$ is also smooth. We conclude that vector bundle charts form an atlas on E. Such atlas is included in a unique smooth structure that we denote $\mathcal{A}_E$. In order to prove that $(E, \mathcal{A}_E)$ is a smooth manifold, we have to show that the atlas topology induced by $\mathcal{A}_E$ is Hausdorff and II-countable. Hausdorffness can be proved exactly as for the tangent bundle and we leave the details to the reader. As for II-countability, note that it follows from the proof of Proposition 1.41 that the topology of M has a countable basis $\mathcal{B}$ consisting of coordinate domains. By choosing from $\mathcal{B}$ those coordinate domains which are contained in some $\mathcal{U} \in \mathcal{C}$, we find an atlas in $\mathcal{A}_E$ consisting of countably many vector bundle charts. So $(E, \mathcal{A}_E)$ is an $(n + m)$-dimensional manifold. Next, we have to show that $\Phi_{\mathcal{U}}$ are diffeomorphisms, hence $(\mathcal{U}, \Phi_{\mathcal{U}})$ are local trivializations of a vector bundle structure. So, let $\mathcal{U} \in \mathcal{C}$, let (U, φ) be a chart such that $U \subseteq \mathcal{U}$, let $\mathcal{R}$ be a frame of V, and let (E_U, φ^E) be the corresponding vector bundle chart on E. The pair $(U \times V, \varphi \times \varphi_{\mathcal{R}})$ is a chart on $\mathcal{U} \times V$ and we have $\Phi_{\mathcal{U}}(E_U) \subseteq U \times V$. In other words, we can consider the coordinate representation

$$\widehat{\Phi}_{\mathcal{U}} : \widehat{U} \times \mathbb{R}^m \to \widehat{U} \times \mathbb{R}^m$$

of $\Phi_{\mathcal{U}}$ in the charts $(E_U, \varphi^E), (U \times V, \varphi \times \varphi_{\mathcal{R}})$. By definition of $\Phi_{\mathcal{U}}$, such coordinate representation is just the identity. This proves that Φ_U is a smooth map and a local diffeomorphism. As it is bijective, it is also a diffeomorphism as desired.

It remains to show that if E possesses two smooth structures, say $\mathcal{A}', \mathcal{A}''$ which turn it into a vector bundle over M with the $(\mathcal{U}, \Phi_{\mathcal{U}})$ as local trivializations, then $\mathcal{A}' = \mathcal{A}''$. This is true, indeed $\Phi_{\mathcal{U}}$ are diffeomorphisms with respect to both smooth structures $\mathcal{A}'$ and $\mathcal{A}''$. It follows that the vector bundle charts constructed above belong to both $\mathcal{A}'$ and $\mathcal{A}''$ (do you see it?). As the vector bundle charts form an atlas on E, $\mathcal{A}'$ and $\mathcal{A}''$ must coincide. $\qquad\square$

Example 9.6 (Tautological Bundle over the Projective Space). Consider the n-dimensional projective space $\mathbb{R}P^n$. For a point $p = [P^0 : \cdots : P^n] \in \mathbb{R}P^n$, denote by $\ell_p \subseteq \mathbb{R}^{n+1}$ the 1-dimensional vector subspace spanned by one, hence any, of the representatives $(P^0, \ldots, P^n)$ of p:

$$\ell_p := \left\{ (P^0, \ldots, P^n) \in \mathbb{R}^{n+1} : p = [P^0 : \cdots : P^n] \right\} \cup \{0\}.$$

The family $\{\ell_p\}_{p \in \mathbb{R}P^n}$ is a rank 1-vector bundle over $\mathbb{R}P^n$ called the *tautological vector bundle*. To see this, we exploit Theorem 9.5. As an open cover of $\mathbb{R}P^n$, fix $\mathcal{C} = \{U_0, \ldots, U_n\}$ the open cover consisting of the coordinate domains of the affine charts on $\mathbb{R}P^n$. For any $i = 0, \ldots, n$, let

$$\Phi_i : \ell_{U_i} \to U_i \times \mathbb{R}$$

be the map given by

$$\Phi_i\big(p, (Q^0, \ldots, Q^n)\big) = (p, Q^i).$$

Then Φ_i is a bijection whose inverse $\Phi_i^{-1} : U_i \times \mathbb{R} \to \ell_{U_i}$ is given by

$$\Phi_i^{-1}\big(p = [P^0 : \cdots : P^n], r\big) = \big(p, (rP^0/P^i, \ldots, rP^n/P^i)\big).$$

We invite the reader to check all the details. For all $i, j = 0, \ldots, n$, the map

$$\Phi_{ij} = \Phi_j \circ \Phi_i^{-1} : (U_i \cap U_j) \times \mathbb{R} \to (U_i \cap U_j) \times \mathbb{R}$$

is given by

$$\Phi_{ij}\big(p = [P^0 : \cdots : P^n], r\big) = \big(p, rP^j/P^i\big)$$

which is clearly (invertible) smooth with smooth inverse $\Phi_{ij}^{-1} = \Phi_{ji}$, i.e., a diffeomorphism. It follows from Theorem 9.5 that ℓ is a vector bundle, as claimed. $\qquad\blacklozenge$

We can compare two vector bundles via *vector bundle maps* between them. Let (E, π, M) and (E', π', M') be vector bundles of rank m and m', with abstract fibers V and V', respectively. Roughly, a vector bundle map between (E, π, M) and (E', π', M') is a smooth map $f : M \to M'$ between the bases, together with a family $\{F_p : E_p \to E'_{f(p)}\}_{p \in M}$ of linear maps smoothly depending on p. The following definition formalizes this idea.

Definition 9.7 (Vector Bundle Map). A *vector bundle map* between (E, π, M) and (E', π', M') (also called a *vector bundle morphism*) is a pair (F, f) consisting of a smooth map $f : M \to M'$ between the bases and a smooth map $F : E \to E'$ between the total spaces such that

(1) the diagram

$$
\begin{array}{ccc}
E & \xrightarrow{\;F\;} & E' \\
{\scriptstyle\pi}\downarrow & & \downarrow{\scriptstyle\pi'} \\
M & \xrightarrow{\;f\;} & M'
\end{array}
\tag{9.5}
$$

commutes; in other words, for any $p \in M$, F maps E_p to $E'_{f(p)}$, and

(2) for every $p \in M$, the restriction

$$
F_p := F|_{E_p} : E_p \to E'_{f(p)}
$$

is a linear map.

In this case, we also say that $F : E \to E'$ is a *vector bundle map covering* $f : M \to M'$. A *vector bundle isomorphism* is a vector bundle map (Φ, ϕ) such that both Φ and ϕ are diffeomorphisms. The vector bundles (E, π, M) and (E', π', M') are *isomorphic* if there is a vector bundle isomorphism (Φ, ϕ) connecting them.

Let $F : E \to E'$ be a vector bundle map covering the smooth map $f : M \to M'$. Take a point $p \in M$, a chart (U, φ) on M around p, and a chart (U', ϕ') on M' around $f(p)$ such that $f(U) \subseteq U'$. Shrinking both U and U' if necessary, we can assume that there are trivializations $\Phi_U : E_U \to U \times V$ and $\Phi_{U'} : E'_{U'} \to U' \times V'$, hence vector bundle charts (E_U, φ^E) and $(E'_{U'}, \varphi'^{E'})$ over (U, φ) and (U', φ'), respectively. Clearly, $F(E_U) \subseteq E'_{U'}$ and the associated coordinate representation

$$
\widehat{F} : \widehat{U} \times \mathbb{R}^m \to \widehat{U}' \times \mathbb{R}^{m'}
$$

is given by

$$\widehat{F}(P, x) = (\widehat{f}(P), B(P)x),$$

where $\widehat{f} : \widehat{U} \to \widehat{U}'$ is the coordinate representation of f in the charts $(U, \varphi), (U', \varphi')$ and $B(P)$ is an $m' \times m$ matrix smoothly depending on P (check the details as an exercise).

Exercise 9.1. Let $(E, \pi, M), (E', \pi', M')$ be vector bundles and let $F : E \to E'$ be a(n *a priori* non-necessarily smooth) map mapping fibers of E to fibers of E'. In other words, there exists a map $f : M \to M'$ such that the diagram (9.5) commutes. Suppose that the restriction $F_p := F|_{E_p} : E_p \to E'_{f(p)}$ is linear for all $p \in M$. Show that, if F is smooth, then f is also smooth, hence (F, f) is a vector bundle map. (**Hint:** *First show that, under the hypothesis in the exercise, for every point $e \in E$, there is a bundle chart (E_U, φ^E) around e and a bundle chart $(E'_{U'}, \varphi'^{E'})$ around $F(e)$ such that $F(E_U) \subseteq E'_{U'}$, then look at what happens on the bases.*)

Example 9.8 (Restriction of a Vector Bundle to a Submanifold). Let (E, π, M) be a vector bundle and let $S \subseteq M$ be a submanifold. Denote

$$E_S := \pi^{-1}(S) = \coprod_{p \in S} E_p.$$

Then (E_S, π, S) (equivalently, the family of vector spaces $\{E_p\}_{p \in S}$) is a vector bundle (over S). This can be easily checked exploiting the Vector Bundle Chart Theorem (and we invite the reader to work out the details). In particular, every trivialization (E_U, Φ_U) of E around a point $p \in S$ induces a trivialization $(E_{U \cap S}, \Phi_U)$ of E_S in the obvious way. The natural inclusion $E_S \hookrightarrow E$ is a vector bundle map. When $S = V$ is an open submanifold, we recover the construction of the vector bundle (E_V, π, V) immediately before Example 9.2. ◆

Example 9.9 (Tangent Map as a Vector Bundle Map). Let $F : M \to N$ be a smooth map between manifolds. Consider the following map between their tangent bundles:

$$dF : TM \to TN, \quad (p, v) \mapsto (F(p), d_p F(v)).$$

Clearly, the diagram

$$
\begin{array}{ccc}
TM & \xrightarrow{\;dF\;} & TN \\
\tau \downarrow & & \downarrow \tau \\
M & \xrightarrow{\;F\;} & N
\end{array}
$$

commutes. If $(U, \varphi = (x^1, \ldots, x^m))$ is a chart on M and $(V, \psi = (y^1, \ldots, y^n))$ is a chart on N such that $F(U) \subseteq V$, then $dF(TU) \subseteq TV$ and the coordinate representation of dF in the standard charts $(TU, T\varphi), (TV, T\psi)$ is given by

$$\widehat{dF}(P; x = (v^1, \ldots, v^m)) = \left(\widehat{F}(P); \frac{\partial \widehat{F}^1}{\partial x^i}(P)v^i, \ldots, \frac{\partial \widehat{F}^n}{\partial x^i}(P)v^i \right)$$

$$= (\widehat{F}(P); J_{\widehat{F}}(P)x),$$

where $\widehat{F}$ is the coordinate representation of F in the charts $(U, \varphi), (V, \psi)$. This shows that dF is a smooth map. As it is linear on fibers, we conclude that it is a vector bundle map covering F. The vector bundle map (dF, F) is called the *tangent map* to F and it encodes partial derivatives of F in a coordinate free manner.   ◆

Example 9.10. Let (E, π, M) be a trivializable vector bundle and let $\Phi : E \to M \times V$ be a global trivialization. Then Φ is a vector bundle isomorphism covering the identity map $\mathrm{id}_M : M \to M$.   ◆

Proposition 9.11. *Let* $(E, \pi, M), (E', \pi', M'), (E'', \pi'', M''), (E''', \pi''', M''')$ *be vector bundles, let* $F : E \to E'$ *and* $G : E' \to E''$ *be vector bundle maps covering the smooth maps* $f : M \to M'$ *and* $g : M' \to M''$, *and let* $\Phi : E \to E'''$ *be a vector bundle isomorphism covering the diffeomorphism* $\phi : M \to M'''$. *Then,*

(1) *the identity* $\mathrm{id}_E : E \to E$ *is a vector bundle isomorphism covering the identity* $\mathrm{id}_M : M \to M$,
(2) *the composition* $G \circ F : E \to E''$ *is a vector bundle map covering the composition* $g \circ f : M \to M''$,
(3) *the inverse* $\Phi^{-1} : E''' \to E$ *is a vector bundle isomorphism covering the inverse* $\phi^{-1} : M''' \to M$.

Proof. Left as Exercise 9.2.   □

Exercise 9.2. Prove Proposition 9.11.

It follows from Proposition 9.11 that *being isomorphic* is an equivalence relation on vector bundles. Additionally, vector bundle isomorphisms from a vector bundle (E, π, M) to itself form a group under composition, called the *group of vector bundle automorphisms* (or, sometimes, the *group of symmetries*) of (E, π, M), and denoted $\mathrm{Diffeo}(E, \pi, M)$. Vector bundle automorphisms covering the identity form a subgroup in $\mathrm{Diffeo}(E, \pi, M)$

sometimes called the *gauge group* of (E, π, M) (particularly in the Physics literature).

Example 9.12. Let M, M', M'', M''' be smooth manifolds, let $F : M \to M'$ and $G : M' \to M''$ be smooth maps, and let $\Phi : M \to M'''$ be a diffeomorphism. It immediately follows from Proposition 3.11 that

(1) the tangent map $(d\,\mathrm{id}_M, \mathrm{id}_M)$ to the identity is the identity vector bundle isomorphism: $d\,\mathrm{id}_M = \mathrm{id}_{TM}$,
(2) the tangent map $(d(G \circ F), G \circ F)$ to the composition is the composition of the tangent maps: $d(G \circ F) = dG \circ dF$,
(3) the tangent map $(d\Phi, \Phi)$ to Φ is a vector bundle isomorphism whose inverse is the tangent map to the inverse: $(d\Phi)^{-1} = d\Phi^{-1}$.

In particular, two smooth manifolds are diffeomorphic if and only if their tangent bundles are isomorphic. ◆

9.2 Sections and Frames

Let (E, π, M) be a rank m vector bundle with abstract fiber V.

Definition 9.13 (Section of a Vector Bundle). A *section* of (E, π, M) is a map $s : M \to TM$ that inverts the vector bundle projection $\pi : E \to M$ on the right, i.e., $\pi \circ s = \mathrm{id}_M$. In other words, a section is the assignment of a vector $s(p)$ in the fiber E_p, for every point $p \in M$. A section of $E_\mathcal{V}$, for some open submanifold $\mathcal{V} \subseteq M$, is also called a *local section* of E.

The space of sections of (E, π, M) is a $C^\infty(M)$-module exactly as in the tangent bundle case (see Example 3.20), and the *zero section* is the section mapping a point $p \in M$ to the zero vector $0 \in E_p$, for all $p \in M$. We now concentrate on smooth sections. Let $(U, \varphi = (x^1, \ldots, x^n))$ be a chart on M and let $(E_U, \varphi^E = (x^1, \ldots, x^n, u^1, \ldots, u^m))$ be a vector bundle chart over (U, φ). A section s of (E, π, M) is smooth provided only the pull-backs $s^*(x^i), s^*(u^\alpha)$ of the coordinates on E_U are smooth, $i = 1, \ldots, n$, $\alpha = 1, \ldots, m$. But $s^*(x^i) = x^i$ for all sections, hence s is smooth provided only the functions $s^*(u^\alpha)$ on U are smooth for some family of vector bundle charts (E_U, φ^E) such that the (U, φ) cover M. In the following we denote $s^\alpha := s^*(u^\alpha)$ and call them the *components* of the section s in the vector bundle chart (E_U, φ^E). We can then rephrase the preceding discussion saying that a section s of (E, π, M) is smooth if and only if its components s^α are smooth in every vector bundle chart of an atlas.

Proposition 9.14. *Smooth sections of a vector bundle* (E, π, M) *form a submodule in the module of all sections.*

Proof. Left as Exercise 9.3. $\qquad\qquad\qquad\qquad\qquad\qquad\qquad\square$

Exercise 9.3. Prove Proposition 9.14.

We denote by $\Gamma(E)$ the $C^\infty(M)$-module of smooth sections of the vector bundle (E, π, M). In the following, by "*section*" we will always mean "*smooth section*", unless otherwise stated.

Example 9.15 (Sections of a Trivial Vector Bundle). Let M be a smooth manifold and let V be a finite-dimensional real vector space. The space of smooth maps $F : M \to V$ is a $C^\infty(M)$-module with the point-wise operations

$$+ : (F_1, F_2) \mapsto F_1 + F_2, \qquad (F_1 + F_2)(p) := F_1(p) + F_2(p),$$
$$\cdot : (f, F) \mapsto fF, \qquad\qquad (fF)(p) := f(p)F(p),$$

where $f \in C^\infty(M)$ and $F, F_1, F_2 : M \to V$ are smooth maps. We denote this module by $C^\infty(M, V)$.

Now let $V_M = M \times V$ be the trivial vector bundle over M with abstract fiber V. Given a section s of V_M we get a smooth map $F_s : M \to V$ by composing with the projection $\mathrm{pr}_V : V_M \to V$ onto the second factor: $F_s := \mathrm{pr}_V \circ s$. It is easy to see that the assignment $s \mapsto F_s$ establishes an isomorphism between the modules $\Gamma(V_M)$ and $C^\infty(M, V)$. The inverse isomorphism maps a smooth map $F \in C^\infty(M, V)$ to the section s_F given by $s_F(p) = (p, F(p)) \in V_M$ for all $p \in M$. We leave the easy details to the reader. $\qquad\qquad\qquad\qquad\qquad\qquad\qquad\qquad\qquad\qquad\blacklozenge$

From now on, in this section, we will assume that we have chosen once for all a frame of V that we use to identify it with $\mathbb{R}^m$. In particular, a local trivialization $(\mathcal{U}, \Phi_\mathcal{U})$ of E will identify $E_\mathcal{U}$ with $\mathcal{U} \times \mathbb{R}^m$.

The module of sections of a vector bundle E does not possess a (finite) frame in general. It does precisely when E is a trivializable vector bundle. To see this, we begin proving that, if E is trivializable, then $\Gamma(E)$ possesses a finite frame. This is basically the content of the following

Proposition 9.16. *Let* (E, π, M) *be a vector bundle with a global trivialization* $\Phi : E \to M \times \mathbb{R}^m$, *let* $(E_1, \ldots, E_m)$ *be the canonical frame of* $\mathbb{R}^m$, *and let* $s_1^\Phi, \ldots, s_m^\Phi$ *be the sections of* E *defined by putting*

$$s_\alpha^\Phi(p) = \Phi^{-1}(p, E_\alpha), \quad \alpha = 1, \ldots, m. \tag{9.6}$$

Then $(s_1^\Phi, \ldots, s_m^\Phi)$ *is a frame of* $\Gamma(E)$.

Proof. It is clear that the s_α^Φ are smooth sections (they are sections and can be written as compositions of smooth maps. Do you see it?). We have to show that they form a frame of $\Gamma(E)$. To do this, first note that, for all $p \in M$, $(s_1^\Phi(p), \ldots, s_m^\Phi(p))$ is a frame of E_p. Indeed, it is the image of the frame $((p, E_1), \ldots, (p, E_m))$ of $p \times \mathbb{R}^m \cong \mathbb{R}^m$ under $\Phi^{-1} : M \times \mathbb{R}^m \to E$. But Φ, hence Φ^{-1}, is a vector bundle isomorphism, hence it maps linearly and bijectively, i.e., isomorphically, fibers of $M \times V$ to fibers of E. As an isomorphism of vector spaces transforms frames into frames, we conclude that $(s_1^\Phi(p), \ldots, s_m^\Phi(p))$ is a frame of E_p as claimed. Now, we are ready to show that the s_α^Φ are linearly independent. To do this, take functions $f^1, \ldots, f^m \in C^\infty(M)$ such that

$$f^\alpha s_\alpha^\Phi = 0.$$

This means that, for all $p \in M$,

$$0 = (f^\alpha s_\alpha^\Phi)(p) = f^\alpha(p)s_\alpha^\Phi(p).$$

As the $s_\alpha^\Phi(p)$ form a frame of E_p, it follows that $f^\alpha(p) = 0$. From the arbitrariness of p, we get $f^\alpha = 0$ for all $\alpha = 1, \ldots, m$. Now we show that the s_α^Φ span $\Gamma(E)$. So let $s \in \Gamma(E)$ and define the vector valued function $F : M \to \mathbb{R}^m$ as the composition

$$M \xrightarrow{s} E \xrightarrow{\Phi} M \times \mathbb{R}^m \longrightarrow \mathbb{R}^m,$$

where the last arrow is the projection onto the second factor. Being the composition of smooth maps, F is a smooth map, hence its components $f^1, \ldots, f^m$ are smooth functions on M. Let $p \in M$ and compute

$$
\begin{aligned}
(f^\alpha s_\alpha^\Phi)(p) &= f^\alpha(p)s_\alpha^\Phi(p) \\
&= f^\alpha(p)\Phi_p^{-1}(p, E_\alpha) \\
&= \Phi_p^{-1}(p, f^\alpha(p)E_\alpha) \\
&= \Phi^{-1}(p, F(p)) \\
&= s(p).
\end{aligned}
$$

This shows that $s = f^\alpha s_\alpha^\Phi$ is a linear combination of the s_α^Φ as desired. $\square$

Example 9.17. Let $\mathbb{R}_M^m = M \times \mathbb{R}^m \to M$ be the trivial vector bundle with abstract fiber $\mathbb{R}^m$. According to Proposition 9.16, the *identical trivialization* id : $\mathbb{R}_M^m \to M \times \mathbb{R}^m$ determines a frame of $\Gamma(\mathbb{R}_M^m)$ that, abusing the notation, we denote $(E_1, \ldots, E_m)$ (as the canonical frame of $\mathbb{R}^m$). The identification

$$\Gamma(\mathbb{R}^m_M) \cong C^\infty(M, \mathbb{R}^m)$$

from Example 9.15 now identifies a section $s = f^\alpha E_\alpha$ of $\mathbb{R}^m_M$ with the vector-valued function $(f_1, \ldots, f_m) : M \to \mathbb{R}^m$. ◆

Proving that, when the module of sections $\Gamma(E)$ of the vector bundle E possesses a finite frame, then E is trivializable is much harder, and we will need some intermediate results of an independent interest. For now, we discuss the relationship between sections and vector bundle maps. Consider two vector bundles $(E, \pi, M), (E', \pi', M)$ over the same base M. We will show that there is a canonical bijection between $C^\infty(M)$-module homomorphisms $\Gamma(E) \to \Gamma(E')$ and vector bundle maps $E \to E'$ covering the identity. The discussion here is very similar to that going from Lemma 7.12 to Proposition 7.14 and does actually generalize the latter. First, we need two lemmas.

Lemma 9.18. *Let $h : \Gamma(E) \to \Gamma(E')$ be a $C^\infty(M)$-linear map, let $p_0 \in M$, and let $s \in \Gamma(E)$. Then the value*

$$h(s)(p_0)$$

of the section $h(s) \in \Gamma(E')$ at p_0 does only depend on (h and) the value $s(p_0)$ of the section s at p_0.

Proof. The proof is formally identical to that of Lemma 7.12 and we only list the necessary steps leaving the details to the reader.

Step I: Homomorphisms between Sections are Local Operators. *For every open subset $\mathcal{U} \subseteq M$, the section $h(s)|_{\mathcal{U}}$ of $\Gamma(E'_{\mathcal{U}})$ does only depend on (h and) the restriction $s|_{\mathcal{U}}$.* This can be proved using a bump function exactly as in Step I of the proof of Lemma 7.12.

Step II: Local Extension Lemma for Sections of a Vector Bundle. *Let $S \subseteq M$ be a submanifold, and let s be a section of the vector bundle E_S. For every point p_0 in S, there exists an open neighborhood $U \subseteq M$ of p_0 and a section $\tilde{s}$ of E such that $\tilde{s}|_{S \cap U} = s|_{S \cap U}$.* This can be proved exactly as in the proof of the Local Extension Lemma 2.47 for functions but using the Gluing Lemma for Smooth Maps.

Step III: Sections Vanishing at a Point. *Let s be a section of E vanishing at a point p, i.e., $s(p) = 0$, then there exist sections $s_1, \ldots, s_m \in \Gamma(E)$ and smooth* *functions $f_1, \ldots, f_m$ such that* (1) *the f_i all vanish at p, i.e., $f_i(p) = 0$, and* (2)

$$s = f_1 s_1 + \cdots + f_m s_m$$

in an open neighborhood of p. Similarly as in Step III of the proof of Lemma 7.12, consider a local trivialization $(E_{\mathcal{U}}, \Phi_{\mathcal{U}})$ of E around p. Applying Proposition 9.16 to the vector bundle $E_{\mathcal{U}}$, we find a frame $(s_1^{\Phi_{\mathcal{U}}}, \ldots, s_m^{\Phi_{\mathcal{U}}})$ of $\Gamma(E_{\mathcal{U}})$. Hence,

$$s|_{\mathcal{U}} = g^\alpha s_\alpha^{\Phi_{\mathcal{U}}}$$

for some smooth functions $g^\alpha \in C^\infty(\mathcal{U})$. As

$$0 = s(p) = g^\alpha(p) s_\alpha^{\Phi_{\mathcal{U}}}(p)$$

and $(s_1^{\Phi_{\mathcal{U}}}(p), \ldots, s_m^{\Phi_{\mathcal{U}}}(p))$ is a frame of E_p (see the proof of Proposition 9.16), we get $g^\alpha(p) = 0$ for all $\alpha = 1, \ldots, m$. Now let $f_1, \ldots, f_m \in C^\infty(M)$ be local extensions of $g^1, \ldots, g^m$ and let $s_1, \ldots, s_m \in \Gamma(E)$ be local extensions of $s_1^{\Phi_{\mathcal{U}}}, \ldots, s_m^{\Phi_{\mathcal{U}}}$ around p (the latter exist by the Local Extension Lemma in Step II). Then $f_1 s_1 + \cdots + f_m s_m$ agrees with $s|_{\mathcal{U}}$, hence with s, in a neighborhood of p as desired.

Step IV: Conclusion. Exactly as in Step IV of the proof of Lemma 7.12.

$\square$

Lemma 9.19. *Let (E, π, M) be a vector bundle, let $p \in M$, and let $e \in E_p$. Then there exists a section $s \in \Gamma(E)$ such that $s(p) = e$.*

Proof. The proof is similar to that of Lemma 7.13. Let $(\mathcal{U}, \Phi_{\mathcal{U}})$ be a local trivialization of E around p, i.e., $p \in \mathcal{U}$, and let $x \in \mathbb{R}^m$ be defined by

$$\Phi_{\mathcal{U}}(e) = (p, x).$$

The map

$$c_x : \mathcal{U} \to \mathcal{U} \times \mathbb{R}^m, \quad q \mapsto (q, x)$$

is a smooth section of the trivial vector bundle $\mathcal{U} \times \mathbb{R}^m$, hence $\tau := \Phi_{\mathcal{U}}^{-1} \circ c_x$ is a smooth section of $E_{\mathcal{U}}$ (do you see it?). We have

$$\tau(p) = \Phi_{\mathcal{U}}^{-1}(c_x(p)) = \Phi_{\mathcal{U}}^{-1}(p, x) = \Phi_{\mathcal{U}}^{-1}(\Phi_{\mathcal{U}}(e)) = e.$$

Unfortunately, τ cannot be extended to a global smooth section of E, in general. However, from the Local Extension Lemma for Sections of a Vector Bundle (Step II in the proof of Lemma 9.18), there exists a global section $s \in \Gamma(E)$ such that s and τ agree around p. In particular, $s(p) = \tau(p) = e$ as desired. $\qquad\qquad\qquad\qquad\qquad\qquad\qquad\qquad\qquad\qquad\qquad\qquad\qquad\square$

The next result will be extremely useful both in this and the following section and, similarly as Propositions 5.8 and 7.14, shows the close connection between algebraic aspects and geometric aspects in Differential Geometry. Begin with two vector bundles $(E, \pi, M), (E', \pi', M')$ over the same base and a $C^\infty(M)$-module homomorphism $h : \Gamma(E) \to \Gamma(E')$. For any point $p \in M$, we want to define a linear map $h_p : E_p \to E'_p$. So, let $e \in E_p$, and let $s \in \Gamma(E)$ be any section such that $s(p) = e$ (it exists in view of Lemma 9.19). Put

$$h_p(e) := h(s)(p) \in E'_p.$$

From Lemma 9.18, $h_p(e)$ does only depend on (h and) e. Hence, h_p is a well-defined map $h_p : E_p \to E'_p$. It is easy to see that h_p is a linear map (we leave the details to the reader, see the discussion preceding Proposition 7.14). We can glue the h_p together in a(n *a priori* non-necessarily smooth) map

$$F_h : E \to E', \quad e \mapsto h_p(e),$$

where $p = \pi(e)$. Clearly, the diagram

commutes.

Theorem 9.20. *The assignment*

$$h \mapsto F_h \qquad\qquad\qquad\qquad\qquad\qquad (9.7)$$

establishes a well-defined bijection between $C^\infty(M)$-linear maps $h : \Gamma(E) \to \Gamma(E')$ and vector bundle maps $E \to E'$ covering the identity.

Proof. The proof is very similar to that of Proposition 7.14 (and actually generalizes the latter; we invite the reader to compare the two discussions and note the similarities). First, we prove that, for every $C^\infty(M)$-linear map $h : \Gamma(E) \to \Gamma(E')$, the map $F_h : E \to E'$ is smooth. To do this, take a chart $(U, \varphi = (x^1, \ldots, x^n))$ on M. Shrinking U if necessary, we can always assume that there are vector bundle charts

$(E_U, \varphi^E = (x^1, \ldots, x^n, u^1, \ldots, u^m))$ and $(E'_U, \varphi^{E'} = (x^1, \ldots, x^n, v^1, \ldots, v^{m'}))$ over (U, φ) on E and E', induced by local trivializations (E_U, Φ_U) and (E'_U, Φ'_U). It is clear that $F_h(E_U) \subseteq E'_U$. The restriction $F_h : E_U \to E'_U$ is smooth if and only if the pull-backs $F_h^*(x^i)$, $F_h^*(v^a)$ are smooth functions on E_U for all $i = 1, \ldots, n$, $a = 1, \ldots, m'$. Now, let $p \in U$ and $e \in E_p$. It is easy to see that $F_h^*(x^i) = x^i$ for all i, and it remains to check that the $F_h^*(v^a)$ are smooth. But the $F_h^*(v^a)$ are linear functions on the fibers of E_U. It follows that there are functions $F_\alpha^a : U \to \mathbb{R}$ such that

$$F_h^*(v^a)(e) = F_\alpha^a(p)u^\alpha(e), \quad e \in E_U, \quad p = \pi(e)$$

(do you see it?). We are thus led to show that the F_α^a are smooth functions. To do this, we fix $p_0 \in U$ and show that the F_α^a are smooth around p_0. So, consider the sections $s_1 := s_1^{\Phi_U}, \ldots, s_m := s_m^{\Phi_U}$ from the proof of Proposition 9.16. Formula (9.6) now means that, for all $p \in U$ and all $\alpha, \beta = 1, \ldots, m$,

$$u^\alpha\left(s_\beta(p)\right) = \delta_\beta^\alpha.$$

It follows that

$$F_\beta^a(p) = F_\alpha^a(p)\delta_\beta^\alpha = F_\alpha^a(p)u^\alpha\left(s_\beta(p)\right) = F_h^*(v^a)\left(s_\beta(p)\right)$$
$$= v^a\left(F_h\left(s_\beta(p)\right)\right).$$

To continue the computation, we need to choose a section $\tilde{s}_\beta \in \Gamma(E)$ such that $\tilde{s}_\beta(p) = s_\beta(p)$. It is convenient to choose $\tilde{s}_\beta$ in a more precise way. Namely, using the Local Extension Lemma for Sections of a Vector Bundle, we choose $\tilde{s}_\beta$ such that it agrees with s_β in an open neighborhood $V \subseteq U$ of p_0. Then

$$F_\beta^a(p) = v^a\left(F_h\left(s_\beta(p)\right)\right) = v^a\left(h(\tilde{s}_\beta)(p)\right)$$

for all $p \in V$. This shows that $F_\beta^a|_V$ is smooth. From the arbitrariness of p_0, the F_β^a are smooth, as desired.

It remains to check that the assignment $h \mapsto F_h$ is one-to-one. To do this, we define its inverse (which is even easier). Given a vector bundle map $F : E \to E'$ covering the identity, we construct a $C^\infty(M)$-linear map $h_F : \Gamma(E) \to \Gamma(E')$ in such a way that

$$F_{h_F} = F \quad \text{and} \quad h_{F_h} = h, \tag{9.8}$$

for all vector bundle maps $F : E \to E'$ covering the identity and all $C^\infty(M)$-linear maps $h : \Gamma(E) \to \Gamma(E')$. For every sections $s \in \Gamma(E)$, we define

$$h_F(s) := F \circ s : M \to E'.$$

Clearly, $h_F(s)$ is a smooth section of E'. We still have to show that

- $h_F : \Gamma(E) \to \Gamma(E')$ is a $C^\infty(M)$-linear map,
- the identities (9.8) hold.

We leave this to the reader as Exercise 9.4. This concludes the proof. $\square$

Exercise 9.4. Complete the proof of Theorem 9.20 showing that

(1) for every vector bundle map $F : E \to E'$ covering the identity, the map $h_F : \Gamma(E) \to \Gamma(E')$ is $C^\infty(M)$-linear,
(2) the assignment $F \mapsto h_F$ inverts $h \mapsto F_h$.

Theorem 9.20 has some important consequences.

Corollary 9.21 (Global Trivializations and Frames of Sections). *Let (E, π, M) be a rank m vector bundle. The assignment $\Phi \mapsto (s_1^\Phi, \dots, s_m^\Phi)$ of Proposition 9.16 establishes a bijection between global trivializations of E and frames of the module $\Gamma(E)$ of sections.*

Proof. We have to invert the assignment $\Phi \mapsto (s_1^\Phi, \dots, s_m^\Phi)$ in the statement. First note that, given a frame $\mathcal{R} = (s_1, \dots, s_q)$ of $\Gamma(E)$, there exists a unique $C^\infty(M)$-linear isomorphism $h_{\mathcal{R}} : \Gamma(E) \to \Gamma(\mathbb{R}_M^q)$ mapping s_α to E_α, $\alpha = 1, \dots, q$ (see Example 9.17 for the frame $(E_1, \dots, E_q)$ of the module $\Gamma(\mathbb{R}_M^q)$ of sections of the trivial vector bundle $\mathbb{R}_M^q$). According to Theorem 9.20, $h_{\mathcal{R}}$ corresponds to a vector bundle map $\Phi_{\mathcal{R}} : E \to \mathbb{R}_M^q$. Actually, $\Phi_{\mathcal{R}}$ is a vector bundle isomorphism whose inverse is the vector bundle map corresponding to $h_{\mathcal{R}}^{-1}$. We conclude that $\Phi_{\mathcal{R}}$ is a global trivialization. Finally, the assignment $\mathcal{R} \mapsto \Phi_{\mathcal{R}}$ inverts the assignment $\Phi \mapsto (s_1^\Phi, \dots, s_m^\Phi)$ of Proposition 9.16. In particular, $q = m$. We leave the details as Exercise 9.5. $\square$

Exercise 9.5. Complete the proof of Corollary 9.21 showing that

(1) the vector bundle map $\Phi_{\mathcal{R}}$ is a vector bundle isomorphism, and
(2) the assignment $\mathcal{R} \mapsto \Phi_{\mathcal{R}}$ inverts the assignment $\Phi \mapsto (s_1^\Phi, \dots, s_m^\Phi)$ of Proposition 9.16.

Corollary 9.22. Let (E, π, M) be a rank m vector bundle and let $s_1, \ldots, s_m$ be sections of E. Then $(s_1, \ldots, s_m)$ is a frame of the module $\Gamma(E)$ if and only if $(s_1(p), \ldots, s_m(p))$ is a frame of the vector space E_p for all $p \in M$.

Proof. Let $(s_1, \ldots, s_m)$ be a frame of $\Gamma(E)$. According to Corollary 9.21, it corresponds to a trivialization $\Phi : E \to M \times \mathbb{R}^m$ of E. The latter identifies the fiber E_p of E over $p \in M$ with $\{p\} \times \mathbb{R}^m$ by mapping $(s_1(p), \ldots, s_m(p))$ to $(E_1, \ldots, E_m)$, the canonical frame (do you see it?). As linear isomorphisms identify frames, it follows that $(s_1(p), \ldots, s_m(p))$ is a frame of E_p. Conversely, let $\mathcal{R}_p := (s_1(p), \ldots, s_m(p))$ be a frame of E_p for all p. Then the map $\Phi : E \to M \times \mathbb{R}^m$ mapping $e \in E_p$ to $(p, \varphi_{\mathcal{R}_p}(e))$ is a trivialization of E (do you see it?). Finally, the trivialization Φ corresponds to a frame $(s_1^\Phi, \ldots, s_m^\Phi)$ of $\Gamma(E)$, but it is clear that, actually, $s_\alpha^\Phi = s_\alpha$ for all $\alpha = 1, \ldots, m$. We conclude that $(s_1, \ldots, s_m)$ is a frame of $\Gamma(E)$ as desired. $\qquad \square$

Remark 9.23 (Local Frames of a Vector Bundle). Let (E, π, M) be a rank m vector bundle and let $(\mathcal{U}, \Phi_\mathcal{U})$ be a local trivialization. As $\Phi_\mathcal{U} : E_\mathcal{U} \to \mathcal{U} \times \mathbb{R}^m$ is a global trivialization of the vector bundle $E_\mathcal{U}$, it corresponds to a frame $(s_1, \ldots, s_m)$ of the $C^\infty(\mathcal{U})$-module $\Gamma(E_\mathcal{U})$. In particular, $(s_1(p), \ldots, s_m(p))$ is a frame of E_p for all $p \in \mathcal{U}$. A frame $(s_1, \ldots, s_m)$ of $\Gamma(E_\mathcal{U})$ is sometimes called a *local frame* of E and this discussion shows that (as it possesses local trivializations) every vector bundle possesses local frames. $\qquad \diamond$

9.3 Constructions with Vector Bundles

There are many natural ways to construct new vector bundles from given vector bundles (over the same base). Roughly, let (E, π, M) be a vector bundle and suppose that there is a *sufficiently natural* "construction" $\mathcal{C}$ in Linear Algebra giving a new finite dimensional vector space $\mathcal{C}(V)$ from a given one V (think, e.g., of the case $\mathcal{C}(V) = V^*$, the dual vector space). Now apply the construction $\mathcal{C}$ to the family $\{E_p\}_{p \in M}$ of fibers of E to get a new family $\{\mathcal{C}(E_p)\}_{p \in M}$ of finite dimensional vector spaces. We do not explain here what we exactly mean by "*sufficiently natural*". We only mention that, in most cases, $\{\mathcal{C}(E_p)\}_{p \in M}$ will be again a vector bundle over M. The same strategy will usually work if $\mathcal{C}$ is a "*construction*" giving a new vector space $\mathcal{C}(V_1, \ldots, V_k)$ from multiple given vector spaces $V_1, \ldots, V_k$ (think, e.g., of the case $\mathcal{C}(V_1, V_2) = \mathrm{Hom}(V_1, V_2)$, the vector space of linear maps $V_1 \to V_2$). In this section, we make this idea rigorous in two cases, which can serve as building blocks for several other examples. Specifically, we consider the cases when $\mathcal{C}$ is "*linear maps*" or "*tensor product*" between 2-vector spaces.

We begin with the *vector bundle of linear maps*. So let (E, π, M), (E', π', M) be two vector bundles over the same base manifold M. We denote by V, V' the abstract fibers and by $m = \dim V$, $m' = \dim V'$ the ranks. Consider the corresponding families $\{E_p\}_{p \in M}$, $\{E'_p\}_{p \in M}$ of vector spaces. For every $p \in M$, the vector space $\mathrm{Hom}(E_p, E'_p)$ of linear maps $E_p \to E'_p$ is a finite dimensional real vector space of dimension mm'.

Theorem 9.24 (The Hom Vector Bundle). *The family of vector spaces*

$$\big\{ \mathrm{Hom}(E_p, E'_p) \big\}_{p \in M}$$

can be given a natural structure of rank mm' vector bundle (to be specified in the proof) with abstract fiber $\mathrm{Hom}(V, V')$.

Proof.   We use the Vector Bundle Chart Theorem (Theorem 9.5). First of all, denote

$$\mathrm{Hom}(E, E') := \coprod_{p \in M} \mathrm{Hom}(E_p, E'_p).$$

Note that, for every point $p_0 \in M$, there are local trivializations $(\mathcal{U}, \Phi_{\mathcal{U}}), (\mathcal{U}, \Phi'_{\mathcal{U}})$ of E, E' over the same open neighborhood $\mathcal{U} \ni p_0$ (this can be achieved by passing to subtrivializations if necessary, do you see it?). In what follows, for every $p \in \mathcal{U}$, we denote by $\Phi_{\mathcal{U},p} : E_p \to V$ (resp. $\Phi'_{\mathcal{U},p} : E'_p \to V'$) the vector space isomorphism induced by $\Phi_{\mathcal{U}}$ (resp. $\Phi'_{\mathcal{U}}$) on the fiber:

$$\Phi_{\mathcal{U},p} := \Phi_{\mathcal{U}} : E_p \to \{p\} \times V \cong V$$

$$(\text{resp. } \Phi'_{\mathcal{U},p} := \Phi'_{\mathcal{U}} : E'_p \to \{p\} \times V' \cong V').$$

Every pair of trivializations as above determines a bijection

$$\Phi_{\mathcal{U}}^{\mathrm{Hom}} : \mathrm{Hom}(E, E')_{\mathcal{U}} \to \mathcal{U} \times \mathrm{Hom}(V, V')$$

as follows. For every $p \in \mathcal{U}$ and every linear map $h : E_p \to E'_p$, we put

$$\Phi_{\mathcal{U}}^{\mathrm{Hom}}(p, h) := (p, \Phi'_{\mathcal{U},p} \circ h \circ \Phi_{\mathcal{U},p}^{-1} : V \to V').$$

By construction, the diagram

commutes. Moreover, for all $p \in \mathcal{U}$, the restriction $\Phi_{\mathcal{U}}^{\mathrm{Hom}} : \mathrm{Hom}(E_p, E_p') \to \{p\} \times \mathrm{Hom}(V, V')$ maps h to $\Phi_{\mathcal{U},p}' \circ h \circ \Phi_{\mathcal{U},p}^{-1}$, hence it is a vector space isomorphism. Finally, let $(\mathcal{U}, \Phi_{\mathcal{U}}), (\mathcal{V}, \Phi_{\mathcal{V}})$ be two local trivializations of E and let $(\mathcal{U}, \Phi_{\mathcal{U}}'), (\mathcal{V}, \Phi_{\mathcal{V}}')$ be two local trivializations of E' over the same open subsets $\mathcal{U}, \mathcal{V}$. We want to compute

$$\Phi_{\mathcal{U}\mathcal{V}}^{\mathrm{Hom}}$$
$$= \Phi_{\mathcal{V}}^{\mathrm{Hom}} \circ \Phi_{\mathcal{U}}^{\mathrm{Hom}\,-1} : (\mathcal{U} \cap \mathcal{V}) \times \mathrm{Hom}(V, V') \to (\mathcal{U} \cap \mathcal{V}) \times \mathrm{Hom}(V, V').$$

So, first denote by $A : \mathcal{U} \cap \mathcal{V} \to \mathrm{GL}(V)$ (resp. $A' : \mathcal{U} \cap \mathcal{V} \to \mathrm{GL}(V')$) the smooth map determined by

$$\Phi_{\mathcal{U}\mathcal{V}}(p, v) = (p, A(p)(v)) \quad (\text{resp. } \Phi_{\mathcal{U}\mathcal{V}}'(p, v') = (p, A'(p)(v')))$$

for all $(p, v) \in (\mathcal{U} \cap \mathcal{V}) \times V$ (resp. $(p, v') \in (\mathcal{U} \cap \mathcal{V}) \times V'$. Then, for all $(p, h) \in (\mathcal{U} \cap \mathcal{V}) \times \mathrm{Hom}(V, V')$, we have

$$
\begin{aligned}
\Phi_{\mathcal{U}\mathcal{V}}^{\mathrm{Hom}}(p, h) &= \Phi_{\mathcal{V}}^{\mathrm{Hom}} \circ \Phi_{\mathcal{U}}^{\mathrm{Hom}\,-1}(p, h) \\
&= \Phi_{\mathcal{V}}^{\mathrm{Hom}}\left(p, \Phi_{\mathcal{U},p}'^{-1} \circ h \circ \Phi_{\mathcal{U},p}\right) \\
&= \left(p, \Phi_{\mathcal{V},p}' \circ \Phi_{\mathcal{U},p}'^{-1} \circ h \circ \Phi_{\mathcal{U},p} \circ \Phi_{\mathcal{V},p}^{-1}\right) \\
&= \left(p, A'(p) \circ h \circ A(p)^{-1}\right).
\end{aligned}
$$

In other words,

$$\Phi_{\mathcal{U}\mathcal{V}}^{\mathrm{Hom}}(p, h) = (p, A^{\mathrm{Hom}}(p)(h)),$$

where $A^{\mathrm{Hom}} : \mathcal{U} \cap \mathcal{V} \to \mathrm{GL}\left(\mathrm{Hom}(V, V')\right)$ is the map given by

$$A^{\mathrm{Hom}}(p)(h) := A'(p) \circ h \circ A(p)^{-1}.$$

As A, A' are smooth maps, it easily follows that A^{Hom} is also a smooth map, and this concludes the proof. □

Example 9.25 (Dual Vector Bundle). When $E' = \mathbb{R}_M$, the trivial vector bundle with abstract fiber $\mathbb{R}$, then the vector bundle $\mathrm{Hom}(E, E') = \mathrm{Hom}(E, \mathbb{R}_M)$ is also called the *dual vector bundle* of E, and it is denoted by E^*. The fibers of E^* are duals to the fibers of E: $E_p^* = (E_p)^*$ for all $p \in M$. For instance, the cotangent bundle T^*M is the dual vector bundle of the tangent bundle: $T^*M = (TM)^*$. ◆

We now discuss sections of the Hom-vector bundle. Namely, we prove that, given two vector bundles E, E' over the same base M, the $C^\infty(M)$-module homomorphisms $h : \Gamma(E) \to \Gamma(E')$ not only are equivalent to vector bundle maps $F : E \to E'$ covering the identity (Theorem 9.20) but they are also equivalent to sections of $\mathrm{Hom}(E, E')$. To see this, we begin remarking that a vector bundle map $F : E \to E'$ covering the identity can be seen as a section $\eta_F \in \Gamma(\mathrm{Hom}(E, E'))$ by putting $\eta_F(p) := F_p : E_p \to E'_p$ for all $p \in M$. The smoothness of η_F easily follows from that of F. It is also clear that the assignment $F \mapsto \eta_F$ establishes a bijection between vector bundle maps $F : E \to E'$ covering the identity and sections of the Hom-vector bundle. Indeed, given a section $\eta \in \Gamma(\mathrm{Hom}(E, E'))$, we can define a vector bundle map $F_\eta : E \to E'$ by putting $F_\eta(e) = \eta(p)(e)$, where $p = \pi(e)$. The smoothness of F_η follows from that of η. Finally, the assignment $\eta \mapsto F_\eta$ inverts the assignment $F \mapsto \eta_F$. We leave the straightforward details to the reader.

Proposition 9.26. *The map*

$$\mathrm{Hom}\left(\Gamma(E), \Gamma(E')\right) \to \Gamma\left(\mathrm{Hom}(E, E')\right), \quad h \mapsto \eta_h := \eta_{F_h} \qquad (9.9)$$

is a well-defined $C^\infty(M)$-module isomorphism (here $F_h : E \to E'$ is the vector bundle map corresponding to $h \in \mathrm{Hom}\left(\Gamma(E), \Gamma(E')\right)$).

Proof.   From Theorem 9.20 and the above discussion, it follows that the map in the statement is a bijection. The $C^\infty(M)$-linearity is easy and it is left as Exercise 9.6.   □

Exercise 9.6. Prove that the map (9.9) is $C^\infty(M)$-linear (hence a $C^\infty(M)$-module isomorphism).

The map (9.9) is explicitly given by

$$\eta_h(p)(e) = h(s)(p), \quad p \in M, \quad e \in E_p,$$

where $s \in \Gamma(E)$ is any section such that $s(p) = e$. The inverse $\eta \mapsto h_\eta$ is given by

$$h_\eta(s)(p) = \eta(p)\left(s(p)\right), \quad s \in \Gamma(E), \quad p \in M.$$

When $E' = \mathbb{R}_M$, we find a canonical $C^\infty(M)$-module isomorphism $\Gamma(E)^* \cong \Gamma(E^*)$. If, additionally, $E = TM$, this recovers Proposition 7.14.

Exercise 9.7. Let (E, π, M) be a vector bundle. Prove that the standard biduality isomorphisms $E_p \cong E_p^{**}$, $p \in M$, determine a vector bundle isomorphism $E \cong E^{**}$. Conclude that there are canonical $C^\infty(M)$-module isomorphisms

$$\Gamma(E) \cong \Gamma(E^{**}) \cong \Gamma(E^*)^*.$$

There is another important construction in Linear Algebra that can be applied to fibers of vector bundles to get a new vector bundle: the *tensor product*. We first need some algebraic preliminaries on the tensor product. We work in the general setting of *tensor products of modules* over an algebra. So let A be a real (associative, commutative, and unital) algebra and let $\mathcal{M}, \mathcal{N}$ be A-modules.

Definition 9.27 (Tensor Product). A *tensor product* of $\mathcal{M}$ and $\mathcal{N}$ is an A-module $\mathcal{T}$ together with an A-bilinear map

$$\otimes : \mathcal{M} \times \mathcal{N} \to \mathcal{T}, \quad (\mu, \nu) \mapsto \mu \otimes \nu$$

satisfying the following *universal property of tensor products*: for any other A-module $\mathcal{P}$ and any other bilinear map $\beta : \mathcal{M} \times \mathcal{N} \to \mathcal{P}$, there exists a unique linear map $\beta_\otimes : \mathcal{T} \to \mathcal{P}$ such that the diagram

$$
\begin{array}{ccc}
\mathcal{M} \times \mathcal{N} & \xrightarrow{\ \beta\ } & \mathcal{P} \\
{\scriptstyle \otimes}\big\downarrow & \nearrow & \\
\mathcal{T} & \exists! \, \beta_\otimes &
\end{array}
\tag{9.10}
$$

commutes.

Exercise 9.8. Let $A = \mathbb{R}$, $\mathcal{M} = \mathbb{R}^m$, and $\mathcal{N} = \mathbb{R}^n$. Put $\mathcal{T} = M(m, n; \mathbb{R})$, the real vector space of $m \times n$ real matrices, and define

$$\otimes : \mathcal{M} \times \mathcal{N} \to \mathcal{T}$$

via

$$x \otimes y = \left(x^i y^j\right)_{j=1,\dots,n}^{i=1,\dots,m} \in M(m, n; \mathbb{R}),$$

for all $x = (x^1, \dots, x^m) \in \mathcal{M} = \mathbb{R}^m$ and $y = (y^1, \dots, y^n) \in \mathbb{R}^n$. Show that $(\mathcal{T}, \otimes)$ is a tensor product of $\mathcal{M}$ and $\mathcal{N}$.

Tensor products exist and are essentially unique according to the following

Theorem 9.28 (Existence and Uniqueness of the Tensor Product). *Let* $\mathcal{M}, \mathcal{N}$ *be A-modules. Then*

(1) *there exists a tensor product* $(\mathcal{T}, \otimes)$ *of* $\mathcal{M}$ *and* $\mathcal{N}$,
(2) *the tensor product is* unique up to a unique isomorphism *in the following sense: let* $(\mathcal{T}_1, \otimes_1), (\mathcal{T}_2, \otimes_2)$ *be tensor products of* $\mathcal{M}$ *and* $\mathcal{N}$, *then there exists a unique isomorphism* $\Phi : \mathcal{T}_1 \to \mathcal{T}_2$ *such that the following diagram commutes*

$$
\begin{array}{ccc}
& \mathcal{M} \times \mathcal{N} & \\
{\scriptstyle \otimes_1} \swarrow & & \searrow {\scriptstyle \otimes_2} \\
\mathcal{T}_1 & \xrightarrow[\exists! \, \Phi]{} & \mathcal{T}_2
\end{array}
\qquad (9.11)
$$

Proof. For exposition purposes, it is convenient, in this proof, to change the usual notation (μ, ν) for an element in $\mathcal{M} \times \mathcal{N}$ into the following: $\mu \widetilde{\otimes} \nu := (\mu, \nu)$. The proof requires some Linear Algebra of modules which we didn't discuss in details. However, the statements that we need are identical to the analogous statements for vector spaces and we can be quick on this material.

For the existence, begin with the set $\widetilde{\mathcal{T}}$ of *formal finite linear combinations*

$$
\sum_i a_i \left(\mu_i \widetilde{\otimes} \nu_i \right)
$$

of elements $\mu_i \widetilde{\otimes} \nu_i$ of $\mathcal{M} \times \mathcal{N}$ with coefficients a_i in A. The very demanding reader can interpret such a linear combination as a map $\mathcal{M} \times \mathcal{N} \to A$ which vanish on all but finitely many elements $\mu_i \widetilde{\otimes} \nu_i \in \mathcal{M} \times \mathcal{N}$. The coefficients a_i are then just the images of the $\mu_i \widetilde{\otimes} \nu_i$ under this map. The usual sum and multiplication by a scalar of linear combinations give to $\widetilde{\mathcal{T}}$ the structure of an A-module (when you interpret a linear combination as a map $\mathcal{M} \times \mathcal{N} \to A$ then the sum and the multiplication by a scalar become the point-wise operations). There is an obvious injection $\mathcal{M} \times \mathcal{N} \hookrightarrow \widetilde{\mathcal{T}}$ mapping (μ, ν) to $\mu \widetilde{\otimes} \nu$ and whose image consists of independent generators of $\widetilde{\mathcal{T}}$. However, this map is not A-bilinear. We can *turn it into a bilinear map* with the following trick: consider the submodule $\mathcal{Q} \subseteq \widetilde{\mathcal{T}}$ spanned by linear combinations of the form

$$
(a\mu + a'\mu') \widetilde{\otimes} \nu'' - a \left(\mu \widetilde{\otimes} \nu'' \right) - a' \left(\mu' \widetilde{\otimes} \nu'' \right)
$$

and

$$
\mu'' \widetilde{\otimes} (b\nu + b'\nu') - b \left(\mu'' \widetilde{\otimes} \nu \right) - b' \left(\mu'' \widetilde{\otimes} \nu' \right),
$$

where $\mu, \mu', \mu'' \in \mathcal{M}$, $\nu, \nu', \nu'' \in \mathcal{N}$, and $a, a', b, b' \in A$. Now, let $\mathcal{T}$ be the *quotient module* $\widetilde{\mathcal{T}}/\mathcal{Q}$. This means that $\mathcal{T}$ consists of equivalence classes with respect to the equivalence relation identifying two elements $T, T' \in \widetilde{\mathcal{T}}$ whenever $T - T' \in \mathcal{Q}$. As $\mathcal{Q}$ is a submodule by definition, the module operations on $\widetilde{\mathcal{T}}$ induce well-defined module operations on $\mathcal{T}$. Note that the projection $\mathrm{pr} : \widetilde{\mathcal{T}} \to \mathcal{T}$ mapping an element T to its equivalence class $[T]$ is A-linear.

We claim that the map

$$\otimes : \mathcal{M} \times \mathcal{N} \to \mathcal{T}, \quad (\mu, \nu) \mapsto \mu \otimes \nu$$

given by the composition

$$\mathcal{M} \times \mathcal{N} \hookrightarrow \widetilde{\mathcal{T}} \xrightarrow{\mathrm{pr}} \mathcal{T}$$

of the inclusion $\mathcal{M} \times \mathcal{N} \hookrightarrow \widetilde{\mathcal{T}}$ followed by the projection $\mathrm{pr} : \widetilde{\mathcal{T}} \to \mathcal{T}$ is A-bilinear. To see this, let $\mu, \mu' \in \mathcal{M}$, $\nu'' \in \mathcal{N}$, and $a, a' \in A$ and compute

$$(a\mu + a'\mu') \otimes \nu'' - a\left(\mu \otimes \nu''\right) - a'\left(\mu' \otimes \nu''\right)$$
$$= \mathrm{pr}\left((a\mu + a'\mu') \widetilde{\otimes} \nu'' - a\left(\mu \widetilde{\otimes} \nu''\right) - a'\left(\mu' \widetilde{\otimes} \nu''\right)\right) = 0_{\mathcal{T}},$$

where we used that $T = (a\mu + a'\mu') \widetilde{\otimes} \nu'' - a\left(\mu \widetilde{\otimes} \nu''\right) - a'\left(\mu' \widetilde{\otimes} \nu''\right) \in \mathcal{Q}$, hence its equivalence class $[T]$ is the null element $0_{\mathcal{T}}$ in the quotient $\mathcal{T} = \widetilde{\mathcal{T}}/\mathcal{Q}$. This shows that $\otimes$ is linear in the first argument. In the same way, it is also linear in the second argument, hence it is bilinear.

We want to show that $(\mathcal{T}, \otimes)$ is a tensor product of $\mathcal{M}$ and $\mathcal{N}$. We have to prove the universal property. So, let $\beta : \mathcal{M} \times \mathcal{N} \to \mathcal{P}$ be any bilinear map. As elements in $\mathcal{M} \times \mathcal{N}$ are independent generators of $\widetilde{\mathcal{T}}$, the map β extends to a unique A-linear map $\widetilde{\beta} : \widetilde{\mathcal{T}} \to \mathcal{P}$ (check this claim as an exercise). Note that $\mathcal{Q} \subseteq \ker \widetilde{\beta}$. Indeed, let $\mu, \mu', \mu'' \in \mathcal{M}$, $\nu, \nu', \nu'' \in \mathcal{N}$, and $a, a', b, b' \in A$, and compute

$$\widetilde{\beta}\left((a\mu + a'\mu') \widetilde{\otimes} \nu'' - a\left(\mu \widetilde{\otimes} \nu''\right) - a'\left(\mu' \widetilde{\otimes} \nu''\right)\right)$$
$$= \beta(a\mu + a'\mu', \nu'') - a\beta(\mu, \nu'') - a'\beta(\mu', \nu'') = 0$$

and

$$\widetilde{\beta}\left(\mu'' \widetilde{\otimes} (b\nu + b'\nu') - b\left(\mu'' \widetilde{\otimes} \nu\right) - b'\left(\mu'' \widetilde{\otimes} \nu'\right)\right)$$
$$= \beta(\mu'', b\nu + b'\nu') - b\beta(\mu'', \nu) - b'\beta(\mu'', \nu') = 0,$$

where we used that β is A-bilinear. As the generators of $\mathcal{Q}$ belong to $\ker \widetilde{\beta}$, we conclude that $\mathcal{Q} \subseteq \ker \widetilde{\beta}$, as claimed. It easily follows from this latter condition that the map

$$\beta_\otimes : \mathcal{T} \to \mathcal{P}, \quad [T] \mapsto \beta_\otimes[T] := \widetilde{\beta}(T)$$

is well defined and A-linear (check the details as an exercise). Moreover, for all $\mu \in \mathcal{M}$ and $\nu \in \mathcal{N}$,

$$\beta_\otimes(\mu \otimes \nu) = \widetilde{\beta}(\mu \widetilde{\otimes} \nu) = \beta(\mu, \nu),$$

showing that the diagram (9.10) commutes. Finally, if $\beta'_\otimes : \mathcal{T} \to \mathcal{P}$ is another linear map such that $\beta'_\otimes \circ \otimes = \beta$, then $\beta_\otimes$ and $\beta'_\otimes$ agree on the image of $\otimes$ which generates $\mathcal{T}$, and, by linearity, they must agree on the all $\mathcal{T}$. This concludes the proof of item (1) in the statement.

For the uniqueness, let $(\mathcal{T}_1, \otimes_1), (\mathcal{T}_2, \otimes_2)$ be two tensor products. As $\otimes_2 : \mathcal{M} \times \mathcal{N} \to \mathcal{T}_2$ is a bilinear map, from the universal property of $(\mathcal{T}_1, \otimes_1)$, it follows that there is a unique linear map $\Phi : \mathcal{T}_1 \to \mathcal{T}_2$ such that the diagram (9.11) commutes. On the other hand, from the universal property of $(\mathcal{T}_2, \otimes_2)$, there is also a (unique) linear map $\Psi : \mathcal{T}_2 \to \mathcal{T}_1$ such that $\Psi \circ \otimes_2 = \otimes_1$. It follows that

$$\Phi \circ \Psi \circ \otimes_2 = \Phi \circ \otimes_1 = \otimes_2 = \mathrm{id}_{\mathcal{T}_2} \circ \otimes_2.$$

But there is only one linear map $F : \mathcal{T}_2 \to \mathcal{T}_2$ such that $F \circ \otimes_2 = \otimes_2$, hence $\Phi \circ \Psi = \mathrm{id}_{\mathcal{T}_2}$. Similarly, $\Psi \circ \Phi = \mathrm{id}_{\mathcal{T}_1}$. So, Φ, Ψ are mutually inverse A-module isomorphisms. This concludes the proof. □

As there is a canonical tensor product $(\mathcal{T}, \otimes)$ (the one defined in the proof of Theorem 9.28) and all other tensor products identify canonically with $(\mathcal{T}, \otimes)$, we call $\mathcal{T}$ *the* tensor product of $\mathcal{M}$ and $\mathcal{N}$ and denote it $\mathcal{M} \otimes \mathcal{N}$, or $\mathcal{M} \otimes_A \mathcal{N}$ if we want to insist on which is the algebra of scalars. The following proposition states the main properties of the tensor product construction.

Proposition 9.29 (Main Properties of the Tensor Product). *Let A be an algebra and let $\mathcal{M}, \mathcal{N}, \mathcal{P}$ be A-modules.*

(1) (associativity) *There exists a unique A-module isomorphism*

$$(\mathcal{M} \otimes \mathcal{N}) \otimes \mathcal{P} \cong \mathcal{M} \otimes (\mathcal{N} \otimes \mathcal{P})$$

identifying $(\mu \otimes \nu) \otimes \pi$ and $\mu \otimes (\nu \otimes \pi)$, for all $\mu \in \mathcal{M}, \nu \in \mathcal{N}$ and $\pi \in \mathcal{P}$.

(2) (unitality) *There are unique A-module isomorphisms*

$$A \otimes \mathcal{M} \cong \mathcal{M} \cong \mathcal{M} \otimes A$$

identifying $1_A \otimes \mu$, μ *and* $\mu \otimes 1_A$, *for all* $\mu \in \mathcal{M}$.

Proof. Left as Exercise 9.9. $\square$

Exercise 9.9. Prove Proposition 9.29.

We will always use Proposition 9.29 to identify $(\mathcal{M} \otimes \mathcal{N}) \otimes \mathcal{P}$ and $\mathcal{M} \otimes (\mathcal{N} \otimes \mathcal{P})$. Accordingly, we will remove the brackets and simply write $\mathcal{M} \otimes \mathcal{N} \otimes \mathcal{P}$. Similarly, the elements $(\mu \otimes \nu) \otimes \pi$ and $\mu \otimes (\nu \otimes \pi)$ which identify under the isomorphism $(\mathcal{M} \otimes \mathcal{N}) \otimes \mathcal{P} \cong \mathcal{M} \otimes (\mathcal{N} \otimes \mathcal{P})$ will be simply denoted $\mu \otimes \nu \otimes \pi$. More generally, if $\mathcal{M}_1, \ldots, \mathcal{M}_k$ are A-modules, by induction on k, any two pairwise nested tensor products of $\mathcal{M}_1, \ldots, \mathcal{M}_k$ identify canonically and, if we understand this identification, we can safely use the symbol $\mu_1 \otimes \cdots \otimes \mu_k$ for any pairwise nested tensor product of elements $\mu_i \in \mathcal{M}_i, i = 1, \ldots, k$.

The following proposition explains in which precise sense *tensor products represent multilinear maps*.

Proposition 9.30 (Tensor Products Represent Multilinear Maps). *Let A be an algebra, and let* $\mathcal{M}_1, \ldots, \mathcal{M}_k$ *and* $\mathcal{P}$ *be A-modules. For any A-multilinear map* $\gamma : \mathcal{M}_1 \times \cdots \times \mathcal{M}_k \to \mathcal{P}$, *there exists a unique A-linear map* $\gamma_\otimes : \mathcal{M}_1 \otimes \cdots \otimes \mathcal{M}_k \to \mathcal{P}$ *such that the diagram*

$$
\begin{array}{ccc}
\mathcal{M}_1 \times \cdots \times \mathcal{M}_k & \xrightarrow{\ \gamma\ } & \mathcal{P} \\
{\scriptstyle \otimes}\downarrow & \nearrow{\scriptstyle \exists!\ \gamma_\otimes} & \\
\mathcal{M}_1 \otimes \cdots \otimes \mathcal{M}_k & &
\end{array}
$$

commutes. The assignment $\gamma \mapsto \gamma_\otimes$ *is an A-linear bijection*

$$\mathrm{Mult}_k(\mathcal{M}_1, \ldots, \mathcal{M}_k; \mathcal{P}) \xrightarrow{\ \cong\ } \mathrm{Hom}(\mathcal{M}_1 \otimes \cdots \otimes \mathcal{M}_k, \mathcal{P})$$

from the module of multilinear maps $\mathcal{M}_1 \times \cdots \times \mathcal{M}_k \to \mathcal{P}$ *to the module of linear maps* $\mathcal{M}_1 \otimes \cdots \otimes \mathcal{M}_k \to \mathcal{P}$.

Proof. For $k = 2$, the first part of the statement is just the universal property of the tensor product. The rest is omitted, but we invite the reader to attempt a full proof. $\square$

We conclude these algebraic preliminaries on the tensor product by discussing the tensor product of modules possessing finite frames.

Proposition 9.31 (Tensor Product and Finite Frames). *Let A be an algebra, and let $\mathcal{M}_1, \ldots, \mathcal{M}_k$ be A-modules possessing finite frames $(\varepsilon_{i_1}^{(1)})_{i_1}, \ldots, (\varepsilon_{i_k}^{(k)})_{i_k}$. Then the tensor product $\mathcal{M}_1 \otimes \cdots \otimes \mathcal{M}_k$ does also possess a finite frame given by*

$$\left(\varepsilon_{i_1}^{(1)} \otimes \cdots \otimes \varepsilon_{i_k}^{(k)} \right)_{i_1, \ldots, i_k}.$$

Proof. We sketch the proof for $k = 2$. So, let $\mathcal{M}, \mathcal{N}$ be A-modules possessing finite frames $(\varepsilon_i)_i, (\varphi_j)_j$. As the latter generate $\mathcal{M}, \mathcal{N}$, it follows that the system $(\varepsilon_i \otimes \varphi_j)_{i,j}$ generates $\mathcal{M} \otimes \mathcal{N}$ (do you see it?). It remains to check that the vectors $\varepsilon_i \otimes \varphi_j$ are linearly independent. So let $a^{ij} \in A$ be scalars such that

$$a^{ij} \varepsilon_i \otimes \varphi_j = 0$$

(where we are using the Einstein convention). As $(\varepsilon_i)_i, (\varphi_j)_j$ are frames, for every l, k, there exists a unique bilinear map $\beta^{lk} : \mathcal{M} \times \mathcal{N} \to A$ such that $\beta^{lk}(\varepsilon_i, \varphi_j) = \delta_i^l \delta_j^k$. Consider the corresponding linear map $\beta_\otimes^{lk} : \mathcal{M} \otimes \mathcal{N} \to A$. We have

$$0 = \beta_\otimes^{lk} \left(a^{ij} \varepsilon_i \otimes \varphi_j \right) = a^{ij} \beta^{lk}(\varepsilon_i, \varphi_j) = a^{ij} \delta_i^l \delta_j^k = a^{lk}.$$

For $k > 2$, use induction. $\qquad \square$

Corollary 9.32 (Tensor Products as Multilinear Maps). *Let A be an algebra, let $\mathcal{M}_1, \ldots, \mathcal{M}_k$ be A-modules possessing finite frames, and let $\mathcal{M}_1^*, \ldots, \mathcal{M}_k^*$ be their dual modules. Then there are canonical A-module isomorphisms*

$$\mathcal{M}_1 \otimes \cdots \otimes \mathcal{M}_k \cong \mathrm{Mult}_k(\mathcal{M}_1^*, \ldots, \mathcal{M}_k^*; A) \tag{9.12}$$

and

$$\mathcal{M}_1^* \otimes \cdots \otimes \mathcal{M}_k^* \cong \mathrm{Mult}_k(\mathcal{M}_1, \ldots, \mathcal{M}_k; A) \tag{9.13}$$

(to be specified in the proof).

Proof. First of all, let $\mathcal{M}$ be an A-module possessing a finite frame, and let $\mathcal{M}^{**}$ be its bidual module. Then the map

$$\iota : \mathcal{M} \to \mathcal{M}^{**},$$

defined by

$$\iota(\mu)(\varphi) := \varphi(\mu), \quad \varphi \in \mathcal{M}^*,$$

is an A-module isomorphism. The isomorphism (9.12) generalizes the latter *biduality isomorphism* ι to the case $k > 1$. Let

$$\iota_k : \mathcal{M}_1 \otimes \cdots \otimes \mathcal{M}_k \to \mathrm{Mult}_k(\mathcal{M}_1^*, \ldots, \mathcal{M}_k^*; A)$$

be the unique linear map such that

$$\iota_k(\mu_1 \otimes \cdots \otimes \mu_k)(\varphi_1, \ldots, \varphi_k) := \varphi_1(\mu_1) \cdots \varphi_k(\mu_k),$$

for all $\mu_i \in \mathcal{M}_i$ and all $\varphi_i \in \mathcal{M}_i^*$, $i = 1, \ldots, k$. We want to show that ι_k is a well-defined A-module isomorphism. We sketch the proof for $k = 2$. So, let $\mathcal{M}, \mathcal{N}$ be A-modules possessing finite frames $(\varepsilon_i)_i, (\varphi_j)_j$. We have defined $\iota_2 : \mathcal{M} \otimes \mathcal{N} \to \mathrm{Mult}_2(\mathcal{M}, \mathcal{N}; A)$ by putting

$$\iota_2(\mu \otimes \nu)(\varphi, \psi) := \varphi(\mu)\psi(\nu) \tag{9.14}$$

for all $\mu \in \mathcal{M}, \nu \in \mathcal{N}, \varphi \in \mathcal{M}^*$, and $\psi \in \mathcal{N}^*$. As the expression in the right hand side of (9.14) is linear in both φ, ψ, it follows that $\iota_2(\mu \otimes \nu)$ is a bilinear map. The expression in the right hand side of (9.14) is also linear in μ, ν, hence, from the universal property of the tensor product, a linear map ι_2 as required does actually exist. We have to show that ι_2 is both injective and surjective. We will need the dual bases $(\varepsilon^i)_i, (\varphi^j)_j$ of $(\varepsilon_i)_i, (\varphi_j)_j$ in $\mathcal{M}^*, \mathcal{N}^*$.

For the injectivity, take an element $T \in \mathcal{M} \otimes \mathcal{N}$. From Proposition 9.31, there are unique scalars $a^{ij} \in A$ such that $T = a^{ij}\varepsilon_i \otimes \varphi_j$. If $T \in \ker \iota_2$, in particular,

$$0 = \iota_2(T)(\varepsilon^l, \varphi^k) = a^{ij}\varepsilon^l(\varepsilon_i)\varphi^k(\varphi_j) = a^{ij}\delta_i^l\delta_j^k = a^{kl}$$

for all k, l. It follows that $T = 0$.

For the surjectivity, let $\beta \in \mathrm{Mult}_2(\mathcal{M}^*, \mathcal{N}^*; A)$. Denote $\beta^{ij} := \beta(\varepsilon^i, \varphi^j)$. A similar computation as above now shows that

$$\beta = \iota_2\left(\beta^{ij}\varepsilon_i \otimes \varphi_j\right).$$

For the isomorphism (9.13), use (9.12) and $\mathcal{M}^{**} \cong \mathcal{M}$. This concludes the proof. $\qquad \square$

For more on tensor products in Algebra, see, e.g., Atiyah and Macdonald (2016).

We are now ready to define the tensor product of vector bundles. So, let (E, π, M) and (E', π', M) be vector bundles with abstract fibers V and V'. Let $m = \dim V$ and $m' = \dim V'$. For every $p \in M$, we can take the vector space $E_p \otimes E'_p = E_p \otimes_{\mathbb{R}} E'_p$. In view of Proposition 9.31, it is a vector space of dimension mm'.

Theorem 9.33 (Tensor Product of Vector Bundles). *The family of vector spaces*

$$\left\{ E_p \otimes E'_p \right\}_{p \in M}$$

can be given a natural structure of rank mm' vector bundle (to be specified in the proof) with abstract fiber $V \otimes V'$.

Proof. Similarly as for the Hom-vector bundle, we use the Vector Bundle Chart Theorem. Denote

$$E \otimes E' := \coprod_{p \in M} E_p \otimes E'_p.$$

Choose local trivializations $(\mathcal{U}, \Phi_{\mathcal{U}}), (\mathcal{U}, \Phi'_{\mathcal{U}})$ of E, E' over the same open subset $\mathcal{U} \subseteq M$. We have a bijection

$$\Phi^{\otimes}_{\mathcal{U}} : (E \otimes E')_{\mathcal{U}} \to \mathcal{U} \times (V \otimes V')$$

defined as follows. For every $p \in \mathcal{U}$, every $e \in E_p$, and every $e' \in E'_p$, put

$$\Phi^{\otimes}_{\mathcal{U}}(p, e \otimes e') := (p, \Phi_{\mathcal{U},p}(e) \otimes \Phi'_{\mathcal{U},p}(e'))$$

and use the universal property of the tensor product to extend $\Phi^{\otimes}_{\mathcal{U}}$ to generic elements in $E_p \otimes E'_p$ (this is possible because the expression $\Phi_{\mathcal{U},p}(e) \otimes \Phi'_{\mathcal{U},p}(e')$ is bilinear in the argument (e, e')). The diagram

$$
\begin{array}{ccc}
(E \otimes E')_{\mathcal{U}} & \xrightarrow{\;\;\Phi^{\otimes}_{\mathcal{U}}\;\;} & \mathcal{U} \times (V \otimes V') \\
& \searrow \qquad \swarrow {\scriptstyle \mathrm{pr}_{\mathcal{U}}} & \\
& \mathcal{U} &
\end{array}
$$

commutes. Moreover, $\Phi^{\otimes}_{\mathcal{U}}$ is clearly an isomorphism on fibers. Finally, let $(\mathcal{U}, \Phi_{\mathcal{U}}), (\mathcal{V}, \Phi_{\mathcal{V}})$ be two local trivializations of E and let $(\mathcal{U}, \Phi'_{\mathcal{U}}), (\mathcal{V}, \Phi'_{\mathcal{V}})$

be two local trivializations of E' over the same open subsets $\mathcal{U}, \mathcal{V}$. We want to compute

$$\Phi^{\otimes}_{\mathcal{U}\mathcal{V}} = \Phi^{\otimes}_{\mathcal{V}} \circ \Phi^{\otimes -1}_{\mathcal{U}} : (\mathcal{U} \cap \mathcal{V}) \times (V \otimes V') \to (\mathcal{U} \cap \mathcal{V}) \times (V \otimes V').$$

First, denote by $A : \mathcal{U} \cap \mathcal{V} \to \mathrm{GL}(V)$ (resp. $A' : \mathcal{U} \cap \mathcal{V} \to \mathrm{GL}(V')$) the smooth map determined by

$$\Phi_{\mathcal{U}\mathcal{V}}(p, v) = (p, A(p)(v)) \quad (\text{resp. } \Phi'_{\mathcal{U}\mathcal{V}}(p, v') = (p, A'(p)(v')))$$

for all $(p, v) \in (\mathcal{U} \cap \mathcal{V}) \times V$ (resp. $(p, v') \in (\mathcal{U} \cap \mathcal{V}) \times V'$). Then,

$$\Phi^{\otimes}_{\mathcal{U}\mathcal{V}}(p, v \otimes v') = \Phi^{\otimes}_{\mathcal{V}} \circ \Phi^{\otimes -1}_{\mathcal{U}}(p, v \otimes v') = (p, A(p)(v) \otimes A'(p)(v')).$$

In other words,

$$\Phi^{\otimes}_{\mathcal{U}\mathcal{V}}(p, T) = (p, A^{\otimes}(p)(T)), \quad (p, T) \in (\mathcal{U} \cap \mathcal{V}) \times (V \otimes V'),$$

where $A^{\otimes}(p) \in \mathrm{GL}(V \otimes V')$ is the unique vector space isomorphism such that $A^{\otimes}(p)(v \otimes v') = A(p)(v) \otimes A(p)(v')$ for all $v \in V$ and $v' \in V$ (such isomorphism exists thanks to the universal property of tensor products). An easy computation in coordinates exploiting Proposition 9.31 now shows that the map $A^{\otimes} : \mathcal{U} \cap \mathcal{V} \to \mathrm{GL}(V \otimes V')$ is smooth and this concludes the proof. □

We now briefly discuss sections of the vector bundle $E \otimes E'$. One can actually show that they are equivalent to elements in the tensor product $\Gamma(E) \otimes \Gamma(E') = \Gamma(E) \otimes_{C^\infty(M)} \Gamma(E')$ of the $C^\infty(M)$-modules $\Gamma(E), \Gamma(E')$. Begin with two sections $s \in \Gamma(E)$ and $s' \in \Gamma(E')$. We can define a section τ of $E \otimes E'$ by putting $\tau(p) = s(p) \otimes s'(p) \in E_p \otimes E'_p$. The map $\Gamma(E) \times \Gamma(E') \to \Gamma(E \otimes E')$ defined in this way is $C^\infty(M)$-bilinear (do you see it?), hence it defines a $C^\infty(M)$-linear map

$$\Gamma(E) \otimes \Gamma(E') \to \Gamma(E \otimes E'), \quad T \mapsto \tau_T. \tag{9.15}$$

Proposition 9.34. *The map (9.15) is a $C^\infty(M)$-module isomorphism.*

Proof. Omitted (for a detailed proof, see, e.g., Conlon, 2008). □

Remark 9.35. There are various natural linear maps connecting tensor products and multilinear maps. An instance is provided by the map ι_k in the proof of Corollary 9.32. Accordingly, there are various vector bundle maps connecting tensor products of vector bundles and vector bundles of multilinear maps. In this remark, we briefly discuss a prototypical example. Begin with an algebra A and two A-modules $\mathcal{M}, \mathcal{N}$. There is a linear map

$$\iota : \mathcal{M}^* \otimes \mathcal{N} \to \operatorname{Hom}(\mathcal{M}, \mathcal{N})$$

uniquely defined by

$$\iota(\varphi \otimes \nu)(\mu) = \varphi(\mu)\nu \tag{9.16}$$

for all $\varphi \in \mathcal{M}^*$, $\nu \in \mathcal{N}$, and $\mu \in \mathcal{M}$. When $\mathcal{M}, \mathcal{N}$ possess finite frames, then ι is an A-module isomorphism (the proof is very similar to that of Corollary 9.32).

Now, given vector bundles $(E, \pi, M), (E', \pi', M)$ over the same base manifold M, the vector space isomorphisms

$$\iota : E_p^* \otimes E_p' \to \operatorname{Hom}(E_p, E_p')$$

glue together to a vector bundle isomorphism $\Gamma(E^* \otimes E') \cong \Gamma(\operatorname{Hom}(E, E'))$. Finally, in view of Propositions 9.20, 9.26, and 9.34, Formula (9.16) does also define a $C^\infty(M)$-module isomorphism $\Gamma(E)^* \otimes \Gamma(E) \cong \operatorname{Hom}(\Gamma(E), \Gamma(E'))$. We invite the reader to check the details. $\diamondsuit$

Remark 9.36 (Differential Forms and Tensor Products). There is a relationship between differential forms and tensor products. Let M be a manifold. According to Corollary 9.32, for all k and any point $p \in M$, there is a vector space isomorphism

$$\operatorname{Mult}_k(T_pM, \ldots, T_pM; \mathbb{R}) \xrightarrow{\cong} \underbrace{T_p^*M \otimes_{\mathbb{R}} \cdots \otimes_{\mathbb{R}} T_p^*M}_{k \text{ times}}.$$

We can restrict these isomorphisms to alternating multilinear maps obtaining $\mathbb{R}$-linear injections

$$\wedge^k T_p^*M = \operatorname{Alt}_k(T_pM, \mathbb{R}) \hookrightarrow \underbrace{T_p^*M \otimes_{\mathbb{R}} \cdots \otimes_{\mathbb{R}} T_p^*M}_{k \text{ times}}.$$

All these injections glue together to an injective vector bundle map

$$\wedge^k T^*M \hookrightarrow \underbrace{T^*M \otimes \cdots \otimes T^*M}_{k \text{ times}}.$$

It follows from Propositions 9.20 and 9.34 that we also have a $C^\infty(M)$-linear injection

$$\Omega^k(M) \cong \Gamma(\wedge^k T^* M) \hookrightarrow \underbrace{\Omega^1(M) \otimes \cdots \otimes \Omega^1(M)}_{k \text{ times}},$$

allowing one to interpret k-forms as tensor products. It is not hard to see that, given 1-forms $\theta_1, \ldots, \theta_k \in \Omega^1(M)$, their wedge product $\theta_1 \wedge \cdots \wedge \theta_k$ identifies with

$$\sum_{\sigma \in S_k} (-)^\sigma \theta_{\sigma(1)} \otimes \cdots \otimes \theta_{\sigma(k)}.$$

For more on differential forms as tensors, see, e.g., Lee (2013). $\diamond$

Chapter 10

Integration on Manifolds

In this chapter, we discuss integration on manifolds. It turns out that, for a coordinate free theory of integration, the correct *integrands*, i.e., object to be integrated on manifolds, are *not* functions but rather differential forms of top degree, up to the choice of an *orientation*. So we will first need to develop a theory of orientations on manifolds and *oriented manifolds*. Once this has been done, the definition of the *definite integral* is transported from open subsets of $\mathbb{R}^n$ to a generic manifold via charts and a new tool known as a *partition of unity*. We will leave out of the discussion the most technical aspects.

10.1 Oriented Manifolds

In standard Calculus courses, we usually learn how to integrate a(n integrable) function $f : \mathcal{U} \to \mathbb{R}$ on (a measurable subset, e.g.) an open subset $\mathcal{U} \subseteq \mathbb{R}^n$ of some standard Euclidean space $\mathbb{R}^n$. Let $t = (t^1, \dots, t^n)$ be standard coordinates on $\mathcal{U}$, and denote by

$$\int_{\mathcal{U}} f(t) \mathrm{d}^n t$$

the integral. Note that if $\mathcal{V} \subseteq \mathbb{R}^n$ is another open subset with standard coordinates denoted $s = (s^1, \dots, s^n)$, and $\Phi : \mathcal{V} \to \mathcal{U}$ is a diffeomorphism, then the following *Coordinate Change Formula* holds:

$$\int_{\mathcal{U}} f(t) \mathrm{d}^n t = \int_{\mathcal{V}} f(\Phi(s)) \left| \det J_\Phi(s) \right| \mathrm{d}^n s.$$

If $\det J_\Phi > 0$, then the above formula simplifies a bit to

$$\int_U f(t)\mathrm{d}^n t = \int_V f(\Phi(s)) \det J_\Phi(s)\mathrm{d}^n s.$$

This suggests that, in a coordinate free theory of integration, the "correct integrands" are objects which locally look like functions $f = f(t)$ but that, under a change of coordinates $\Phi : s \mapsto t$, change as follows:

$$f(t) \mapsto (\Phi^* f)(s) \det J_\Phi(s), \tag{10.1}$$

at least when the coordinate change is *orientation preserving*, i.e., it satisfies $\det J_\Phi > 0$.

It is not hard to see that (10.1) is exactly the *Coordinate Change Formula* for a differential n-form on an n-dimensional manifold. Namely, let M be an n-dimensional manifold, let $(U, \varphi = (x^1, \ldots, x^n))$, $(U, \psi = (y^1, \ldots, y^n))$ be two charts on M with the same coordinate domain U, and let $\omega \in \Omega^n(M)$ be an n-form on M. Then there exists a smooth function $f : U \to \mathbb{R}^n$ such that

$$\omega|_U = f dx^1 \wedge \cdots \wedge dx^n$$

but $dx^i = \frac{\partial x^i}{\partial y^j} dy^j$, hence

$$\omega|_U = f \frac{\partial x^1}{\partial y^{j_1}} \cdots \frac{\partial x^n}{\partial y^{j_n}} dy^{j_1} \wedge \cdots \wedge dy^{j_n} = f \det \left(\frac{\partial x}{\partial y}\right) dy^1 \wedge \cdots \wedge dy^n,$$

where

$$\frac{\partial x}{\partial y} = \left(\frac{\partial x^i}{\partial y^j}\right)_{j=1,\ldots,n}^{i=1,\ldots,n}$$

is the Jacobian matrix of the transition map, and the last step follows from the skew-symmetry of the wedge product (do you see it?).

This discussion shows that, in order to be able to define a meaningful theory of integration on arbitrary n-dimensional manifolds, we need at least to take care of the following two aspects:

(1) we have to work with an atlas whose transition maps are all orientation preserving,
(2) we have to integrate differential n-forms (rather than functions).

In this section, we take care of the first aspect. This is done introducing the notions of *orientable* and *oriented manifold* (see the following).

Let M be an n-dimensional manifold. In general, M will not possess an atlas whose transition maps are all orientation preserving. For this reason, we introduce the following

Definition 10.1 (Orientable Manifold). Two charts (U, φ) and (V, ψ) on M *define the same local orientation* if either

- $U \cap V = \varnothing$ or
- $U \cap V \neq \varnothing$ and, additionally, the transition map $\psi \circ \varphi^{-1} : \varphi(U \cap V) \to \psi(U \cap V)$ is an *orientation preserving diffeomorphism*, i.e., $\det J_{\psi \circ \varphi^{-1}} > 0$.

An *oriented atlas* on M is an atlas $\mathcal{A}$ such that any two charts in $\mathcal{A}$ define the same local orientation. An n-dimensional manifold M is *orientable* if it possesses an oriented atlas (otherwise it is *non-orientable*). Two oriented atlases $\mathcal{A}, \mathcal{B}$ on an orientable manifold M *define the same orientation* if the union $\mathcal{A} \cup \mathcal{B}$ is an orientable atlas as well (otherwise $\mathcal{A}, \mathcal{B}$ *define opposite orientations*).

Example 10.2. The standard atlas on $\mathbb{R}^n$ is clearly oriented. It follows that $\mathbb{R}^n$ is an orientable manifold. For the same reason, any open submanifold of $\mathbb{R}^n$ is orientable. ◆

Example 10.3. Consider the circle S^1 and the stereographic atlas $\{(U_+, \varphi_+), (U_-, \varphi_-)\}$ on it. It is easy to see that the transition map between the two stereographic charts is given by

$$\varphi_- \circ \varphi_+^{-1} : \mathbb{R} \setminus \{0\} \to \mathbb{R} \setminus \{0\}, \quad t \mapsto t^{-1}$$

(the reader who solved Exercise 1.3 should have already found this formula in the general case of the n-dimensional sphere, see also Example 10.18). Hence, the Jacobian $J_{\varphi_- \circ \varphi_+^{-1}}$ is the 1×1 matrix whose only element is

$$\frac{d}{dt} t^{-1} = -t^{-2} < 0.$$

We conclude that the stereographic atlas is *not* oriented. However, if we replace the chart (U_-, φ_-) by the (compatible) chart $(U_-, -\varphi_-)$, then, clearly, the new transition map $-\varphi_- \circ \varphi_+^{-1}$ is $t \mapsto -t^{-1}$ which is orientation preserving (do you see it?). In other words, the atlas $\{(U_+, \varphi_+), (U_-, -\varphi_-)\}$ is oriented, and the circle is an orientable manifold. ◆

Exercise 10.1. Let $M_1, \ldots, M_k$ be smooth manifolds and let $\mathcal{A}_1, \ldots, \mathcal{A}_k$ be oriented atlases on them. Consider the product manifold $M_1 \times \cdots \times M_k$, and show that the product atlas $\mathcal{A}^\times$ determined by $\mathcal{A}_1, \ldots, \mathcal{A}_k$ as in the discussion preceding Exercise 1.14 is oriented as well. Conclude that the product of orientable manifolds is orientable as well.

In the following, we characterize orientable manifolds in a simple way which will allow us to make more examples of orientable manifolds (see Theorem 10.7). Now, let M be an orientable n-dimensional manifold.

Proposition 10.4. *"Defining the same orientation" is an equivalence relation on the set of all oriented atlases on M.*

Proof. Left as Exercise 10.2. □

Exercise 10.2. Prove Proposition 10.4.

Definition 10.5 (Orientation and Oriented Manifold). An oriented atlas $\mathcal{A}_+$ on an orientable manifold M is *maximal* if it is not properly contained in any other orientable atlas or, equivalently, if it already contains all atlases defining the same orientation. A maximal oriented atlas on M is an *orientation* of M. An *oriented manifold* is an orientable manifold equipped with an orientation.

Proposition 10.6. *Let M be an orientable manifold. Then*

(1) *every oriented atlas $\mathcal{A}$ on M is contained in a unique maximal oriented atlas, necessarily determining the same orientation as $\mathcal{A}$,*
(2) *two oriented atlases are contained in the same maximal oriented atlas if and only if they define the same orientation.*

Proof. Left as Exercise 10.3. □

Exercise 10.3. Prove Proposition 10.6.

Let $(M, \mathcal{A}_+)$ be an oriented manifold, $\mathcal{A}_+$ being the orientation, i.e., the chosen maximal oriented atlas. In this situation, we will adopt the following terminology: $\mathcal{A}_+$ is called the *positive orientation* and any chart in $\mathcal{A}_+$ is called *positively oriented* (or simply *positive*).

The orientability of an n-dimensional manifold M can be characterized in terms of differential n-forms. Remember that any such form can be seen as a section of the vector bundle $\wedge^n T^* M \to M$ of n-covectors. Moreover, the latter is actually a *line bundle*, i.e., a rank 1 vector bundle, that, from now on, we will denote by $\mathrm{Vol}(M) \to M$ and call the (line) *bundle of volume forms*. The reason for this terminology will be clarified in the next section (see Proposition 10.17 Point (3), see also Example 10.18). We are now ready to state the main theorem of this section.

Theorem 10.7 (Orientability and Volume Forms). *An n-dimensional smooth manifold M is orientable if and only if the line bundle $\mathrm{Vol}(M)$ of volume forms is trivializable or, equivalently, there exists a nowhere zero differential n-form on M. Any trivialization of $\mathrm{Vol}(M)$, equivalently any nowhere zero differential n-form on M, determines an orientation on M in a canonical way (to be specified in the proof).*

By a *nowhere zero differential n-form* on M in the above statement, we mean a differential n-form $\omega \in \Omega^n(M)$ such that $\omega_p \neq 0$ for all $p \in M$. We also call such ω a *volume form* on M. Note that, in view of Corollary 9.22, a volume form is also a frame of the $C^\infty(M)$-module $\Omega^n(M)$, so, from Corollary 9.21, the line bundle $\mathrm{Vol}(M)$ is trivializable if and only if there exists a volume form, and, even more, trivializations are in one-to-one correspondence with volume forms, which explains the statement. For the proof of Theorem 10.7, we will need a technical tool called a *partition of unity* which will also be useful in the next section and we introduce here.

Let M be a manifold, and let $C = \{\mathcal{U}\}$ be an open cover of M. A *partition of unity subordinate to C* is a family $\{f_\mathcal{U}\}_{\mathcal{U} \in C}$ of smooth functions $f_\mathcal{U} \in C^\infty(M)$ parameterized by C with the following properties:

(1) $0 \leq f_\mathcal{U}(p) \leq 1$ for all $\mathcal{U} \in C$ and all $p \in M$,
(2) $\mathrm{supp}\, f_\mathcal{U} \subseteq \mathcal{U}$ for all $\mathcal{U} \in C$,
(3) for all $p \in M$, there exists an open neighborhood $\mathcal{V} \subseteq M$ of p such that $\mathcal{V}$ intersects $\mathrm{supp}\, f_\mathcal{U}$ for only finitely many $\mathcal{U} \in C$,
(4) for all $p \in M$, $\sum_{\mathcal{U} \in C} f_\mathcal{U}(p) = 1$.

Note that, in view of Property (3), the sum in Property (4) has only finitely many non-trivial terms, hence it is actually a well-defined finite sum. Partitions of unities are often useful in Differential Geometry to "*transport properties from the local to the global setting*". We state the existence of partitions of unity without a proof in the following

Proposition 10.8. *For any open cover C of M, there exists a partition of unity subordinate to C.*

Proof. Omitted, but see Lee 2013. □

We are now ready to prove Theorem 10.7.

Proof of Theorem 10.7. Begin with a volume form $\omega \in \Omega^n(M)$. For every chart $(U, \varphi = (x^1, \ldots, x^n))$, we have

$$\omega|_U = f dx^1 \wedge \cdots \wedge dx^n,$$

for some function $f \in C^\infty(U)$. As $\omega_p \neq 0$ for all $p \in M$, we must have $f(p) \neq 0$ for all $p \in U$. Let $p_0 \in U$ be such that $f(p_0) > 0$ (resp. $f(p_0) < 0$). Then, by continuity, $f > 0$ (resp. $f < 0$) in a whole neighborhood of p_0 (do you agree?). Let $\mathcal{A}_\omega$ be the collection of all charts $(U, \varphi = (x^1, \ldots, x^n))$ such that $f > 0$. We claim that $\mathcal{A}_\omega$ is an oriented atlas on M. To see this, first note that it is an atlas; indeed for any point $p \in M$, let $(V, \psi = (y^1, \ldots, y^n))$ be any chart around p. Then

$$\omega|_V = g dy^1 \wedge \cdots \wedge dy^n$$

for some $g \in C^\infty(U)$ with $g \neq 0$. If $g(p) > 0$, then $g > 0$ in a whole neighborhood $U \subseteq V$ of p and the subchart (U, ψ) is a chart in $\mathcal{A}_\omega$ around p. On the other hand, if $g(p) < 0$, then $g < 0$ in a whole neighborhood $U \subseteq V$ of p. In this case, consider the chart

$$(U, \varphi = (x^1, \ldots, x^n) := (-y^1, y^2, \ldots, y^n)).$$

Then

$$\omega|_U = g|_U dy^1 \wedge \cdots \wedge dy^n = -g|_U dx^1 \wedge \cdots \wedge dx^n$$

As $-g|_U > 0$, it follows that (U, φ) is again a chart in $\mathcal{A}_\omega$ around p. So the charts in $\mathcal{A}_\omega$ cover M and $\mathcal{A}_\omega$ is an atlas as claimed. Next, we have to show that it is an oriented atlas. So let $(U, \varphi = (x^1, \ldots, x^n)), (V, \psi = (y^1, \ldots, y^n))$ be two charts in $\mathcal{A}_\omega$. Either $U \cap V = \varnothing$ or $U \cap V \neq \varnothing$, in which case, from the coordinate change formula for n-forms,

$$\omega|_{U \cap V} = f dx^1 \wedge \cdots \wedge dx^n = f \det\left(\frac{\partial x}{\partial y}\right) dy^1 \wedge \cdots \wedge dy^n, \qquad (10.2)$$

where $f > 0$ because $(U, \varphi) \in \mathcal{A}_\omega$. But also $(V, \psi) \in \mathcal{A}_\omega$, hence $f \det(\partial x/\partial y) > 0$. So $\det(\partial x/\partial y) > 0$. As $\det(\partial x/\partial y)$ is the Jacobian matrix of the transition map between $(V, \psi), (U, \psi)$, we conclude that the two charts define the same local orientation. It follows from their arbitrariness that $\mathcal{A}_\omega$ is an oriented atlas as claimed.

We now assume that M is orientable and show that it possesses a volume form. So let $\mathcal{A}$ be an oriented atlas on M. Denote by $\mathcal{C}_\mathcal{A}$ the open cover of M consisting of coordinate domains of charts in $\mathcal{A}$:

$$C_{\mathcal{A}} := \{U : (U, \varphi) \in \mathcal{A}\}.$$

From Proposition 10.8, there exists a partition of unity $\{f_U\}_{(U,\varphi)\in\mathcal{A}}$ subordinate to $C_{\mathcal{A}}$. We use it to construct a volume form as follows. For any chart $(U, \varphi = (x^1, \ldots, x^n))$ in $\mathcal{A}$, consider the coordinate n-form on U:

$$\omega_{\text{coord}} := dx^1 \wedge \cdots \wedge dx^n \in \Omega^n(U).$$

Actually, ω_U is a volume form on U. However, in general, it cannot be extended to a volume form on M, i.e., there is no volume form, not even just a form, $\omega \in \Omega^n(M)$ such that $\omega_{\text{coord}} = \omega|_U$. Nonetheless, we can consider the form $\omega_U \in \Omega^n(M)$ defined by its values as follows:

$$(\omega_U)_p = \begin{cases} f_U(p)(\omega_{\text{coord}})_p & \text{if } p \in U \\ 0 & \text{if } p \in M \setminus \text{supp } f_U. \end{cases}$$

As $\text{supp } f_U \subseteq U$ (by definition of partition of unity), ω_U is well defined, and it is smooth by the Gluing Lemma for Differential Forms (do you see it?). Finally, consider the n-form

$$\omega := \sum_{(U,\varphi)\in\mathcal{A}} \omega_U.$$

To be more precise, ω is defined by its values as follows: $\omega_p = \sum_{(U,\varphi)\in\mathcal{A}}(\omega_U)_p$. As p belongs to $\text{supp } f_U$ for only finitely many (U, φ) (by definition of partition of unity), this is a finite sum. By the Gluing Lemma again, ω is a well-defined differential n-form (check the details as an exercise). It remains to check that ω is a volume form. To do this, let $p \in M$, and let

$$(U_i, \varphi_i = (x^1_{(i)}, \ldots, x^n_{(i)})), \quad i = 0, \ldots, k,$$

be the only (finitely many) charts such that $\text{supp } f_{U_i} \ni p, i = 0, \ldots, k$. Then

$$\omega_p = \sum_{i=0}^{k}(\omega_{U_i})_p$$

$$= \sum_{i=0}^{k} f_{U_i}(p) d_p x^1_{(i)} \wedge \cdots \wedge d_p x^n_{(i)}$$

$$= \sum_{i=0}^{k} f_{U_i}(p) \det\left(\frac{\partial x_{(i)}}{\partial x_{(0)}}(p)\right) d_p x^1_{(0)} \wedge \cdots \wedge d_p x^n_{(0)}. \tag{10.3}$$

But

$$f_{U_i}(p) \geq 0 \quad \text{for all } i, \quad \text{moreover} \sum_{j=0}^{k} f_{U_j}(p) = 1$$

and

$$\det\left(\frac{\partial x_{(i)}}{\partial x_{(0)}}(p)\right) > 0$$

because it is the Jacobian of the transition map between two charts of an oriented atlas. Hence the coefficient

$$\sum_{i=0}^{k} f_{U_i}(p) \det\left(\frac{\partial x_{(i)}}{\partial x_{(0)}}(p)\right)$$

in (10.3) is greater then 0 and $\omega_p \neq 0$. It follows from the arbitrariness of p that ω is a volume form as desired. $\qquad\square$

Example 10.9. Let $F : \mathbb{R}^n \to \mathbb{R}$ be a smooth function. Suppose that 0 is a regular value of F so that F is a submersion and $Z(F) = F^{-1}(0) \subseteq \mathbb{R}^n$ is a submanifold of dimension $n - 1$. We want to use Theorem 10.7 to show that $Z(F)$ is an orientable manifold. In other words, we want to show that $Z(F)$ possesses a volume form. To do this, consider the *coordinate volume form on $\mathbb{R}^n$*

$$\omega_{\text{coord}} := dt^1 \wedge \cdots \wedge dt^n \in \Omega^n(\mathbb{R}^n)$$

and the vector field

$$X := \sum_{i=1}^{n} \frac{\partial F}{\partial t^i} \frac{\partial}{\partial t^i} \in \mathfrak{X}(\mathbb{R}^n).$$

Finally, consider the following $(n-1)$-form on $Z(F)$:

$$\eta := (\iota_X \omega_{\text{coord}})|_{Z(F)} \in \Omega^{n-1}(Z(F)).$$

The form η is actually a volume form on $Z(F)$. To see this, first note that, for all $P \in Z(F)$, the value X_P of the vector field X at P is

$$X_P = \sum_{i=1}^{n} \frac{\partial F}{\partial t^i}(P) \frac{\partial}{\partial t^i}\Big|_P.$$

As $0 \neq d_P F = \frac{\partial F}{\partial t^i}(P) dpt^i$, it follows that $X_P \neq 0$. Next, a tangent vector $v = v^i \frac{\partial}{\partial t^i}|_P \in T_P \mathbb{R}^n$ is tangent to $Z(F)$ if

$$0 = d_P F(v) = v(F) = v^i \frac{\partial F}{\partial t^i}(P).$$

So, if we use the coordinate frame $\left(\frac{\partial}{\partial t^1}|_P, \ldots, \frac{\partial}{\partial t^n}|_P \right)$ to identify $T_P \mathbb{R}^n$ with the vector space $\mathbb{R}^n$ itself, the tangent space $T_P Z(F)$ identifies with the subspace of vectors which are orthogonal to X_P with respect to the standard scalar product on $\mathbb{R}^n$ (see Example 4.14 for a specific instance). Finally, choose a frame $(V_1, \ldots, V_{n-1})$ of $T_P Z(F)$. As $X_P \neq 0$ and the V_i are orthogonal to X_P (with respect to an appropriate scalar product), then $(X_P, V_1, \ldots, V_{n-1})$ is a frame of $T_P M$. As ω_{coord} is a volume form, we have $(\omega_{\text{coord}})_P(X_P, V_1, \ldots, V_{n-1}) \neq 0$, otherwise $(\omega_{\text{coord}})_P$ would be zero (do you agree?). Hence,

$$0 \neq (\omega_{\text{coord}})_P(X_P, V_1, \ldots, V_{n-1}) = (\iota_X \omega_{\text{coord}})_P(V_1, \ldots, V_{n-1})$$
$$= \eta_P(V_1, \ldots, V_{n-1}),$$

i.e., $\eta_P \neq 0$. It follows from the arbitrariness of P that η is a volume form, as claimed, and $Z(F)$ is orientable. ◆

Example 10.10 (The Standard Orientation of the Sphere). It follows from Example 10.9 that the n-dimensional sphere S^n is an orientable manifold. If we present it as usual as the zero locus of the smooth function

$$F : \mathbb{R}^{n+1} \to \mathbb{R}, \quad (t^1, \ldots, t^{n+1}) \mapsto (t^1)^2 + \cdots (t^n)^2 - 1,$$

we see from Example 10.9 that the n-form

$$\omega_{S^n} = \frac{1}{2}(-)^{n+1}(\iota_X \omega_{\text{coord}})|_{S^n},$$

where

$$X = \sum_{i=1}^{n} \frac{\partial F}{\partial t^i} \frac{\partial}{\partial t^i} = 2t^i \frac{\partial}{\partial t^i},$$

is a volume form on S^n, called the *standard volume form* on S^n. In its turn, ω_{S^n} determines an orientation of S^n called the *standard orientation*. ◆

Example 10.11. Not all smooth manifolds are orientable. For instance, the projective plane $\mathbb{R}P^2$ is non-orientable. To see this, consider the affine charts $(U_i, \varphi_i = (x_i, y_i))$ on $\mathbb{R}P^2$, $i = 0, 1, 2$. We will actually need only the first two of them. Remember that $\widehat{U}_0 = \widehat{U}_1 = \mathbb{R}^2$. A direct computation (that the reader has already done if they solved Exercise 1.4) shows that

$$\varphi_0(U_0 \cap U_1) = \varphi_1(U_0 \cap U_1) = \left\{(t, s) \in \mathbb{R}^2 : t \neq 0\right\},$$

and the transition map

$$\varphi_1 \circ \varphi_0^{-1} : \varphi_0(U_0 \cap U_1) \to \varphi_1(U_0 \cap U_1)$$

is given by

$$\varphi_1 \circ \varphi_0^{-1}(s, t) = \left(\frac{1}{s}, \frac{t}{s}\right).$$

This means that

$$x_1 = \frac{1}{x_0} \quad \text{and} \quad y_1 = \frac{y_0}{x_0}$$

on $U_0 \cap U_1$, where both sides are well defined. Hence, the determinant of the Jacobian of the transition map is

$$\det \begin{pmatrix} \frac{\partial x_1}{\partial x_0} & \frac{\partial x_1}{\partial y_0} \\ \frac{\partial y_1}{\partial x_0} & \frac{\partial y_1}{\partial y_0} \end{pmatrix} = \det \begin{pmatrix} -\frac{1}{x_0^2} & 0 \\ -\frac{y_0}{x_0^2} & \frac{1}{x_0} \end{pmatrix} = -\frac{1}{x_0^3}$$

which is positive where $x_0 < 0$ and negative where $x_0 > 0$. Now, assume by contradiction that $\mathbb{R}P^2$ is orientable. Then, from Theorem 10.7, there exists a volume form $\omega \in \Omega^2(\mathbb{R}P^2)$ and

$$\omega|_{U_0} = f_0 dx_0 \wedge dy_0$$

for some nowhere zero function $f_0 \in C^\infty(U_0)$. As the coordinate representation $\widehat{f}_0 = f_0 \circ \varphi_0^{-1}$ of f_0 is a smooth nowhere zero function on $\mathbb{R}^2$, we conclude that either $f_0 > 0$ or $f_0 < 0$, i.e., f_0 has constant sign. Similarly,

$$\omega|_{U_1} = f_1 dx_1 \wedge dy_1$$

for some nowhere zero, constant sign, smooth function $f_1 \in C^\infty(U_1)$. But, from the coordinate change formula (10.2), on $U_0 \cap U_1$ we have

$$f_0 = f_1 \det \frac{\partial(x_1, y_1)}{\partial(x_0, y_0)} = -\frac{f_1}{x_0^3}$$

whose sign changes. This is a contradiction. So $\mathbb{R}P^2$ cannot be orientable.

♦

We conclude this section defining *orientation preserving diffeomorphisms*. So, let M, M' be orientable and oriented manifolds (of the same dimension).

Definition 10.12 (Orientation Preserving Diffeomorphism). An *orientation preserving diffeomorphism* between M and M' is a diffeomorphism $\Phi : M \to M'$ such that, for any positive chart (U, φ) on M, the chart $(\Phi(U), \varphi \circ \Phi^{-1})$ is positive as well (equivalently, the coordinate representation of Φ in any two positive charts has positive Jacobian determinant).

10.2 Integral of a Differential Form

Let M be a smooth orientable and oriented manifold. In this section, we define the integral $\int_M \omega$ of a differential n-form ω on M. Just as not all functions are integrable on an open subset of $\mathbb{R}^n$, not all n-forms ω are integrable on M. A necessary condition for ω to be integrable is that it is only non-zero on a "sufficiently small" subset of M. In order to make this idea more precise, we introduce the notion of a *compactly supported differential form*. Given a k-form $\omega \in \Omega^k(M)$, the *support* of ω is the closure

$$\operatorname{supp} \omega := \overline{\{p \in M : \omega_p \neq 0\}}$$

(recall that the closure $\overline{A}$ of a subset A of a topological space is the smallest closed subset containing A). For $k = 0$, the latter definition obviously agrees with that of support of a function discussed in Section 2.1.

Definition 10.13 (Compactly Supported Differential Form). A *compactly supported differential k-form* on M is a differential k-form $\omega \in \Omega^k(M)$ whose support $\operatorname{supp} \omega$ is a compact subset of M. A *compactly supported differential form* is a (possibly non-homogeneous) differential form $(\omega_k)_{k \in \mathbb{Z}} \in \Omega^\bullet(M)$ all whose components ω_k are compactly supported.

Compactly supported differential k-forms on M form a vector subspace in $\Omega^k(M)$ denoted $\Omega_c^k(M) \subseteq \Omega^k(M)$. Similarly, compactly supported differential forms form a graded vector subspace $\Omega_c^\bullet(M) := \bigoplus_{k\in\mathbb{Z}} \Omega_c^k(M) \subseteq \Omega^\bullet(M)$ in the graded algebra of differential forms (see Exercise 10.4 for more information on the algebraic properties of compactly supported differential forms). Note that, if M is already compact, then $\Omega_c^\bullet(M) = \Omega^\bullet(M)$.

Exercise 10.4. Prove that the subspace $\Omega_c^\bullet(M) \subseteq \Omega^\bullet(M)$ of compactly supported differential forms is a *graded ideal* in the graded algebra $\Omega^\bullet(M)$, i.e., not only $\Omega_c^\bullet(M)$ is a graded vector subspace, but additionally, for every $\omega \in \Omega_c^\bullet(M)$ and every $\rho \in \Omega^\bullet(M)$, we have

$$\omega \wedge \rho \in \Omega_c^\bullet(M).$$

We are now almost ready to define the integral of a (compactly supported) differential form $\omega \in \Omega_c^n(M)$. We begin discussing the case when $\mathrm{supp}\,\omega$ is entirely contained in the coordinate domain U of some positively oriented chart (U, φ). In this case, the pull-back $\varphi^{-1*}\omega := \varphi^{-1*}\omega|_U$ is a compactly supported n-form on $\widehat{U}$, hence

$$\varphi^{-1*}\omega = f\,dt^1 \wedge \cdots \wedge dt^n$$

for some compactly supported smooth function $f = f(t^1,\ldots,t^n) \in C^\infty(\widehat{U})$. Put

$$\int_{(U,\varphi)} \omega := \int_{\widehat{U}} f(t)\mathrm{d}^n t \in \mathbb{R}, \qquad (10.4)$$

where the integral in the right-hand side is either the Riemann or the Lebegue integral on $\mathbb{R}^n$. As f is smooth and its support is compact, hence closed and bounded, then f is a limited function, so it is integrable, and (10.4) is a well-defined real number. The integral defined in (10.4) is actually independent of the coordinate map φ as explained in the following

Lemma 10.14. *Let* $(U, \varphi), (U, \varphi')$ *be two positive charts with the same coordinate domain* U, *and let* $\omega \in \Omega_c^n(M)$ *be a compactly supported differential n-form with* $\mathrm{supp}\,\omega \subseteq U$. *Then*

$$\int_{(U,\varphi)} \omega = \int_{(U,\varphi')} \omega.$$

Proof. Denote $\widehat{U} := \varphi(U)$ and $\widehat{U}' := \varphi'(U)$. Moreover, let

$$\Phi := \varphi \circ \varphi'^{-1} : \widehat{U}' \to \widehat{U}$$

be the transition map. In order to avoid confusion, we denote by $(t^1,\ldots,t^n)$ the standard coordinates on $\widehat{U}$, and by $(s^1,\ldots,s^n)$ the standard coordinates on $\widehat{U}'$. Then, if

$$\varphi^{-1*}\omega = f dt^1 \wedge \cdots \wedge dt^n,$$

we have

$$\varphi'^{-1*}\omega = \Phi^* \left(\varphi^{-1*}\omega\right) = \Phi^* \left(f dt^1 \wedge \cdots \wedge dt^n\right)$$
$$= \Phi^*(f) \det J_\Phi \, ds^1 \wedge \cdots \wedge ds^n.$$

Therefore,

$$\int_{(U,\varphi')} \omega = \int_{\widehat{U}'} f(\Phi(s)) \det J_\Phi(s) d^n s$$
$$= \int_{\widehat{U}'} f(\Phi(s)) \left|\det J_\Phi(s)\right| d^n s$$
$$= \int_{\widehat{U}} f(t) d^n t$$
$$= \int_{(U,\varphi)} \omega,$$

where we used that (U, φ) and (U, φ') are both positive charts, hence $\det J_\Phi > 0$. $\qquad\square$

In view of Lemma 10.14, we can simply write

$$\int_U \omega$$

instead of $\int_{(U,\varphi)} \omega$ (when we have a positive chart (U, φ) and $\omega \in \Omega_c^n(M)$ such that $\operatorname{supp} \omega \subseteq U$) and, in what follows, we will always do so.

Exercise 10.5. Let U be the coordinate domain of a positive chart and let $V \subseteq U$ be a non-empty open subset, so that V is itself the coordinate domain of a positive chart (just take the corresponding subchart). Moreover, let $\omega \in \Omega_c^n(M)$ be a compactly supported n-form with $\operatorname{supp} \omega \subseteq V \subseteq U$. Show that

$$\int_V \omega = \int_U \omega.$$

Now let $\omega \in \Omega_c^n(M)$ be a compactly supported differential n-form on M. The support of ω is compact, hence there are finitely many positive charts $(U_1, \varphi_1), \ldots, (U_k, \varphi_k)$ whose coordinate domains $U_1, \ldots, U_k$ cover it, i.e., $\operatorname{supp} \omega \subseteq U_1 \cup \cdots \cup U_k$. The union $\mathcal{U} := U_1 \cup \cdots \cup U_k$ is an open submanifold (containing $\operatorname{supp} \omega$). Choose a partition of unity $\{f_i = f_{U_i}\}_{i=1,\ldots,k}$ on $\mathcal{U}$ subordinate to the open cover $\{U_i\}_{i=1,\ldots,k}$ of $\mathcal{U}$. For every $i = 1, \ldots, k$, the n-form $f_i \omega := f_i \omega|_{\mathcal{U}}$ is a compactly supported n-form on $\mathcal{U}$. Indeed,

$$\operatorname{supp} f_i \omega \subseteq \operatorname{supp} f_i \cap \operatorname{supp} \omega$$

(do you see it?). So $\operatorname{supp} f_i \omega$ is a closed subset in the compact subspace $\operatorname{supp} \omega$, hence it is compact itself (see Exercise 1.13). Moreover, $\operatorname{supp} f_i \omega \subseteq U_i$. We remark that $f_i \omega$ can also be understood as a compactly supported n-form on M (rather than on $\mathcal{U}$) by extending it as the zero form outside $\mathcal{U}$. Indeed, as $\operatorname{supp} f_i \omega \subseteq U_i$ is compact, it is also closed in M (see Exercise 1.13 again), and the latter extension correctly defines a compactly supported n-form on M (whose support is contained into U_i) by the Gluing Lemma for Differential Forms. Put

$$\int_M \omega := \sum_{i=1}^{k} \int_{U_i} f_i \omega \in \mathbb{R}. \tag{10.5}$$

Proposition 10.15. *The real number $\int_M \omega$ is independent of the choice of the open cover $\{U_i\}_{i=1,\ldots,k}$ of $\operatorname{supp} \omega$ by coordinate domains and of the partition of unity $\{f_i\}_{i=1,\ldots,k}$, i.e., if $\{V_j\}_{j=1,\ldots,m}$ is another open cover of $\operatorname{supp} \omega$ by coordinate domains of positive charts and $\{g_j = f_{V_j}\}_{j=1,\ldots,m}$ is a partition of unity of $\mathcal{V} := V_1 \cup \cdots \cup V_m$ subordinate to $\{V_j\}_{j=1,\ldots,m}$, then*

$$\sum_{j=1}^{m} \int_{V_j} g_j \omega = \sum_{i=1}^{k} \int_{U_i} f_i \omega.$$

Proof. We remark preliminarily that, if ω_1, ω_2 are compactly supported n-forms with $\operatorname{supp} \omega_1, \operatorname{supp} \omega_2 \subseteq U$ for some coordinate domain U of a positive chart, and $a_1, a_2 \in \mathbb{R}$, then $a_1 \omega_1 + a_2 \omega_2$ is also a compactly supported n-form with $\operatorname{supp}(a_1 \omega_1 + a_2 \omega_2) \subseteq U$ and

$$\int_U (a_1 \omega_1 + a_2 \omega_2) = a_1 \int_U \omega_1 + a_2 \int_U \omega_2 \tag{10.6}$$

(prove the latter formula as an exercise).

Now, f_i form a partition of unity, in particular $\sum_{i=1}^{k} f_i = 1$. Hence,

$$\sum_{j=1}^{m} \int_{V_j} g_j \omega = \sum_{j=1}^{m} \int_{V_j} \sum_{i=1}^{k} f_i g_j \omega$$

$$= \sum_{j=1}^{m} \sum_{i=1}^{k} \int_{V_j} f_i g_j \omega \qquad \text{(Formula (10.6))}$$

$$= \sum_{j=1}^{m} \sum_{i=1}^{k} \int_{U_i \cap V_j} f_i g_j \omega \qquad \text{(Exercise 10.5)}$$

$$= \sum_{i=1}^{k} \sum_{j=1}^{m} \int_{U_i \cap V_j} g_j f_i \omega$$

$$= \sum_{i=1}^{k} \sum_{j=1}^{m} \int_{U_i} g_j f_i \omega \qquad \text{(Exercise 10.5)}$$

$$= \sum_{i=1}^{k} \int_{U_i} \sum_{j=1}^{m} g_j f_i \omega \qquad \text{(Formula (10.6))}$$

$$= \sum_{i=1}^{k} \int_{U_i} f_i \omega \qquad \text{(the g_j form a partition of unity),}$$

where we put $\int_{U_i \cap V_j} f_i g_j \omega = 0$ whenever $U_i \cap V_j = \varnothing$. $\qquad \square$

Definition 10.16 (Integral of a Compactly Supported n-Form). The real number $\int_M \omega$ is called the *integral* of the compactly supported differential form ω on the oriented manifold M.

Proposition 10.17 (Elementary Properties of the Integral). *Let M be an orientable and oriented manifold of dimension n, and let $\int_M : \Omega_c^n(M) \to \mathbb{R}$, $\omega \mapsto \int_M \omega$ be the integral defined by the positive orientation on M. Then,*

(1) *the integral $\int_M : \Omega_c^n(M) \to \mathbb{R}$ is a surjective $\mathbb{R}$-linear map,*
(2) *if M' is another oriented n-dimensional manifold, and $\Phi : M \to M'$ is an orientation preserving diffeomorphism, then, for every $\omega \in \Omega_c^n(M')$*

(the pull-back $\Phi^*(\omega)$ is also compactly supported: $\Phi^*(\omega) \in \Omega_c^n(M)$ and),

$$\int_M \Phi^*(\omega) = \int_{M'} \omega,$$

(3) *let M be compact, and let $\omega \in \Omega^n(M)$ be a volume form determining the positive orientation, then $\int_M \omega > 0$.*

Proof. Left as Exercise 10.6. □

Exercise 10.6. Prove Proposition 10.17.

Exercise 10.7. Let $M_1, \ldots, M_k$ be smooth oriented manifolds of dimensions $n_1, \ldots, n_k$. Denote by $\mathrm{pr}_i : M_1 \times \cdots \times M_k \to M_i$ the projection onto the i-th factor. Moreover, for all $i = 1, \ldots, k$, let $\omega_i \in \Omega_c^{n_i}(M_i)$ be a compactly supported n_i-form on M_i. Show that

$$\omega^\times := \mathrm{pr}_1^*(\omega_1) \wedge \cdots \wedge \mathrm{pr}_k^*(\omega_k) \in \Omega^{n_1 + \cdots + n_k}(M_1 \times \cdots \times M_k)$$

is also a compactly supported differential form and that

$$\int_{M_1 \times \cdots \times M_k} \omega^\times = \prod_{i=1}^k \int_{M_i} \omega_i,$$

where the orientation on $M_1 \times \cdots \times M_k$ is the one induced by those of the factors.

Example 10.18 (Volume of Spheres). As an example, we now compute the integral of the standard volume form on the sphere (with respect to the standard orientation). Recall that the n-dimensional sphere S^n is a compact, orientable manifold, hence $\Omega_c^n(S^n) = \Omega^n(S^n)$. Moreover, the standard volume form ω_{S^n} determines the standard orientation of S^n (see Example 10.10)). In order to compute $\int_{S^n} \omega_{S^n}$, we first show that the stereographic chart $(U_+, \varphi_+ = (X_+^1, \ldots, X_+^n))$ (resp. $(U_-, \varphi_- = (X_-^1, \ldots, X_-^n))$) is positive (resp. negative) with respect to the standard orientation.

To do this, first note that

$$\omega_{S^n} = (-)^{n+1} \left(t^i \iota_{\partial/\partial t^i} \omega_{\text{coord}} \right) \big|_{S^n}$$

$$= \left(\sum_{i=1}^{n+1} (-)^{n-i} t^i dt^1 \wedge \cdots \wedge \widehat{dt^i} \wedge \cdots \wedge dt^{n+1} \right) \Big|_{S^n}$$

$$= \sum_{i=1}^{n+1} (-)^{n-i} T^i dT^1 \wedge \cdots \wedge \widehat{dT^i} \wedge \cdots \wedge dT^{n+1}$$

$$= \sum_{j=1}^{n} (-)^{n-j} T^j dT^1 \wedge \cdots \wedge \widehat{dT^j} \wedge \cdots \wedge dT^n \wedge dT^{n+1}$$

$$\quad - T^{n+1} dT^1 \wedge \cdots \wedge dT^n, \tag{10.7}$$

where we denoted $T^i := t^i \big|_{S^n}$. It is now easy to see that

$$T^j \big|_{U_+} = \frac{2X_+^j}{\Sigma_+ + 1}, \quad \text{for } j = 1, \dots, n,$$

and

$$T^{n+1} \big|_{U_+} = \frac{\Sigma_+ - 1}{\Sigma_+ + 1}$$

(as those readers who solved Exercises 1.1 and 1.3(1) should know, see also Example 1.36) where we put $\Sigma_+ = \|X_+\|^2 = (X_+^1)^2 + \cdots + (X_+^n)^2$. It follows that

$$dT^j \big|_{U_+} = \frac{2}{\Sigma_+ + 1} dX_+^j - \frac{2X_+^j}{(\Sigma_+ + 1)^2} d\Sigma_+ \tag{10.8}$$

for all $j = 1, \dots, n$, and

$$dT^{n+1} \big|_{U_+} = \frac{2}{(\Sigma_+ + 1)^2} d\Sigma_+.$$

As the wedge product is graded commutative, the second summand of (10.8) does not contribute to the first summand of (10.7) (when restricted to U_+) and we get

$$\left(\sum_{j=1}^{n} (-)^{n-j} T^j dT^1 \wedge \cdots \wedge \widehat{dT^j} \wedge \cdots \wedge dT^n \wedge dT^{n+1} \right) \bigg|_{U_+}$$

$$= \sum_{j=1}^{n} (-)^{n-j} 2^{n+2} \frac{X_+^j}{(\Sigma_+ + 1)^{n+2}} dX_+^1 \wedge \cdots \wedge \widehat{dX_+^j} \wedge \cdots \wedge dX_+^n \wedge d\Sigma_+.$$

$$(10.9)$$

Now

$$d\Sigma_+ = 2 X_+^1 dX_+^1 + \cdots + 2 X_+^n dX_+^n,$$

but, from the wedge product being graded commutative again, only the j-th summand contribute in the j-th summand of (10.9), i.e.,

$$\left(\sum_{j=1}^{n} (-)^{n-j} T^j dT^1 \wedge \cdots \wedge \widehat{dT^j} \wedge \cdots \wedge dT^n \wedge dT^{n+1} \right) \bigg|_{U_+}$$

$$= \sum_{j=1}^{n} \frac{(-)^{n-j} 2^{n+2} (X_+^j)^2}{(\Sigma_+ + 1)^{n+2}} dX_+^1 \wedge \cdots \wedge \widehat{dX_+^j} \wedge \cdots \wedge dX_+^n \wedge dX_+^j$$

$$= \sum_{j=1}^{n} \frac{2^{n+2} (X_+^j)^2}{(\Sigma_+ + 1)^{n+2}} dX_+^1 \wedge \cdots \wedge dX_+^n$$

$$= \frac{2^{n+2} \Sigma_+}{(\Sigma_+ + 1)^{n+2}} dX_+^1 \wedge \cdots \wedge dX_+^n.$$

Similarly,

$$- \left(T^{n+1} dT^1 \wedge \cdots \wedge dT^n \right) \bigg|_{U_+} = \frac{2^n (\Sigma_+ - 1)^2}{(\Sigma_+ + 1)^{n+2}} dX_+^1 \wedge \cdots \wedge dX_+^n.$$

Therefore,

$$\omega_{S^n}|_{U_+} = \left(\sum_{j=1}^{n} (-)^{n-j} T^j dT^1 \wedge \cdots \wedge \widehat{dT^j} \wedge \cdots \wedge dT^n \wedge dT^{n+1} \right.$$

$$\left. - T^{n+1} dT^1 \wedge \cdots \wedge dT^n \right) \bigg|_{U_+}$$

$$= \frac{2^{n+2} \Sigma_+}{(\Sigma_+ + 1)^{n+2}} dX_+^1 \wedge \cdots \wedge dX_+^n$$

$$+ \frac{2^n (\Sigma_+ - 1)^2}{(\Sigma_+ + 1)^{n+2}} dX_+^1 \wedge \cdots \wedge dX_+^n$$

$$= \left(\frac{2^{n+2}\Sigma_+}{(\Sigma_+ + 1)^{n+2}} + \frac{2^n(\Sigma_+ - 1)^2}{(\Sigma_+ + 1)^{n+2}} \right) dX_+^1 \wedge \cdots \wedge dX_+^n$$

$$= \frac{2^n}{(\Sigma_+ + 1)^n} dX_+^1 \wedge \cdots \wedge dX_+^n. \tag{10.10}$$

As $\omega_{S^n}|_{U_+}$ is a positive multiple of the coordinate form, we conclude that (U_+, φ_+) is a positively oriented chart as announced. A very similar computation shows that

$$\omega_{S^n}|_{U_-} = -\frac{2^n}{(\Sigma_- + 1)^n} dX_-^1 \wedge \cdots \wedge dX_-^n,$$

where $\Sigma_- = \|X_-\|^2 = (X_-^1)^2 + \cdots + (X_-^n)^2$. Hence, (U_-, φ_-) is a negatively oriented chart. We can define a positively oriented chart $(U_-, \psi_- = (Y_-^1, \ldots, Y_-^n))$ by putting

$$Y_-^1 = -X_-^1 \quad \text{and} \quad Y_-^k = X_-^k \quad \text{for all } k = 2, \ldots, n$$

so that $\Sigma_- = \|Y_-\|^2 = (Y_-^1)^2 + \cdots + (Y_-^n)^2$ and

$$\omega_{S^n}|_{U_-} = \frac{2^n}{(\Sigma_- + 1)^n} dY_-^1 \wedge \cdots \wedge dY_-^n.$$

In order to compute $\int_{S^n} \omega_{S^n}$, choose a partition of unity $\{f = f_{U_+}, g = f_{U_-} = 1 - f\}$ subordinate to the open cover (by coordinate domains of positive charts) $\{U_+, U_-\}$ so that

$$\int_{S^n} \omega_{S^n} = \int_{U_+} f\omega_{S^n} + \int_{U_-} (1 - f)\omega_{S^n}. \tag{10.11}$$

Now, from (10.10),

$$\varphi_+^{-1*}(f\omega_{S^n}) = \varphi_+^{-1*}(f) \cdot \varphi_+^{-1*}(\omega_{S^n}) = \frac{2^n \widehat{f_{U_+}}}{(\|t\|^2 + 1)^n} dt^1 \wedge \cdots \wedge dt^n,$$

where we denoted by $\widehat{f_{U_+}} = f \circ \varphi_+^{-1}$ the coordinate representation of f in the chart (U_+, φ_+). Similarly,

$$\psi_-^{-1*}((1 - f)\omega_{S^n}) = \psi_-^{-1*}(1 - f) \cdot \varphi_+^{-1*}(\omega_{S^n})$$

$$= \frac{2^n(1 - \widehat{f_{U_-}})}{(\|t\|^2 + 1)^n} dt^1 \wedge \cdots \wedge dt^n,$$

where $\widehat{f_{U_-}} = f \circ \psi_-^{-1}$. Hence, from (10.11),

$$\int_{S^n} \omega_{S^n} = \int_{\mathbb{R}^n} \frac{2^n \left(1 + \widehat{f_{U_+}}(t) - \widehat{f_{U_-}}(t)\right)}{(\|t\|^2 + 1)^n} d^n t.$$

Now the integral $\int_{\mathbb{R}^n} \frac{2^n}{(\|t\|^2+1)^n} d^n t$ is a well-known (finite) integral that amounts to

$$\int_{\mathbb{R}^n} \frac{2^n}{(\|t\|^2 + 1)^n} d^n t = \frac{2\pi^{\frac{n+1}{2}}}{\Gamma\left(\frac{n+1}{2}\right)},$$

where Γ is the *gamma function* (which agrees with the factorial of $q - 1$ on $q \in \mathbb{N}$). Hence,

$$\int_{S^n} \omega_{S^n} = \frac{2\pi^{\frac{n+1}{2}}}{\Gamma\left(\frac{n+1}{2}\right)} + 2^n \int_{\mathbb{R}^n} \frac{\widehat{f_{U_+}}(t)}{(\|t\|^2 + 1)^n} d^n t$$

$$- 2^n \int_{\mathbb{R}^n} \frac{\widehat{f_{U_-}}(s)}{(\|s\|^2 + 1)^n} d^n s.$$

The last two summands do actually cancel for any choice of the partition of unity (we leave the proof of this last fact to the reader as Exercise 10.8) and we conclude that

$$\int_{S^n} \omega_{S^n} = \frac{2\pi^{\frac{n+1}{2}}}{\Gamma\left(\frac{n+1}{2}\right)}. \tag{10.12}$$

For instance,

$$\int_{S^1} \omega_{S^1} = 2\pi, \quad \text{and} \quad \int_{S^2} \omega_{S^2} = 4\pi.$$

The positive number (10.12) is sometimes called the *volume* of the (unit) n-dimensional sphere (not to be confused with the volume of the ball!). ◆

Exercise 10.8. Let $\widehat{f}_{U_+}, \widehat{f}_{U_-}$ be as in Example 10.18. Show that

$$\int_{\mathbb{R}^n} \frac{\widehat{f}_{U_+}(t)}{(\|t\|^2 + 1)^n} d^n t = \int_{\mathbb{R}^n} \frac{\widehat{f}_{U_-}(s)}{(\|s\|^2 + 1)^n} d^n s. \qquad (10.13)$$

(**Hint:** *In the right-hand side of* (10.13), *perform the change of variables*

$$s = (\psi_- \circ \varphi_+^{-1})(t).$$

Note that you will need to compute the determinant of the Jacobian of the transition map $\psi_- \circ \varphi_+^{-1}$: *do this using the coordinate change formula for differential n-forms!*)

Remark 10.19. It is possible to define the integral over a non-orientable n-dimensional manifold, but, to do this, one needs to change the space of *integrands*. Namely, differential n-forms do not work anymore in the non-orientable case, and one needs to pass to a new line bundle satisfying the appropriate coordinate change formula, namely possessing a local frame on every coordinate domain, such that the local frames corresponding to different charts differ by an overall factor equal to the *absolute value* of the determinant of the Jacobian of the transition map. Such a line bundle does actually exist and, because of its specific coordinate change formula, one can indeed correctly define the integral of its compactly supported sections. Discussing the details of this construction goes beyond the scopes of these notes. The interested reader may consult, e.g., Nicolaescu (2021). ◇

10.3   Stokes Theorem

We conclude this chapter by briefly discussing how the integral interacts with the de Rham differential. Let M be a smooth manifold. First of all, we remark that the de Rham differential $d : \Omega^\bullet(M) \to \Omega^\bullet(M)$ preserves compactly supported differential forms. To see this, note that, given a differential k-form $\omega \in \Omega^k(M)$, the complement $M \smallsetminus \operatorname{supp} \omega$ of the support of ω consists of those points $p \in M$ with the property that there exists an open neighborhood $\mathcal{U}$ of p in M such that $\omega|_{\mathcal{U}} = 0$ (do you see it? If not, prove the latter fact as an exercise). This easily implies that

$$\operatorname{supp} d\omega \subseteq \operatorname{supp} \omega.$$

Indeed if $p \in M \smallsetminus \operatorname{supp} \omega$ and $\mathcal{U}$ is an open neighborhood of p such that $\omega|_{\mathcal{U}} = 0$, then $(d\omega)|_{\mathcal{U}} = d(\omega|_{\mathcal{U}}) = 0$, i.e., $p \in M \smallsetminus \operatorname{supp} d\omega$. In particular, if $\operatorname{supp} \omega$ is compact, then $\operatorname{supp} d\omega$ is also compact (it is a closed subset in a compact space), as announced. We are now ready to state and prove the main result of this section.

Theorem 10.20 (Stokes Theorem). *Let M be an orientable and oriented n-dimensional manifold, and let $\eta \in \Omega_c^{n-1}(M)$ be a compactly supported $(n-1)$-form on M (so that $d\eta$ is a compactly supported n-form). Then*

$$\int_M d\eta = 0. \tag{10.14}$$

Proof. We begin with some preliminary remarks. Let M be a smooth n-dimensional manifold. For any point $p \in M$, there exists a chart (U, φ) around p with the property that $\widehat{U} \subseteq \mathbb{R}^n$ is a *multi-rectangle*, i.e., an open subset of the form

$$\widehat{U} = (a_1, b_1) \times \cdots \times (a_n, b_n) \subseteq \mathbb{R}^n,$$

with $a_i, b_i \in \mathbb{R}$ and $a_i < b_i$, for all $i = 1, \dots, n$. This easily follows from the fact that multi-rectangles form a basis for the topology of $\mathbb{R}^n$ (do you see it?). Any such chart will be called a *coordinate multi-rectangle*.

Now, let M and η be as in the statement. Cover $\operatorname{supp} \eta$ by finitely many coordinate domains of positively oriented coordinate multi-rectangles $(U_1, \varphi_1), \dots, (U_k, \varphi_k)$. By the compactness of $\operatorname{supp} \eta$, this is always possible. As $\operatorname{supp} d\eta \subseteq \operatorname{supp} \eta$, the U_i do also cover $\operatorname{supp} d\eta$. Consider the following finite open cover of M: $\mathcal{C} := \{M \smallsetminus \operatorname{supp} \eta\} \cup \{U_i\}_{i=1,\dots,k}$, and choose a partition of unity $\{g = f_{M \smallsetminus \operatorname{supp} \eta}\} \cup \{f_i = f_{U_i}\}_{i=1,\dots,k}$ subordinate to $\mathcal{C}$ so that

$$g + \sum_{i=1}^{k} f_i = 1$$

and

$$\eta = \left(g + \sum_{i=1}^{k} f_i\right)\eta = g\eta + \sum_{i=1}^{k} f_i\eta.$$

But $g\eta = 0$. Indeed, if $p \in \operatorname{supp} \eta$, then $g(p) = 0$ (as $\operatorname{supp} g \subseteq M \smallsetminus \operatorname{supp} \eta$), while if $p \notin \operatorname{supp} \eta$, then $\eta_p = 0$. In any case, $(g\eta)_p = g(p)\eta_p = 0$. Hence,

$$\eta = \sum_{i=1}^{k} f_i\eta$$

and

$$\int_M d\eta = \int_M d\left(\sum_{i=1}^k f_i\eta\right) = \sum_{i=1}^k \int_M d(f_i\eta),\qquad(10.15)$$

where we used both the linearity of the de Rham differential and the linearity of the integral. For all $i = 1,\ldots,k$, we have that $\operatorname{supp}(f_i\eta)$ is contained in the coordinate domain U_i of the coordinate multi-rectangle (U_i, φ_i). From (10.15), in order to prove (10.14), we can assume that $\operatorname{supp}(d\eta) \subseteq \operatorname{supp}\eta \subseteq U$ for some positively oriented multi-rectangle (U, φ). In this case,

$$\int_M d\eta = \int_U d\eta.$$

Now,

$$\varphi^{-1*}\eta = \sum_{i=1}^n (-)^{i-1} f_i dt^1 \wedge \cdots \wedge \widehat{dt^i} \wedge \cdots \wedge dt^n$$

for some smooth compactly supported functions $f_i \in C^\infty(\widehat{U})$. Note that, being compactly supported, the functions f_i can be extended to smooth functions on the whole $\mathbb{R}^n$, denoted F_i, declaring the latter to be zero outside $\widehat{U}$ so that $\operatorname{supp} F_i = \operatorname{supp} f_i$. In the following, we will freely do so. We have

$$\varphi^{-1*}(d\eta) = d(\varphi^{-1*}\eta)$$

$$= \sum_{i=1}^n (-)^{i-1} df_i \wedge dt^1 \wedge \cdots \wedge \widehat{dt^i} \wedge \cdots \wedge dt^n$$

$$= \sum_{i,j=1}^n (-)^{i-1} \frac{\partial f_i}{\partial t^j} dt^j \wedge dt^1 \wedge \cdots \wedge \widehat{dt^i} \wedge \cdots \wedge dt^n. \qquad(10.16)$$

As the wedge product is graded commutative, the only terms contributing to (10.16) are those with $i = j$, i.e.,

$$\varphi^{-1*}(d\eta) = \sum_{i=1}^n (-)^{i-1} \frac{\partial f_i}{\partial t^i} dt^i \wedge dt^1 \wedge \cdots \wedge \widehat{dt^i} \wedge \cdots \wedge dt^n$$

$$= \sum_{i=1}^n \frac{\partial f_i}{\partial t^i} dt^1 \wedge \cdots \wedge dt^n.$$

Therefore, if $\widehat{U} = (a_1, b_1) \times \cdots \times (a_n, b_n)$, we get

$$\int_M d\eta = \int_U d\eta$$

$$= \int_{(a_1,b_1)\times\cdots\times(a_n,b_n)} \sum_{i=1}^{n} \frac{\partial f_i}{\partial t^i}(t)\mathrm{d}^n t$$

$$= \sum_{i=1}^{n} \int_{(a_1,b_1)\times\cdots\times(a_n,b_n)} \frac{\partial f_i}{\partial t^i}(t)\mathrm{d}^n t$$

$$= \sum_{i=1}^{n} \int_{[a_1,b_1]\times\cdots\times[a_n,b_n]} \frac{\partial F_i}{\partial t^i}(t)\mathrm{d}^n t$$

$$= \sum_{i=1}^{n} \int_{a_1}^{b_1} \cdots \widehat{\int_{a_i}^{b_i}} \cdots \int_{a_n}^{b_n} \left(\int_{a_i}^{b_i} \frac{\partial F_i}{\partial t^i}(t)\mathrm{d}t^i \right) \mathrm{d}t^1 \cdots \widehat{\mathrm{d}t^i} \cdots \mathrm{d}t^n,$$

where, in the last step, we used the Fubini Theorem. But, from the Fundamental Theorem of Calculus, for all $i = 1, \ldots, n$,

$$\int_{a_i}^{b_i} \frac{\partial F_i}{\partial t^i}(t)\mathrm{d}t^i = F_i|_{t_i=b_i} - F_i|_{t_i=a_i} = 0,$$

as $\mathrm{supp}\, F_i = \mathrm{supp}\, f_i \subseteq \widehat{U} = (a_1, b_1) \times \cdots \times (a_n, b_n)$ so that $F_i(P) = 0$ for all P on the boundary of the closed multi-rectangle $[a_1, b_1] \times \cdots \times [a_n, b_n]$. This concludes the proof.   □

Remark 10.21. Let M be a compact, connected, orientable, and oriented manifold of dimension n. Then the Stokes Theorem says that the image of the linear map $d : \Omega^{n-1}(M) \to \Omega^n(M)$ is contained in the kernel of the linear map $\int_M : \Omega^n(M) \to \mathbb{R}$. It follows from the Homomorphism Theorem that $\int_M$ induces a well-defined linear map

$$I_M : \frac{\Omega^n(M)}{\mathrm{im}\,(d : \Omega^{n-1}(M) \to \Omega^n(M))} \to \mathbb{R}, \quad [\omega] \mapsto I_M[\omega] := \int_M \omega,$$

where we denoted by $[\omega]$ the equivalence class of $\omega \in \Omega^n(M)$ modulo the vector subspace $\mathrm{im} : \Omega^{n-1}(M) \to \Omega^n(M)$. The domain of I_M is exactly the n-th homogeneous piece $H^n_{\mathrm{dR}}(M)$ of the de Rham cohomology space $H_{\mathrm{dR}}(M)$ (see Section 8.3). So there is a canonical linear map

$$I_M : H^n_{\mathrm{dR}}(M) \to \mathbb{R}.$$

It can be proved that I_M is actually a vector space isomorphism. In other words, given $\omega \in \Omega^n(M)$, the integral $\int_M \omega$ vanishes exactly when ω is an exact n-form: $\omega = d\eta$ for some $\eta \in \Omega^{n-1}(M)$ (see, e.g., Michor, 2008, for more details). ◇

Remark 10.22. There is a more general class of spaces than smooth manifolds where one can perform differential and integral calculus, namely smooth manifolds with boundary. They are spaces M locally modeled on open subsets of

$$\mathbb{R}_+^n := \left\{ P = (P^1, \ldots, P^n) \in \mathbb{R}^n : P^n \geq 0 \right\}$$

(instead of simply open subsets of $\mathbb{R}^n$). As in the usual case of smooth manifolds, one defines *charts with boundary* and requires the charts to cover M and the transition maps to be diffeomorphisms between open subsets of $\mathbb{R}_+^n$. Points of M that are mapped to the boundary

$$\partial\mathbb{R}_+^n := \left\{ P = (P^1, \ldots, P^n) \in \mathbb{R}^n : P^n = 0 \right\} \subseteq \mathbb{R}_+^n$$

of $\mathbb{R}_+^n$ by some chart with boundary are called *boundary points* and form a smooth submanifold (without boundary) of dimension $n - 1$ in M, denoted ∂M, and called the *boundary* of M. One can define smooth functions, vector fields, and differential forms on manifolds with boundary in a very similar way as we did for plain manifolds. One can also define orientations. The boundary of an orientable and oriented manifold with boundary M is also orientable and inherits an orientation in a canonical way. The Stokes Theorem has a powerful version for manifolds with boundary stating that, given an orientable and oriented n-dimensional manifold with boundary M and a compactly supported differential $(n-1)$-form η on M, then

$$\int_M d\eta = \int_{\partial M} \eta|_{\partial M}. \tag{10.17}$$

For more details on manifolds with boundary and the Stokes Theorem in the general form (10.17), the reader may consult, e.g., Lee (2013). ◇

Bibliography

Arnol'd Vladimir I., *Geometric Methods in the Theory of Ordinary Differential Equations* (translated from the Russian by Szücs J.; Translation edited by Levi M.). Grundlehren der Mathematischen Wissenschaften [Fundamental Principles of Mathematical Sciences], Vol. 250, Springer-Verlag, New York-Berlin, 1983.

Arnold Vladimir I., *Ordinary Differential Equations* (translated from the third Russian edition by Cooke R.). Springer Textbook, Springer-Verlag, Berlin, 1992.

Atiyah Michael F. and Macdonald Ian G., *Introduction to Commutative Algebra* (student economy edition). Addison-Wesley Series in Mathematics, Westview Press, Boulder, CO, 2016.

Bott R. and Tu L. W., *Differential Forms in Algebraic Topology*. Graduate Text in Mathematics, Vol. 82, Springer-Verlag, New York-Berlin, 1982.

Cartan É., Sur certaines expressions différentielles et le probléme de Pfaff, *Ann. Sci. École Norm. Sup.* (3) **16**, 239–332 (1899).

Conlon L., *Differentiable Manifolds* (reprint of the 2001 2nd edn.). Modern Birkhäuser Classics, Birkhäuser Boston Inc., Boston, MA, 2008.

Harris J., *Algebraic Geometry: A First Course*. Graduate Texts in Mathematics, Vol. 133, Springer-Verlag, New York, 1992.

Hartman P., *Ordinary Differential Equations* (corrected reprint of the 1982 2nd edn. [Brikhauser, Boston, MA], with a foreword by Bates P.). Classics in Applied Mathematics, Vol. 38, Society for Industrial and Applied Mathematics (SIAM), Philadelphia, PA, 2002.

Kosniowski C., *A First Course in Algebraic Topology*, Cambridge University Press, Cambridge-New York, 1980.

Lang S., *Undergraduate Analysis* (2nd edn.). Undergraduate Texts in Mathematics, Springer-Verlag, New York, 1997.

Lee J. M., *Introduction to Topological Manifolds* (2nd edn.). Graduate Texts in Mathematics, Vol. 202, Springer, New York, 2011.

Lee J. M., *Introduction to Smooth Manifolds* (2nd edn.). Graduate Texts in Mathematics, Vol. 218, Springer, New York, 2013.

Manetti M., *Topology* (translated from the 2014 Italian edition by Simon G. Chiossi). Unitext, La Mat. per il 3+2, Vol. 91, Springer, Cham, 2015.

Michor P. W., *Topics in Differential Geometry*. Graduate Studies in Mathematics, Vol. 93, American Mathematical Society, Providence, RI, 2008.

Nakahara M., *Geometry, Topology and Physics* (2nd edn.). Graduate Student Series in Physics, Institute of Physics, Bristol, 2003.

Nestruev J., *Smooth Manifolds and Observables* (2nd edn.). Graduate Texts in Mathematics, Vol. 220, Springer, Cham, 2020.

Nicolaescu L. I., *Lectures on the Geometry of Manifolds* (3rd edn.). World Scientific Publishing Co. Pte. Ltd., Hackensack, NJ, 2021.

Index

www.ingramcontent.com/pod-product-compliance
Lightning Source LLC
Chambersburg PA
CBHW050544190326
41458CB00007B/1913